Hydrogen Production from Renewable Resources and Wastes

This book provides readers with a comprehensive overview of the processes and technologies utilized for producing hydrogen from renewable sources. It discusses common methods like gasification, pyrolysis, and liquefaction, along with novel methods like water thermochemical splitting, biophotolysis, biological water-gas shift reaction, and fermentation processing. The application of various renewable sources, including wind, solar, and geothermal energy, is covered in detail.

- Introduces water splitting conversion processes for hydrogen production in detail
- Uniquely provides different pyrolysis, gasification, and liquefaction processes for hydrogen generation
- Covers different biomass and waste sources for producing hydrogen
- Discusses biochemical methods for converting biomass to hydrogen
- Provides the application of renewable energy sources in hydrogen production

Part of the multivolume Handbook of Hydrogen Production and Applications, this stand-alone book guides researchers and academics in chemical, environmental, energy, and related areas of engineering interested in the development and implementation of hydrogen production technologies.

Mohammad Reza Rahimpour is a Distinguished Professor of Chemical Engineering at Shiraz University, Iran. He earned a PhD in chemical engineering at Shiraz University, in collaboration with the University of Sydney, Australia, in 1988. He leads a pioneering research group specializing in fuel processing technology, with a particular focus on the catalytic conversion of both fossil fuels, such as natural gas, and renewable fuels, like bio-oils derived from lignin, into valuable energy sources. Throughout his academic career, Dr. Rahimpour has specialized in hydrogen production technologies, including steam methane reforming, water electrolysis, photocatalytic hydrogen production, and hydrogen production from both renewable sources and fossil fuels. His extensive expertise and innovative research significantly contribute to advancements in hydrogen production and its practical applications.

Mohammad Amin Makarem earned a PhD in chemical engineering at Shiraz University. His research interests include gas separation and purification, nanofluids, microfluidics, catalyst synthesis, reactor design, and green energy.

Parvin Kiani earned a degree in chemical engineering at Shiraz University. Her research has focused on gas separation, clean energy, and catalyst synthesis.

Hydrogen Production from Renewable Resources and Wastes

Edited by
Mohammad Reza Rahimpour,
Mohammad Amin Makarem and Parvin Kiani

CRC Press
Taylor & Francis Group
Boca Raton London New York

CRC Press is an imprint of the
Taylor & Francis Group, an **informa** business

Designed cover image: © Shutterstock, Scharfsinn

First edition published 2025
by CRC Press
2385 NW Executive Center Drive, Suite 320, Boca Raton FL 33431

and by CRC Press
4 Park Square, Milton Park, Abingdon, Oxon, OX14 4RN

CRC Press is an imprint of Taylor & Francis Group, LLC

Library of Congress Cataloging-in-Publication Data
Names: Rahimpour, Mohammad Reza, 1961- editor. | Makarem, Mohammad Amin, editor. | Kiani, Parvin, editor.
Title: Hydrogen production from renewable resources and wastes / edited by Mohammad Reza Rahimpour, Mohammad Amin Makarem and Parvin Kiani.
Description: First edition. | Boca Raton : CRC Press, 2024. |
Series: Handbook of hydrogen production and applications; 2 |
Includes bibliographical references and index.
Identifiers: LCCN 2024020661 (print) | LCCN 2024020662 (ebook) |
ISBN 9781032465609 (hardback) | ISBN 9781032465616 (paperback) |
ISBN 9781003382270 (ebook)
Subjects: LCSH: Hydrogen as fuel. | Environmental engineering. | Renewable energy sources.
Classification: LCC TP359.H8 H95325 (print) | LCC TP359.H8 (ebook) |
DDC 665.8/1—dc23/eng/20240501
LC record available at https://lccn.loc.gov/2024020661
LC ebook record available at https://lccn.loc.gov/2024020662

ISBN: 978-1-032-46560-9 (hbk)
ISBN: 978-1-032-46561-6 (pbk)
ISBN: 978-1-003-38227-0 (ebk)

DOI: 10.1201/9781003382270

Typeset in Times
by codeMantra

Contents

SECTION I Water Splitting Conversion Processes

SECTION II Biomass and Wastes Thermochemical Conversion Processes

Preface

As the world intensifies its focus on sustainable energy solutions, hydrogen emerges as a key player due to its versatility and eco-friendly attributes. The Handbook of Hydrogen Production and Applications represents a comprehensive and authoritative exploration of the pivotal role hydrogen plays in addressing the global energy landscape and environmental challenges. This ambitious seven-volume project delves into every facet of hydrogen, ranging from its production using both nonrenewable and renewable resources, purification, and separation processes, to its storage, transportation, and various applications across diverse technologies. With an emphasis on fostering a deeper understanding of hydrogen utilization, particularly in fuel cells, this handbook aims to serve as an invaluable resource for researchers, engineers, and policymakers striving to advance the forefront of clean energy technology. Each volume within this compendium is dedicated to a specific aspect, collectively forming a comprehensive guide that illuminates the multifaceted dimensions of hydrogen production and its myriad applications.

Volume 2, *Hydrogen Production from Renewable Resources and Wastes*, continues this exploration by focusing on the dynamic and evolving landscape of hydrogen production from sustainable sources. Section I, "Water Splitting Conversion Processes," delves into the fundamental methods of hydrogen production through electrolysis and water thermochemical splitting. These processes, leveraging the abundant resource of water, lay the foundation for renewable hydrogen generation, aligning with the global pursuit of green energy solutions.

Moving to Section II, titled "Biomass and Wastes Thermochemical Conversion Processes," the volume explores innovative approaches to harnessing hydrogen from renewable resources and wastes. Processes such as pyrolysis for the conversion of lignin and sewage sludge into hydrogen showcase the potential of waste materials to contribute to clean energy production. The importance of lignin utilization and the gasification of sewage sludge further underscore the diverse applications of thermochemical conversion methods in sustainable hydrogen production.

Section III, "Biomass Biochemical Conversion Processes," delves into biological pathways for hydrogen production from biomass. From biophotolysis processes to biological water-gas shift reactions and fermentation processes, this section explores the intricate biochemical conversions that can yield hydrogen from renewable sources. Highlighting the intersection of biology and energy, these processes offer eco-friendly alternatives in the pursuit of sustainable hydrogen production.

The volume concludes with Section IV, exploring "Other Renewable Resources for Hydrogen Production." This section investigates the application of solar, hydropower, and geothermal energy in hydrogen production, emphasizing the potential of diverse renewable sources in shaping the future of clean energy. From harnessing solar energy to utilizing the power of tides, waves, water flow, and falls, this section provides a comprehensive overview of the renewable energy landscape in the context of hydrogen production.

In this volume, we aim to offer a comprehensive resource for researchers, engineers, and policymakers seeking to understand and contribute to the evolving field of hydrogen production from renewable resources and wastes. By delving into water splitting conversion processes, biomass and waste thermochemical conversion, biochemical conversion, and other renewable sources, this volume aims to provide insights that contribute to the broader goals of sustainable energy. We extend our sincere gratitude to all contributors who have brought their expertise to this project, inviting readers to embark on a captivating journey through the pages of this volume. Together, let us explore the intricate world of hydrogen production from renewable resources, paving the way for a cleaner and more sustainable energy future.

Mohammad Reza Rahimpour
Mohammad Amin Makarem
Parvin Kiani

Contributors

Aisha H. Al-Moubaraki
Department of Chemistry
College of Science
University of Jeddah
Jeddah, Saudi Arabia

Carmen María Álvez-Medina
Department of Applied Physics
University of Extremadura
Badajoz, Spain

Mohamed Al Zarooni
American University of Ras Al
    Khaimah
Ras Al-Khaimah, United Arab Emirates

Tolga Ayzit
Izmir Institute of Technology
Urla, Izmir, Turkey

Amizon Azizan
Universiti Teknologi MARA
Selangor, Malaysia

Alper Baba
Izmir Institute of Technology
Urla, Izmir, Turkey

Adrián Bocho-Roas
Department of Applied Physics
University of Extremadura
Badajoz, Spain

Beatriz Ledesma Cano
Department of Applied Physics
University of Extremadura
Badajoz, Spain

Ajay K. Dalai
Department of Chemical Engineering
University of Saskatchewan
Saskatoon, Canada

Sara Faiz
American University of Ras Al
    Khaimah
Ras Al-Khaimah, United Arab Emirates

Zeynab Farzizada
Khazar University
Baku, Azerbaijan

Pedram Fatehi
Department of Chemical Engineering
Lakehead University
London, Canada

Sadia Fida
National University of Sciences and
    Technology (NUST)
Islamabad, Pakistan

Greeshma Gadikota
School of Civil and Environmental
    Engineering
Cornell University
and
Robert Fredrick Smith School of
    Chemical and Biomolecular
    Engineering
Cornell University
Ithaca, New York, USA

Agshin Garashli
Khazar University
Baku, Azerbaijan

Yulin Hu
Faculty of Sustainable Design
    Engineering
University of Prince Edward Island
Charlottetown, Canada

Rafidah Jalil
Forest Research Institute Malaysia
(FRIM)
Selangor, Malaysia

Michal Jeremias
Department of Gasification and
Synthesis Gas Processing
VTT Technical Research Centre of
Finland Ltd.
Espoo, Finland

Yang Jia
School of Civil and Environmental
Engineering
Cornell University
Ithaca, New York, USA

Kang Kang
Biorefining Research Institute
Lakehead University
London, Canada

Mohammad Hasan Khademi
Department of Chemical Engineering
College of Engineering
Isfahan University
Isfahan, Iran

Parvin Kiani
Shiraz University
Shiraz, Fars, Iran

Janusz Kozinski
Biorefining Research Institute
Lakehead University
London, Canada

Mohammad Latifi
Department of Chemical Engineering
Polytechique Montréal
Montréal, Canada

Akanksh Mamidala
School of Civil and Environmental
Engineering
Cornell University
Ithaca, New York, USA

Rasoul Moradi
Khazar University
Baku, Azerbaijan

Sina Mosallanezhad
Shiraz University
Shiraz, Fars, Iran

Sonil Nanda
Department of Engineering
University of Saskatchewan
Saskatoon, Canada

Sergio Nogales-Delgado
Department of Applied Physics
University of Extremadura
Badajoz, Spain

Uday Kumar Nutakki
American University of
Ras Al Khaimah
Ras Al-Khaimah, United Arab Emirates

Prince Ochonma
Robert Fredrick Smith School of
Chemical and Biomolecular
Engineering
Cornell University
Ithaca, New York, USA

Mohamed Abdalla Omar
American University of
Ras Al Khaimah
Ras Al-Khaimah, United Arab Emirates

Alper Özmumcu
Izmir Institute of Technology
Urla, Izmir, Turkey

Mohammad Reza Rahimpour
Department of Chemical Engineering
Shiraz University
Shiraz, Iran

Gholamreza Roohollahi
Department of Chemical Engineering
Polytechique Montréal
Montréal, Canada

Arash Sadeghi
Department of Chemical Engineering
Shiraz University
Shiraz, Iran

Fatemeh Salahi
Shiraz University
Shiraz, Fars, Iran

Zeshan Sheikh
National University of Sciences and
    Technology (NUST)
Islamabad, Pakistan

Vineet Singh Sikarwar
Division of Plasma Chemical
    Technologies
Institute of Plasma Physics of the Czech
    Academy of Sciences
Prague, Czech Republic

Abhishek N. Srivastava
Department of Civil Engineering
National Institute of Technology Calicut
    Kozhikode
Kerala, India

Zia Ullah
National University of Sciences and
    Technology (NUST)
Islamabad, Pakistan

Fatemeh Zarei-Jelyani
Shiraz University
Shiraz, Fars, Iran

Mohammad Zarei-Jelyani
Shiraz University
Shiraz, Fars, Iran

Reviewer Acknowledgments

The editors feel obliged to appreciate the dedicated reviewers (listed below) who were involved in reviewing and commenting on the submitted chapters and whose cooperation and insightful comments were very helpful in improving the quality of the chapters and books in this series.

Dr. Fatemeh Haghighatjoo
Department of Chemical Engineering
Shiraz University
Shiraz, Iran

Dr. Maryam Meshksar
Department of Chemical Engineering
Shiraz University
Shiraz, Iran

Ms. Soheila Zandi Lak
Department of Chemical Engineering
Shiraz University
Shiraz, Iran

Mr. Sina Mosallanejad
Department of Chemical Engineering
Shiraz University
Shiraz, Iran

Section I

Water Splitting Conversion Processes

1 Hydrogen Production through Electrolysis

Aisha H. Al-Moubaraki

1.1 INTRODUCTION

Over the past few decades, the global population's growing demand for energy, coupled with changes in human lifestyles, has led to an unprecedented need for energy. While fossil fuels have provided a significant source of energy, their depletion has pushed industries and governments to search for alternative sources of fuel and develop new processes for energy production. Moreover, using fossil fuels leads to the emission of greenhouse gases (GHG), SO_x, and NO_x that have a detrimental impact on the environment. These emissions are linked to global warming, climate change, and a decline in biodiversity (Mortensen et al., 2011; Van Ruijven and Van Vuuren, 2009; Moravvej et al., 2021).

The increasing energy demand and the limitations of nonrenewable sources have led to a focus on the utilization of renewable energy resources, and hydrogen has emerged as the most promising candidate (Karimi et al., 2022; Abe et al., 2019). Its unique characteristics, such as a wide range of flammability, high ignition temperature, low ignition energy, fast flame speed, and high diffusivity, have led to its use in various industrial processes, including hydrocarbon upgrading, ammonia and methanol production, and the Fischer-Tropsch technology (Sharma and Ghoshal, 2015). Consequently, the advancement of innovative production techniques for hydrogen has become the primary research emphasis in recent decades (Kovač et al., 2021; Makarem et al., 2022).

One promising method of producing dense and environmentally friendly hydrogen is through electrolysis, which involves water splitting into oxygen and hydrogen. However, the efficiency of hydrogen production through electrolysis is limited by the insufficient hydrogen release rate and excessive power usage (Kumar and Himabindu, 2019), leading to the electrochemical water electrolysis method accounting for only 4% of global hydrogen gas production. Additionally, the use of noble and expensive electrocatalysts such as platinum and ruthenium-based compounds has hindered large-scale commercial purposes (Zhang et al., 2020; Seh et al., 2017; Jaramillo et al., 2007; Wang et al., 2016; Tian et al., 2014). Thus, it is crucial to find economical and extremely effective non-precious metal electrocatalysts for both the hydrogen evolution reaction (HER) and the oxygen evolution reaction (OER). Recent studies have shown that composites, metal-organic frameworks (MOFs), and transition-metal-based compounds such as hydroxides (Rosen et al., 2013), metal oxides (Gerken et al., 2011), sulfides (Faber et al., 2014), nitrides (Chen et al., 2015), phosphides (Popczun et al., 2014), and selenides (Suntivich et al., 2011)

DOI: 10.1201/9781003382270-2

3

have demonstrated excellent electrocatalytic characteristics for both the HER and OER processes (Anwar et al., 2021).

Water electrolysis was a commonly used method for hydrogen production before the advent of steam methane reforming (Basye and Swaminathan, 1997). Steam methane reforming is a widely accepted method for the large-scale generation of CO, H_2, and CO_2. The CO_2 stream produced can be of high purity, making it suitable for carbon capture and storage (CCS). However, there is a suggestion that water electrolysis may become a more cost-effective option if renewable power prices continue to decrease, as this process only needs water as an input. According to a report, water electrolysis can generate 380,000 kg of H_2 annually with a consumption rate of 53.4 kWh/kg H_2 (Nikolaidis and Poullikkas, 2017; Younas et al., 2022).

1.2 ELECTROLYSIS CONCEPT

Electrolysis is a chemical decomposition process that occurs in an electrolyzer by passing an electric current through an ion-containing solution (Grimes et al., 2008). Aqueous methanol can be electrolyzed to produce hydrogen at low cost and voltage, although it emits carbon dioxide, which is a GHG. The electrolysis efficiency for hydrogen generation depends on the feedstock used (Sasikumar et al., 2008). However, since methanol is derived from biomass, this technique is time-consuming and requires a lot of energy. Electrolysis of water is a promising method for hydrogen production since it allows pure hydrogen and oxygen production at different electrodes without external separation processes. Nevertheless, this method has a considerable energy consumption compared to other techniques, with a commercial water electrolyzer consuming approximately 4.5–5 kWh/Nm³. However, the utilization of renewable energy sources like wind and photovoltaics (PV electrolysis) could increase the energy efficiency of water electrolysis (Kothari et al., 2008; Younas et al., 2022).

1.3 WATER ELECTROLYSIS ENERGY CONSUMPTION

The energy consumed in water electrolysis is represented by the cell voltage (U). Water has a theoretical decomposition voltage of 1.23 V, which is the voltage required for gas evolution. The HER has a current efficiency of up to 100%, and 2 F of electricity is needed to generate 1 mol H_2 based on Faraday's law. Therefore, the theoretical energy consumption required to generate 1 m³ H_2 is 2.94 kWh/m³H_2. However, due to ohmic voltage drop and significant overpotential, practical cell voltages in industrial cells range from 1.8 to 2.6 V, which are significantly higher compared to the theoretical decomposition voltage. The gas evolution reaction does not occur until 1.65–1.7 V. At a practical cell voltage of 2.0 V, the energy consumption increases to 4.78 kWh/m³H_2, resulting in a low energy efficiency of 61.6% for hydrogen production (Wang et al., 2014):

$$U = U^{\theta} + |\eta_a| + |\eta_c| + i * \Sigma R \tag{1.1}$$

In this equation, the cell voltage is a key indicator of energy consumption. It is determined by various factors, including the cathode potential for HER (E_c), the anode potential for OER (E_a), the total ohmic resistance ($\sum R$), the current density (i), the theoretical decomposition voltage (U^θ), and the anode and cathode overpotentials (η_a and η_c). As shown in equation (1.1), the water electrolysis cell voltage is the sum of U^θ, the ohmic voltage drop ($i_*\sum R$), and the reaction overpotential (η). *Therefore, improving the water electrolysis process involves reducing U^θ, η, or $i_*\sum R$* (Wang et al., 2014).

1.3.1 Theoretical Decomposition Voltage

The process of water electrolysis is a result of the HER occurring on the cathode and the OER on the anode. The reaction and standard equilibrium electrode potential (E^θ) at 1 atm and 25°C can be expressed as follows (Wang et al., 2014):

In acid electrolyte

$$\text{Cathode:} 2H^+ + 2e = H_2 \quad E_c^\theta = 0.00 \text{ V} \tag{1.2}$$

$$\text{Anode:} H_2O = 2H^+ + \tfrac{1}{2}O_2 + 2e \quad E_a^\theta = 1.23 \text{ V} \tag{1.3}$$

In alkaline electrolyte

$$\text{Cathode:} H_2O + 2e = H_2 + 2OH^- \quad E_c^\theta = -0.83 \text{ V} \tag{1.4}$$

$$\text{Anode:} 2OH^- = H_2O + \tfrac{1}{2}O_2 + 2e \quad E_a^\theta = 0.40 \text{ V} \tag{1.5}$$

Total reaction is

$$H_2O = H_2 + \tfrac{1}{2}O_2 \quad U^\theta = 1.23 \text{ V} \tag{1.6}$$

Theoretical decomposition voltage (U^θ) is:

$$U^\theta = E_a^\theta - E_c^\theta$$

The equation for water electrolysis involves the anode potential for OER (E_a), the cathode potential for HER (E_c), and the thermodynamic parameter of theoretical decomposition voltage (U^θ), which has a fixed value of 1.23 V at 25°C and 1 atm. In a traditional electrolytic system, the U^θ value remains constant at a specific temperature and is not influenced by solution composition, electrode materials, pH, stirring, and other factors (Wang et al., 2014).

1.3.2 Overpotential

To accelerate the water electrolysis speed, it is necessary to overcome the energy barrier with overpotentials (η) for both cathode and anode reactions. The relationship between overpotential (η) and current density (i) is described by the Tafel equation (equation 1.7) in this context. When the theoretical decomposition voltage of 1.23 V is applied to the electrolytic cell, water electrolysis is kinetically reversible, and the hydrogen and oxygen evolution reactions occur at a slower rate (Wang et al., 2014).

$$\eta = a + b \log i \tag{1.7}$$

The Tafel equation consists of two constants: a and b, with a representing the overpotential at 1 A/cm^2. The value of a is determined by the surface structure and intrinsic properties of electrode materials. The metals used as cathodes for HER can be classified into three groups based on their overpotential. Metals with a high overpotential include Cd, Hg, Tl, Pb, Sn, Zn, and others. Metals with a medium overpotential include Fe, Ni, Co, Cu, W, Au, Ag, and so on. Metals with low overpotential include Pd and Pt (Wang et al., 2014).

Pt is considered the most efficient catalyst for HER due to its high exchange current density and optimal hydrogen adsorption energy. Typically, the activity of HER in alkaline media is lower than that in acidic media, as the sluggish water dissociation step significantly decreases the rate of reaction by two to three orders of magnitude. Despite this, alkaline electrolysis is more suitable for industrial applications. In order to create effective electrocatalysts with superior alkaline HER performance, catalysts should possess traits such as hydrogen species binding and water dissociation capabilities (Strmcnik et al., 2013; Wang et al., 2021).

1.3.3 OHMIC VOLTAGE DROP

The total ohmic resistance (ΣR) is a significant factor that contributes to the high energy usage of water electrolysis. It can be expressed as follows (Wang et al., 2014):

$$\Sigma R = R_e + R_m + R_b + R_c \tag{1.8}$$

The equation given describes the total ohmic resistance (ΣR) of water electrolysis, which comprises electrolyte resistance (R_e), bubble resistance (R_b), membrane resistance (R_m), and circuit resistance (R_c). The reduction of R_m and R_c can be achieved by optimizing wire connection and membrane production processes. Both R_c and R_m are constant values in the electrolytic cell. R_e is dependent on the solution composition and distance between the electrodes. It can be reduced by using more conductive salts like NaOH or KOH. However, an excessive amount of conductive salts may lead to cell corrosion and membrane damage. Moreover, the presence of bubbles covering the electrode surface could interfere with the electric field and lead to high bubble resistance (R_b) (Wang et al., 2014).

1.4 ELECTROLYSIS TECHNOLOGIES

1.4.1 ALKALINE WATER ELECTROLYSIS (AWE)

The AWE process, which was first proposed by Troostwijk and Diemann in 1789, is considered to be the most widely used and well-established method for producing hydrogen from water (Balat, 2008; Carmo et al., 2013; Ursua et al., 2011). This process typically operates at temperatures ranging from 40°C to 90°C, and it has an efficiency range of 70%–80%. KOH and NaOH, with concentrations ranging from 20% to 30%, are commonly utilized in AWE electrolytes (Rashid et al., 2015; Sapountzi et al., 2017). To prevent the mixing of the gases produced at the respective electrodes, the anode loop and cathode are separated by an asbestos membrane, and the anodic and cathodic electrodes are constructed from nickel (Ni). During the AWE process, water is electrolyzed on the cathode, and hydroxyl ions (OH^-) migrate toward the anode side. Figure 1.1 provides a schematic representation of AWE (Anwar et al., 2021).

Although the AWE technique has been widely used for hydrogen production, it has some disadvantages, such as low operating pressure and temperature and confined current densities of around 0.4 A/cm², which can limit its energy efficiency. Additionally, the ratio of water consumption to hydrogen production in AWE is approximately 11.5:1. To address these issues, a new approach is being investigated that utilizes anion exchange membranes constructed from polymers with anionic conductivity, instead of diaphragm membranes (Kumar and Himabindu, 2019; Sandeep et al., 2017; Hwang et al., 2015; Chakrabarty et al., 2008; Ohmori et al., 2007).

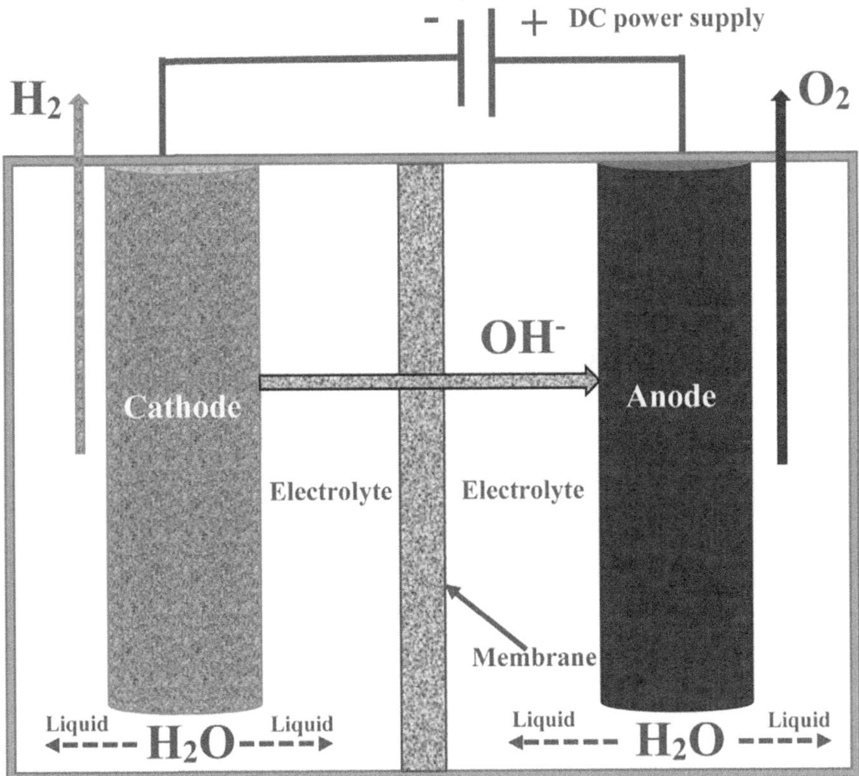

FIGURE 1.1 Schematic diagram of the AWE technique (Anwar et al., 2021).

The fundamental process for producing hydrogen from AWE involves the generation of oxygen at the anode and hydrogen at the cathode. The diaphragm used in the process separates the product gases, and both OH⁻ and water can penetrate it (Carmo et al., 2013). The electrodes are submerged in an electrolyte, which is usually a liquid solution of KOH or NaOH. Cobalt, iron, and nickel are common electrode materials, with nickel being the most widely used due to its low cost and excellent electrocatalytic activity (Zeng and Zhang, 2010). The utilization of AWE has several advantages, including pure hydrogen energy production utilizing non-precious electrocatalysts, and it is an advanced technology that allows for high scalability through the stacking of cells (Sapountzi et al., 2017; Carmo et al., 2013). Nevertheless, the use of AWE can cause corrosion due to the electrolytes, and gas can permeate through the diaphragm, affecting the overall process dynamics (Carmo et al., 2013; Anwar et al., 2021).

1.4.2 SOLID-OXIDE ELECTROLYSIS (SOE)

The SOE is a technology employed for water steam electrolysis, and it acts in reverse on the process of the solid oxide fuel cell (SOFC) (Hauch et al., 2008). Figure 1.2 shows the SOE cell principle (Wang et al., 2014).

SOE is a technology that has gained significant attention for its ability to transform electrical energy into chemical energy to efficiently produce ultra-pure hydrogen. It was initially introduced in 1980 by Donitz and Erdle and is still in the developmental stage with ongoing research (Carmo et al., 2013; Ursua et al., 2011; Seetharaman et al., 2013; Vermeiren et al., 1998). SOE works at a high-temperature range of 500–1000°C, which is comparable to a nuclear reactor's output temperature. Solid ceramic electrolytes are used in SOE, making it compact and responsive like a polymer electrolyte membrane cell. However, high-temperature requirements mean that heat input is also necessary. Moreover, SOE can operate at high pressures of up to 3 MPa, making it commercially feasible for regular hydrogen production. Strontium-doped lanthanum magnetite and nickel-zirconia are commonly used as cathodic materials in solid oxide electrolysis. Zirconia is usually used as an anode material. The cathode and anode have exchange current densities of 2×10^{-1} and

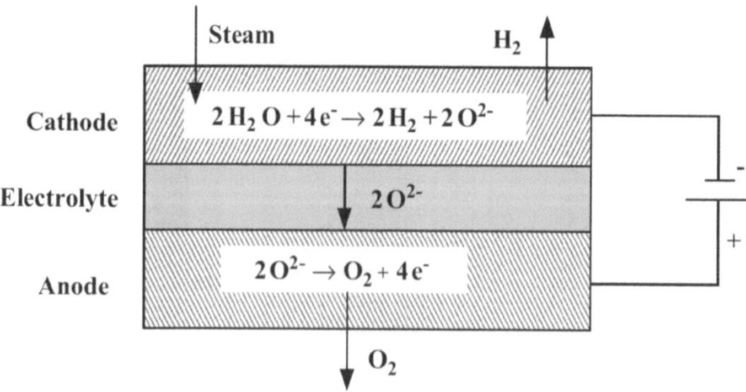

FIGURE 1.2   SOE cell illustration (Wang et al., 2014).

5.3×10^{-1} A/cm^2, respectively. However, due to the cooling requirement, the water consumption in this process is 83.3 times greater than the amount of hydrogen produced (Pinsky et al., 2020; Anwar et al., 2021; Chan and Xia, 2002; Sapountzi et al., 2017; Rashid et al., 2015).

SOFCs commonly use zirconia doped with yttria (YSZ) as the conventional solid electrolyte due to its excellent conductivity and stability, although scandia-doped zirconia, LaGaO$_3$-based, and ceria-based materials are utilized as well (Jiang et al., 2007; Yu et al., 2012; Etsell and Flengas, 1970; Badwal et al., 2000; Hirano et al., 2000; Zhu et al., 2006). Porous electrode materials are often used to enhance gas diffusion and transport rates, these materials must possess high catalytic activity and be resistant to degradation. Ni/YSZ is the most commonly used cathode material for the HER, while titanate/ceria, samaria-doped ceria (SDC), and $(La_{0.75}Sr_{0.25})_{0.95}Mn_{0.5}Cr_{0.5}O_3$(LSCM)/YSZ are also reported as alternatives (Hong et al., 2005; O'Brien et al., 2005; Hong et al., 2008; Ebbesen et al., 2011; Osada et al., 2006; Marina et al., 2007; Yang and Irvine, 2008). For anode materials, lanthanum strontium manganite (LSM)-YSZ, which is also utilized in SOFC, is considered the most suitable option, while $Ce_{0.6}Gd_{0.4}O_{1.8}$ (CG4) and $Ln_2NiO_{4+\gamma}$ (Ln = La, Nd, Pr) are alternatives (Liang et al., 2009; Hauch et al., 2008; Chen and Ai, 2012; Marina et al., 1999; Chauveau et al., 2010; Chauveau et al., 2011). Previous investigations have thoroughly examined the crucial materials for the solid electrolyte, cathode, and anode (Hauch et al., 2008; Ni et al., 2008; Laguna-Bercero, 2012). SOE cell stacks usually use a tubular design for greater mechanical strength or a planar design for improved manufacturability (Ni et al., 2008). The planar SOE cell stacks offer several advantages, including distributing gas species more uniform and easier production of mass as reported by Hino et al. (2004) and Wang et al. (2014).

Currently, extensive research has focused on developing ceramic proton-conducting materials for use in SOE fabrication. Compared to O$_2-$ conductor materials, these materials have better ionic conductivity and higher efficiency. However, SOE technology suffers from drawbacks such as electrode degradation and the use of ceramic electrolytes at temperatures exceeding 1,000°C, which reduce its strength and stability. In contrast, proton exchange membrane electrolysis (PEME) technology employs a solid ceramic membrane operating at temperatures ranging from 500°C to 1,000°C, producing better quality hydrogen than SOE (Sapountzi et al., 2017; Rashid et al., 2015; Anwar et al., 2021).

1.4.3 PROTON EXCHANGE MEMBRANE ELECTROLYSIS (PEME)

PEME has the potential to convert renewable energy sources into pure hydrogen gas with high efficiency. In 1966, PEME was first developed as an alternative to AWE to overcome its limitations. It is also known as polymer electrolyte membrane electrolysis. In contrast to the AWE system, PEME utilizes a polymer electrolyte to prevent the mixing of liquids (Sapountzi et al., 2017; He et al., 2021; Baykara, 2018; Grubb, 1959; Grubb and Niedrach, 1960). PEME operates in a low-temperature range of 20°C–100°C and converts liquid water into hydrogen and oxygen (Rashid et al., 2015). High-pressure operation (up to 40MPa) is typical in PEME systems, which reduces energy demand due to compression. The proton, a charge carrier, is allowed to pass through the membrane in the PEME, while other gases are prevented from transferring. The current density of PEME can be high, ranging from 0.6 to 3 A/cm^2, owing to its low ionic resistance (Carmo et al., 2013; Holladay et al., 2009; Siracusano et al., 2018). Furthermore, similar to AWE, the total consumed water in PEME is approximately 11.5 times higher than the total hydrogen generated, and cooling requirements through the process are minimal (Pinsky et al., 2020; Anwar et al., 2021).

1.4.3.1 Electrocatalysts for PEME

In PEME of water, noble metal-based electrocatalysts like Pt/Pd-based catalysts are widely utilized as cathodes for HER and RuO_2/IrO_2 catalysts as anodes for OER (Xu and Scott, 2010; Giacomini et al., 2003; Rozain et al., 2016a, Rozain et al., 2016b, Yin et al., 2018). Although Pt is well-known for its excellent HER activity and great stability in acidic environments, highly dispersed carbon-supported Pt-based materials have recently been considered the benchmark catalysts for HER in the PEME of water (Xu and Scott, 2010). However, due to their high cost, current research has focused on decreasing the electrocatalyst's price and operational costs while improving their particular effectiveness and resistance (Kumar and Himabindu, 2019).

Alternative electrode materials such as Rh, Rb, Pd, Au, and their oxides, including RhO_2, RbO_2, and PbO_2, have been investigated for their potential use in PEME. Nevertheless, these substances are costly compared to those used in alkaline water electrolysis (Kumar and Himabindu, 2019; Liu et al., 2018; Terezo and Pereira, 1999; Wu et al., 2011; Matz and Calatayud, 2017).

Numerous studies have investigated alternative electrocatalysts for the HER. According to Hinnemann et al. (2005), MoS_2 was found to be an acceptable catalyst for HER. Recently, Sarno and Ponticorvo (2019) synthesized and experimentally studied RuS_2@MoS_2 electrocatalyst to enhance hydrogen generation rates and current density. The synthesized RuS_2@MoS_2 catalyst demonstrated remarkable properties (Kumar and Himabindu, 2019).

Currently, Pd-based electrocatalysts are being researched as a substitute for Pt-based catalysts in the HER, owing to their exceptional electrocatalytic performance for various oxidation and reduction reactions (Lam et al., 2015; Sarkar and Peter, 2018; Escudero et al., 2002). Pd carbon nanotubes (Pd/CNTs) have been examined as well, although the obtained results have not shown a significant difference in comparison with Pt/CNTs (Millet et al., 2009; Grigoriev et al., 2011). To enhance the electrocatalytic activity and stability, doped carbon nanoparticles with hetero atoms (N, P, S, and B) have been widely utilized as a support material to transport electrons for noble metal electrodes in the HER and OER (Wu et al., 2014; Liu et al., 2014; Terrones et al., 2002; Yeh et al., 2018; Zhou et al., 2016). To lower the cost of HER catalysts even further, alternative carbon-supported electrocatalysts composed of low-cost and abundant earth materials such as A-Ni-C, Ni_2P/CNTs, Mo_2C/CNTs, Co-doped WO_2/C nanowires, FeS_2/CNTs, and CoFe nanoalloys encapsulated in N-doped graphene have been thoroughly investigated as possible replacements for Pt in HER (Chen et al., 2013; Fan et al., 2016; Wang et al., 2015; Barman and Nanda, 2018; Wu et al., 2015; Pan et al., 2015). Nevertheless, there are limited reports of these Pt-free HER catalysts supported on carbon and evaluated in PEM water electrolysis.

Overall, various electrocatalysts have been created and researched as alternatives to Pt-based catalysts for the HER in PEME of water (Kumar and Himabindu, 2019).

1.4.4 Carbon Assisted Water Electrolysis (CAWE)

CAWE was suggested as a substitute technique for generating hydrogen by Coughlin and Farooque three decades ago (Coughlin and Farooque, 1979). In CAWE, hydrogen is still generated on the cathode, while the anodic reaction is substituted by equation (1.9) (Wang et al., 2014):

$$\text{Anode: } C + 2H_2O(l) = CO_2(g) + 4H^+ + 4e^- \, E_a^\theta = 0.21 \text{ V} \tag{1.9}$$

$$\text{Total Reaction: } C + 2H_2O(l) = CO_2(g) + 2H_2(g) \, U^\theta = 0.21 \text{ V} \tag{1.10}$$

Unlike regular water electrolysis, where oxygen is evolved at the anode, carbon is oxidized to CO_2 in the anodic reaction of CAWE. The energy consumption and theoretical decomposition voltage for CAWE are lower than those of regular water electrolysis. The theoretical decomposition voltage is only 0.21 V, and the energy required for producing hydrogen is 40.2 kJ/mol. However, the actual operating voltage is higher than the theoretical value because of the reaction overpotential and ohmic voltage drop, and coal, graphite, and activated carbon are commonly used as anodic carbon sources. According to Coughlin and Farooque (Coughlin and Farooque, 1979, 1980a, b), when coal is used as a carbon source, the real operating voltage for practical hydrogen production is typically 0.8–1.0 V, which leads to a 33%–50% reduction in energy consumption compared to regular water electrolysis. Although the use of coal as a carbon source can reduce the cost of hydrogen production, its complex composition and structure, which contain S, O, N, and metal elements in various forms (e.g., 3OH, -S-, –COOH, -NH-, MS_y, and MO_x) (Aho and Pirkonen, 1995; Liya et al., 1999; Wei et al., 2008; Chen et al., 2012), limit the reactivity of carbon during the anodic reaction. In particular, the benzene rings and long carbon chains in coal are difficult to oxidize, which further restricts carbon reactivity. The best carbon material for hydrogen production through CAWE is nanocarbon (BP2000) with a high surface area (Seehra and Bollineni, 2009). At an operating voltage of 1.12 V, this carbon material exhibited a ten-fold increase in the HER compared to carbon GX203. Besides, adding $FeSO_4$ liquid catalyst further reduced the cell voltage to 0.72 V without decreasing the hydrogen evolution rate (Wang et al., 2014).

1.4.5 Microbial Electrolysis

The microbial electrolysis cell (MEC) technology is able to produce hydrogen by using organic matter, such as renewable biomass and wastewater. This technique is similar to that of microbial fuel cells (MFCs), but it operates in reverse. In 2005, two separate research organizations, Penn State University and Wageningen University, Netherlands, introduced the first MEC (Kadier et al., 2016b; Liu et al., 2005). MECs generate chemical energy and produce hydrogen from organic matter by converting

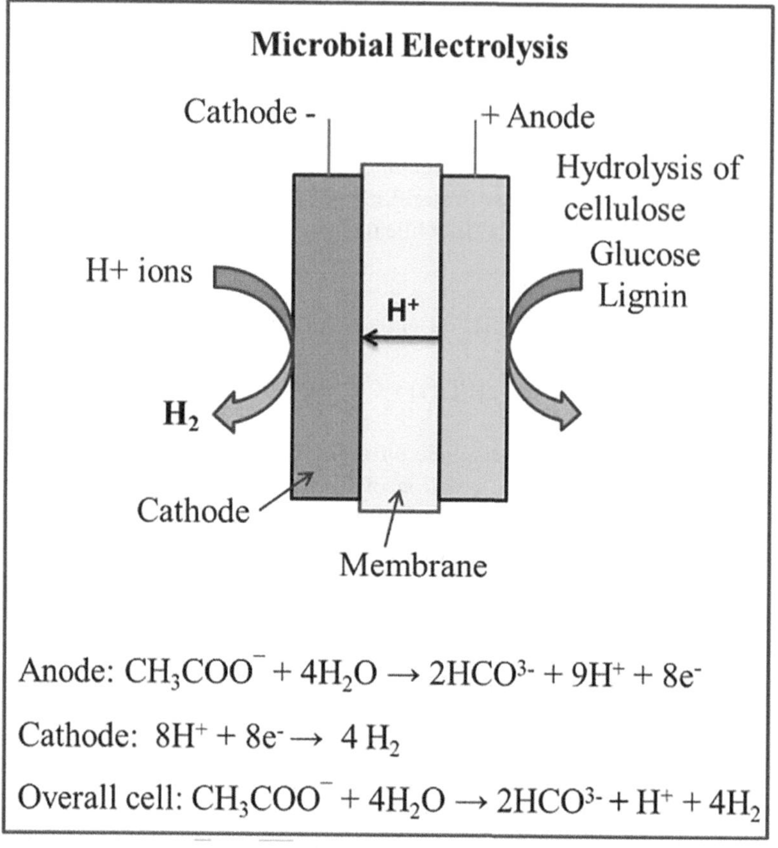

FIGURE 1.3　Microbial electrolysis illustration (Kumar and Himabindu, 2019).

electrical energy using an electric current. On the anode side of the microbial electrolysis process, microbes oxidize the substrate and produce CO_2, electrons, and protons. MECs generate chemical energy and produce hydrogen from organic matter by converting electrical energy, which they combine with electrons to produce hydrogen. Figure 1.3 illustrates the principle of MEC. However, the produced electrochemical potential in the oxidation period inside the anode of MEC is not sufficient to produce the reduced voltage needed for the HER at the cathode, necessitating an additional voltage (0.2–1.0 V). Compared to water electrolysis, MECs require a small amount of external voltage (Rathinam et al., 2018). Nevertheless, there are a number of hurdles that need to be addressed before this technology can be commercialized, including issues with hydrogen generation rate, high internal resistance, electrode materials, and complex design (Kadier et al., 2016a; Kumar and Himabindu, 2019).

1.4.6　Additives in Electrolytes

Recently, a growing interest has been seen in utilizing ionic activators in electrolytes for HERs. Many of these activators are comprised of ethylenediamine-based metal chloride complexes ($[M(en)_3]Cl_x$, $M = Co$, Ni, etc.) and Na_2MoO_4 or Na_2WO_4 (Kaninski et al., 2004; Kaninski et al., 2007; Kaninski et al., 2011a; Maksic et al., 2011; Kaninski et al., 2011b; Stojić et al., 2003; Nikolic et al., 2010; Tasic et al., 2011). Metal composites are deposited on the surface of the cathode during electrolysis, and they show enhanced catalytic activity for HERs than composites deposited ex-situ. Stojić et al. (2003) utilized tris(ethylenediamine)cobalt(III) chloride complex ($[Co(en)_3]Cl_3$) or tris(trimethylenediamine)cobalt(III) chloride complex ($[Co(tn)_3]Cl_3$) as ionic activators in KOH solutions to catalyze HER, saving energy up to 10% in comparison with nonactivated electrolytes. A mixture of Na_2WO_4 and $[Co(en)_3]Cl_3$ as an ionic activator has been utilized by Nikolic et al. (2010), which reduced the needed energy by 15%. The procedure for enhanced water electrolysis was attributed to the cooperative influence of cobalt and tungsten on the surface of the electrode, resulting from the high specific surface area. Tasic et al. (2011) employed Na_2MoO_4 and $[Ni(en)_3]Cl_2$ as ionic activators and further confirmed that the superior catalytic characteristics were not only due to the high surface area but also the true catalytic effect (Wang et al., 2014).

Ionic liquids have high conductivity and do not react with metal electrodes. $BMIBF_4$ was employed by De Souza et al. (2006) as a conductive electrolyte for the process of water electrolysis to produce hydrogen. The catalytic performance of the electrodes remained unaffected, and the mechanism of the HER was consistent across various metal electrodes. Optimization was performed on types of ionic liquid, electrode materials, and concentration in water, leading to the achievement of the highest efficiency of approximately 99% in a solution of 90 vol% water and 10 vol% $BMIBF_4$. Among the tested electrodes, the Mo electrode showed better efficiency, stability, and lower cost than the Pt electrode. A molecular electrocatalyst ($[Ni(P_2N_2)_2](BF_4)_2$) in a concentrated acidic ionic liquid/water-electrolyte to facilitate the reduction of protons to H_2 has been developed by Pool et al. (2012). Table 1.1 presents a comparison of various technologies employed for water electrolysis (Wang et al., 2014).

1.5 COUPLING PHOTOVOLTAIC PANELS TO WATER ELECTROLYZERS

One approach for producing renewable hydrogen that has the potential for commercial viability is to combine photovoltaic panels with electrolyzers in solar-hydrogen hybrid systems (Ngoh and Njomo, 2012; Bhattacharyya et al., 2017) (Figure 1.4). The energy produced by the panels is utilized to power the electrolysis process in these systems. Nevertheless, these systems' current efficiency is not cost-effective when compared to hydrogen obtained from fossil fuels. While an exhibit model was created that attained a maximum solar-to-hydrogen efficiency of 31% by combining two proton exchange membrane electrolyzers with high-performance triple-junction solar cells in sequence, the authors noted that this efficiency would likely be lower in a commercial-sized setup (Jia et al., 2016; Burton et al., 2021).

An alternative strategy to combine photovoltaic panels with an electrolyzer involved a regulator in the electrical circuit to maintain a constant voltage supply to the electrolyzer. Nevertheless, the method needs the photovoltaic panels to perform

TABLE 1.1

Advantages and Disadvantages of Various Electrolysis Technologies of Water (Kumar and Himabindu, 2019)

Electrolysis Process	Advantages	Disadvantages
Alkaline electrolysis	Well-established technology, non-noble electrocatalysts, low-cost technology, the energy efficiency is (70%–80%) Commercialized	Low current densities, formation of carbonates on the electrode decreases the performance of the electrolyzer, low purity of gases, low operational pressure (3–30 bar), low dynamic operation
Solid oxide electrolysis	Higher efficiency (90%–100%), non-noble electrocatalysts, high working pressure	Laboratory stage, large system design, low durability
Microbial electrolysis	Uses different organic wastewaters	Under development, low hydrogen production rate, and low purity of hydrogen
Proton exchange membrane electrolysis	High current densities, compact system design, and quick response, greater hydrogen production rate with, high purity of gases (99.99%), higher energy efficiency (80%–90%), high dynamic operation	New and partially established, high cost of components, acidic environment, low durability, commercialization is in the near term

at a reduced power capacity to consistently produce hydrogen (Ganeshan et al., 2016). Other efforts have been made to improve the photovoltaic panel's efficiency and electrolyzer coupling, such as aligning the photovoltaic panels' power output with the essential input power of the electrolysis cells (Gibson and Kelly, 2010; Rahim et al., 2015). However, the efficiency of the system and the impact of solar irradiance variability on photovoltaic panel output were not addressed by the authors of these studies. The efficiency of currently available solar-hydrogen hybrid systems utilized for producing renewable hydrogen is mainly low, with the coupling efficiency typically less than 12% and in some cases as low as 2.3%. This low efficiency contributes to higher production costs, which limit the commercial viability of these systems (Yilanci et al., 2009; Jia et al., 2016; Abdin et al., 2015; Zini and Tartarini, 2009). While studies have proposed potential techniques to increase the efficiency of solar-hydrogen hybrid systems, currently available systems that are commercially viable are not yet competitive in terms of cost with hydrogen derived from fossil fuels (Burton et al., 2021).

Developing photovoltaic panels with low costs and great efficiency and electrolyzers is critical to enhancing the efficiency and cost competitiveness of solar-hydrogen systems (Jia et al., 2016). Achieving maximum efficiency in solar-hydrogen hybrid systems requires electrolyzers to use the maximum power output of photovoltaic panels while they are operating at optimal efficiency (Yang et al., 2015). In renewable energy systems, the amount of hydrogen produced is often dependent on the

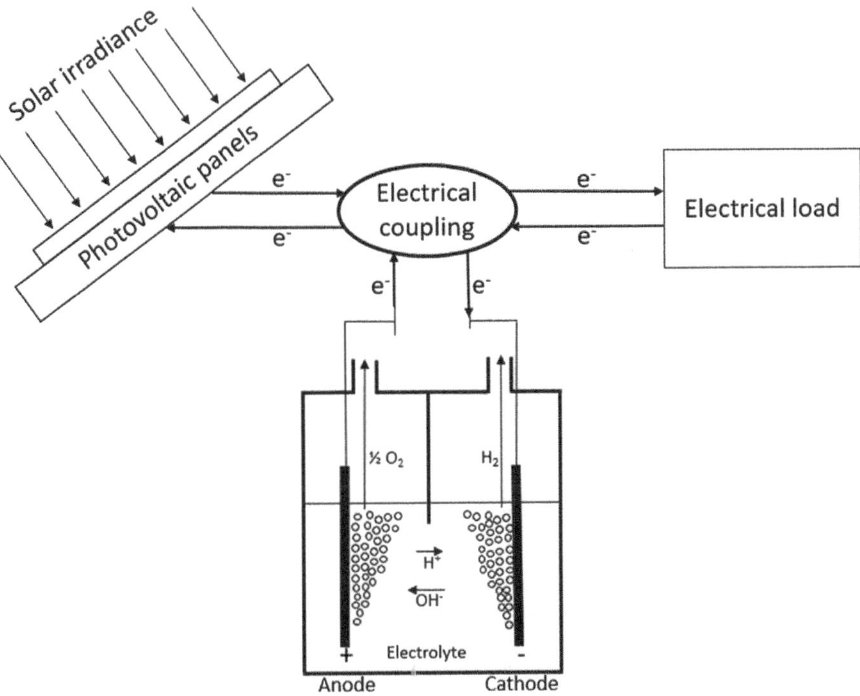

FIGURE 1.4 A straightforward diagram illustrating the integration of photovoltaic panels with an electrolyzer and an electrical load (Burton et al., 2021).

input energy, which can vary. Therefore, it is important to integrate the operating parameters of renewable power systems with the needed power for the electrolyzers (Jia et al., 2016; Ursua et al., 2011). Therefore, there is a need for a system that couples electrolyzers with renewable energy sources effectively (Burton et al., 2021).

1.6 PARAMETERS AFFECTING THE ELECTROLYSIS PROCESS

1.6.1 Temperature

Enhanced performance of electrolyzers can be achieved by operating them at temperatures around 80°C. Such high temperatures provide thermodynamic benefits by reducing the total electricity demand with an increase in thermal energy input. The electrochemical activity and conductivity of the membrane improve, leading to a decrease in electricity demand during electrolysis (Rashid et al., 2015; Selamet et al., 2013; Bazarah et al., 2022).

1.6.2 The Effects of Magnetic Fields and High Voltage Electric Fields

Research has shown that applying a magnetic field to the electrolysis process can have a positive impact on efficiency. By inducing a magnetic field, the hydrogen adsorption from the electrodes is enhanced, leading to a reduction in bubble size and preventing the gases from building up on the electrodes (Matsushima et al., 2013; Koza et al., 2009; Bidin et al., 2017b; Lin and Hourng, 2014; Liu et al., 2019a; Lin et al., 2012). This increase in hydrogen desorption is thought to be due to extra forces imposed on the gases and increased tensional strain on covalent and hydrogen bonds (Koza et al., 2009; Bidin et al., 2017b). The hydrogen evolution process can be enhanced by an induced magnetic field with decreasing resistance produced from gas buildup within the electrodes and creating further active sites on them (Koza et al., 2011; Lin et al., 2012). Additionally, Koza et al. (2011) discovered that the magnetic field resulted in depolarization of the electrodes and that its orientation had no effect on its effectiveness (Burton et al., 2021).

It was shown by several studies that the induced magnetic field can enhance the electrolysis efficiency by creating an upward force on the gases due to the combination of the Lorentz force and buoyancy (Lin et al., 2012; Kaya et al., 2017). The effectiveness of magnetizing different electrode types was found to depend on the magnetic fields' distribution, with ferromagnetic materials being the most effective. According to Wang et al. (2013), the electrolysis process efficiency may be enhanced by inducing a magnetic field, which applies rotational force to the H_2O molecule due to the interaction of its charge with the Lorentz force. The magnetic field reduces the friction produced by hydrogen bonding, resulting in an increase in hydrogen production. Studies have shown that a magnetic field can lead to increased hydrogen production when the distance between electrodes is decreased. Moreover, such studies have revealed that the magnetic field reduces the concentration of electrolytes, which can reduce the alkaline electrolysis's corrosive nature and operating costs associated with electrode corrosion (Lin et al., 2012, 2017a, b; Kaya et al., 2017). Novel studies have also explored the use of magnetic fields with hydrophobic coatings on electrodes (Liu et al., 2019a), foam electrodes (Liu et al., 2019b), or pulsating electric fields (Liu et al., 2019a), all of which have reported a boost in hydrogen generation because of their improved flow rate exiting the system (Burton et al., 2021).

1.6.3 ULTRASOUND FIELD

Although a promising technique for clean and safe hydrogen generation, sonoelectrolysis, which combines ultrasound with electrolysis, has received limited experimental investigation for hydrogen production using water sonolysis. Henglein and Kormann (1985) conducted one of the early experimental studies under oxygen, argon, and oxygen-argon mixtures using ultrasound at 300 kHz and 35 W/cm² (Merabet and Kerboua, 2022).

Pollet and his colleagues are considered to be the pioneers of sonoelectrolysis, having conducted numerous studies to explain how ultrasound can lower the overpotential and ohmic voltage, resulting in higher rates of hydrogen generation (Zore et al., 2021; Pollet, 2012; Islam et al., 2020). For instance, Wang and Chen (2009) exposed water-alkaline solutions of varying NaOH concentrations (0.1, 0.5, and 1 M) to a Ti electrode plated with RuO_2 and IrO_2. Their research found that the efficiency of hydrogen generation was improved by 5%–18%, while energy consumption was reduced by 10%–25% at low electrolyte concentrations and high current densities. Similarly, Symes (2011) and Zadeh (2014) carried out similar studies using mild acidic (H_2SO_4) and alkaline (NaOH and KOH) solutions and various electrode materials (e.g., glassy carbon, platinum, 316 stainless steel, industrial carbon, and nickel-based electrodes) at 20 kHz. The findings of both investigations indicated that hydrogen production was enhanced due to increased mass transport, electrode cleaning, and degassing. Zadeh (2014) showed that the use of ultrasound with the electrolysis process increases H_2 production by 14% and 25% for NaOH and KOH electrolytes, respectively. Recently, Kerboua et al. (2021b) studied the sonolytic pathway for hydrogen generation using a combination of modeling and experimental methods. The study involved varying the acoustic frequencies from 20 to 488 kHz. The experimental findings showed that the maximum electrical-to-acoustic energy conversion efficiency was 61.25% at a frequency of 20 kHz (Merabet and Kerboua, 2022).

In contrast to experimental studies, there are numerous works in the literature that focus on modeling and simulating the sonochemical hydrogen production process. Most of these studies use the "single bubble model," which considers a single bubble as the representative unit of the solution being sonicated (Merouani et al., 2015a; Kerboua et al., 2021a; Merouani et al., 2016; Merouani et al., 2015b; Kerboua and Hamdaoui, 2018a, b). Merouani et al. (2015a) proposed a computational model to show the process of hydrogen generation in water sonication by combining the dynamics of bubbles with chemical kinetics and incorporating 25 reversible chemical reactions occurring in the collapse phase. Upon exposure to ultrasonic waves, water generates hydrogen through water sonolysis, which involves the creation, growth, and violent collapse of tiny bubbles. The production of hydrogen using sonochemistry is affected not only by the chemical mechanism involved but also by various parameters such as the sonication pathway, ultrasound irradiation modes, and configurations of ultrasonic wave sources. These parameters have been extensively investigated in the literature (Torres-Palma and Serna-Galvis, 2018; Mason, 2003; Sancheti and Gogate, 2017). Merouani and Hamdaoui (2020) conducted a review of available research on hydrogen production using ultrasound, examining the factors that affect it. Similarly, Torres-Palma and Serna-Galvis (2018) studied the effect of various operating parameters on hydrogen generation through sonochemical reactions. Many studies have investigated how operating parameters affect sonochemical hydrogen generation (Merabet and Kerboua, 2022).

1.6.4 THE EFFECTS OF LIGHT ENERGY

Prior research has suggested using a green laser during electrolysis can boost hydrogen production through the induction of an electric field (Bidin et al., 2014, 2017b). However, the narrow width of the applied force has been identified as a limiting factor. To address this, it has been suggested that additional lasers or a laser with a wider beam could be used to enhance hydrogen production even further. Combining a magnetic field with the green laser has been shown to increase hydrogen production by ninefold (Bidin et al., 2017b). However, energy consumption by the green laser has not been compared to the extra potential energy generated by the process. In addition, a study showed that both sunlight and concentrated sunlight can increase hydrogen production, with concentrated sunlight leading to a greater improvement of approximately 53% compared to normal sunlight's 31% enhancement (Bidin et al., 2017a). However, using a green laser to simulate collimated sunlight may yield different results than natural collimated sunlight. Enhancement in hydrogen generation is attributed to light energy and is thought to be because of the electric field induced by the green laser that polarizes water molecules in a way that is conducive to electrolysis by placing water molecules in an orientation favorable to electrolysis (Bidin et al., 2017b; Burton et al., 2021).

1.6.5 The Effects of Pulsating the Applied Electric Field

Studies that explored the impact of pulsing the applied electric field on water electrolysis efficiency have also produced mixed results, similar to those investigating the effect of external magnetic fields. However, the majority of the research studies that investigated the impact of varying the intensity and frequency of the electric field through pulsing demonstrated that it enhanced the efficiency of hydrogen production (Demir et al., 2018; Vilasmongkolchai et al., 2016; Vincent et al., 2018; Lin et al., 2012; Huang, 2013; Shimizu et al., 2006). These studies revealed various frequencies, waveforms, and applied power intensities that led to improved hydrogen production in the specific setups they were using. Nevertheless, these variations in results were expected since the application of pulsing electric fields was tested on different arrangements of systems (Burton et al., 2021).

Demir et al. (2018) and Vilasmongkolchai et al. (2016) suggested that the efficiency enhancement was due to a resting time that allowed gas bubbles to escape from the electrodes, decreasing ohmic resistance resulting from the accumulation of gas, and increasing the active electrode area. Meanwhile, Vincent et al. (2018) suggested that by using pulsed power, the formation of the electrodynamic diffusion layer (EDL) can be prevented, resulting in a smaller EDL and less electrical impedance, therefore decreasing the power needed to perform the reaction. This theory is consistent with Shimizu et al. (2006) previous findings that ultrashort pulses hinder electric double-layer formation, reducing overall power consumption. Although both theories may explain the enhanced efficiency with a pulsing electric field, investigating the molecular mechanisms that may reduce electrolysis efficiency is interesting. Using sunlight as an energy source to increase generation production without additional input energy is a promising approach. By integrating sunlight receivers into photovoltaic panels to introduce light into the electrolysis chamber, the efficiency of solar-hydrogen hybrid systems could be improved through this approach. Solar-hydrogen hybrid systems are advantageous because both system components work simultaneously, even when there are variations in solar irradiance, which is challenging. Despite the limited studies that have investigated the effects of light energy on water electrolysis, the findings discussed above exhibit the potential for the use of light energy to boost electrolysis effectiveness, necessitating further research into the possible benefits of this approach (Burton et al., 2021).

1.7 CONCLUSION

The production of hydrogen through water electrolysis is a promising avenue for clean energy. The efficiency of the process has been the subject of numerous studies, and several methods have been proposed to enhance it. One such method is the use of pulsing electric fields, which has been shown to reduce power consumption and increase the active area on electrodes. Another approach is the application of light energy, which has the potential to increase the efficiency of electrolysis and complement solar-hydrogen hybrid systems. However, more research is needed to fully understand the molecular mechanisms behind these methods and to optimize their implementation.

Furthermore, the choice of electrolyte plays a crucial role in the efficiency of water electrolysis. The use of alkaline electrolytes is common, but acidic electrolytes have also shown promise. The use of proton exchange membranes has been proposed as an alternative to traditional electrolytes, as they offer advantages such as improved selectivity and reduced gas crossover. Additionally, the use of electrocatalysts can greatly enhance the efficiency of water electrolysis, and several materials, such as Pt, Ni, and Co, have been investigated.

In conclusion, water electrolysis holds great potential for clean and sustainable hydrogen production. While several methods have been proposed to enhance the efficiency of the process, more research is needed to fully optimize their implementation and understand their molecular mechanisms. The choice of electrolyte and electrocatalyst also plays important roles in the efficiency of water electrolysis, and their optimization can further improve the process.

REFERENCES

Abdin, Z., Webb, C. & Gray, E. 2015. Solar hydrogen hybrid energy systems for off-grid electricity supply: A critical review. *Renewable and Sustainable Energy Reviews*, 52, 1791–1808.

Abe, J. O., Popoola, A., Ajenifuja, E. & Popoola, O. M. 2019. Hydrogen energy, economy and storage: Review and recommendation. *International Journal of Hydrogen Energy*, 44, 15072–15086.

Aho, M. J. & Pirkonen, P. M. 1995. Effects of pressure, gas temperature and CO2 and O2 partial pressures on the conversion of coal-nitrogen to NO, N2O and NO2. *Fuel*, 74, 1677–1681.

Anwar, S., Khan, F., Zhang, Y. & Djire, A. 2021. Recent development in electrocatalysts for hydrogen production through water electrolysis. *International Journal of Hydrogen Energy*, 46, 32284–32317.

Badwal, S., Ciacchi, F. & Milosevic, D. 2000. Scandia-zirconia electrolytes for intermediate temperature solid oxide fuel cell operation. *Solid State Ionics*, 136, 91–99.

Balat, M. 2008. Possible methods for hydrogen production. *Energy Sources, Part A: Recovery, Utilization, and Environmental Effects*, 31, 39–50.

Barman, B. K. & Nanda, K. K. 2018. CoFe nanoalloys encapsulated in N-doped graphene layers as a Pt-free multifunctional robust catalyst: Elucidating the role of Co-alloying and N-doping. *ACS Sustainable Chemistry & Engineering*, 6, 12736–12745.

Basye, L. & Swaminathan, S. 1997. Hydrogen production costs—A survey. Sentech, Inc., Bethesda, MD.

Baykara, S. Z. 2018. Hydrogen: A brief overview on its sources, production and environmental impact. *International Journal of Hydrogen Energy*, 43, 10605–10614.

Bazarah, A., Majlan, E. H., Husaini, T., Zainoodin, A., Alshami, I., Goh, J. & Masdar, M. S. 2022. Factors influencing the performance and durability of polymer electrolyte membrane water electrolyzer: A review. *International Journal of Hydrogen Energy*, 47, 35976–35989.

Bhattacharyya, R., Misra, A. & Sandeep, K. 2017. Photovoltaic solar energy conversion for hydrogen production by alkaline water electrolysis: Conceptual design and analysis. *Energy Conversion and management*, 133, 1–13.

Bidin, N., Azni, S. R., Bakar, M. A. A., Munap, D. H. F. A., Salebi, M. F., Razak, S. N. A., Sahidan, N. S. & Sulaiman, S. N. A. 2017a. The effect of sunlight in hydrogen production from water electrolysis. *International Journal of Hydrogen Energy*, 42, 133–142.

Bidin, N., Azni, S. R., Islam, S., Abdullah, M., Ahmad, M. F. S., Krishnan, G., Johari, A. R., Bakar, M. A. A., Sahidan, N. S. & Musa, N. 2017b. The effect of magnetic and optic field in water electrolysis. *International Journal of Hydrogen Energy*, 42, 16325–16332.

Bidin, N., Razak, S. N. A., Azni, S. R., Nguroho, W., Mohsin, A. K., Abdullah, M., Krishnan, G. & Bakhtiar, H. 2014. Effect of green laser irradiation on hydrogen production. *Laser Physics Letters*, 11, 066001.

Burton, N., Padilla, R., Rose, A. & Habibullah, H. 2021. Increasing the efficiency of hydrogen production from solar powered water electrolysis. *Renewable and Sustainable Energy Reviews*, 135, 110255.

Carmo, M., Fritz, D. L., Mergel, J. & Stolten, D. 2013. A comprehensive review on PEM water electrolysis. *International Journal of Hydrogen Energy*, 38, 4901–4934.

Chakrabarty, B., Ghoshal, A. & Purkait, M. 2008. Preparation, characterization and performance studies of polysulfone membranes using PVP as an additive. *Journal of Membrane Science*, 315, 36–47.

Chan, S. & Xia, Z. 2002. Polarization effects in electrolyte/electrode-supported solid oxide fuel cells. *Journal of Applied Electrochemistry*, 32, 339–347.

Chauveau, F., Mougin, J., Bassat, J.-M., Mauvy, F. & Grenier, J.-C. 2010. A new anode material for solid oxide electrolyser: The neodymium nickelate Nd2NiO4+ δ. *Journal of Power Sources*, 195, 744–749.

Chauveau, F., Mougin, J., Mauvy, F., Bassat, J.-M. & Grenier, J.-C. 2011. Development and operation of alternative oxygen electrode materials for hydrogen production by high temperature steam electrolysis. *International Journal of Hydrogen Energy*, 36, 7785–7790.

Chen, K. & Ai, N. 2012. Enhanced electrochemical performance and stability of (La, Sr) MnO3-(Gd, Ce) O2 oxygen electrodes of solid oxide electrolysis cells by palladium infiltration. *International Journal of Hydrogen Energy*, 37, 1301–1310.

Chen, P., Xu, K., Fang, Z., Tong, Y., Wu, J., Lu, X., Peng, X., Ding, H., Wu, C. & Xie, Y. 2015. Metallic Co4N porous nanowire arrays activated by surface oxidation as electrocatalysts for the oxygen evolution reaction. *Angewandte Chemie*, 127, 14923–14927.

Chen, W.-F., Wang, C.-H., Sasaki, K., Marinkovic, N., Xu, W., Muckerman, J. T., Zhu, Y. & Adzic, R. 2013. Highly active and durable nanostructured molybdenum carbide electrocatalysts for hydrogen production. *Energy & Environmental Science*, 6, 943–951.

Chen, Y., Mastalerz, M. & Schimmelmann, A. 2012. Characterization of chemical functional groups in macerals across different coal ranks via micro-FTIR spectroscopy. *International Journal of Coal Geology*, 104, 22–33.

Coughlin, R. W. & Farooque, M. 1979. Hydrogen production from coal, water and electrons. *Nature*, 279, 301–303.

Coughlin, R. W. & Farooque, M. 1980a. Consideration of electrodes and electrolytes for electrochemical gasification of coal by anodic oxidation. *Journal of Applied Electrochemistry*, 10, 729–740.

Coughlin, R. W. & Farooque, M. 1980b. Electrochemical gasification of coal-simultaneous production of hydrogen and carbon dioxide by a single reaction involving coal, water, and electrons. *Industrial & Engineering Chemistry Process Design and Development*, 19, 211–219.

Demir, N., Kaya, M. F. & Albawabiji, M. S. 2018. Effect of pulse potential on alkaline water electrolysis performance. *International Journal of Hydrogen Energy*, 43, 17013–17020.

De Souza, R. F., Padilha, J. C., Gonçalves, R. S. & Rault-Berthelot, J. 2006. Dialkylimidazolium ionic liquids as electrolytes for hydrogen production from water electrolysis. *Electrochemistry Communications*, 8, 211–216.

Ebbesen, S. D., Høgh, J., Nielsen, K. A., Nielsen, J. U. & Mogensen, M. 2011. Durable SOC stacks for production of hydrogen and synthesis gas by high temperature electrolysis. *International Journal of Hydrogen Energy*, 36, 7363–7373.

Escudero, M., Hontanon, E., Schwartz, S., Boutonnet, M. & Daza, L. 2002. Development and performance characterisation of new electrocatalysts for PEMFC. *Journal of Power Sources*, 106, 206–214.

Etsell, T. & Flengas, S. N. 1970. Electrical properties of solid oxide electrolytes. *Chemical Reviews*, 70, 339–376.

Faber, M. S., Dziedzic, R., Lukowski, M. A., Kaiser, N. S., Ding, Q. & Jin, S. 2014. High-performance electrocatalysis using metallic cobalt pyrite (CoS2) micro-and nano-structures. *Journal of the American Chemical Society*, 136, 10053–10061.

Fan, L., Liu, P. F., Yan, X., Gu, L., Yang, Z. Z., Yang, H. G., Qiu, S. & Yao, X. 2016. Atomically isolated nickel species anchored on graphitized carbon for efficient hydrogen evolution electrocatalysis. *Nature communications*, 7, 10667.

Ganeshan, I. S., Manikandan, V., Sundhar, V. R., Sajiv, R., Shanthi, C., Kottayil, S. K. & Ramachandran, T. 2016. Regulated hydrogen production using solar powered electroly-ser. *International Journal of Hydrogen Energy*, 41, 10322–10326.

Gerken, J. B., Mcalpin, J. G., Chen, J. Y., Rigsby, M. L., Casey, W. H., Britt, R. D. & Stahl, S. S. 2011. Electrochemical water oxidation with cobalt-based electrocatalysts from pH 0-14: The thermodynamic basis for catalyst structure, stability, and activity. *Journal of the American Chemical Society*, 133, 14431–14442.

Giacomini, M. T., Balasubramanian, M., Khalid, S., Mcbreen, J. & Ticianellia, E. 2003. Characterization of the activity of palladium-modified polythiophene electrodes for the hydrogen oxidation and oxygen reduction reactions. *Journal of the Electrochemical Society*, 150, A588.

Gibson, T. L. & Kelly, N. A. 2010. Predicting efficiency of solar powered hydrogen generation using photovoltaic-electrolysis devices. *International Journal of Hydrogen Energy*, 35, 900–911.

Grigoriev, S., Mamat, M., Dzhus, K., Walker, G. & Millet, P. 2011. Platinum and palladium nano-particles supported by graphitic nano-fibers as catalysts for PEM water electroly-sis. *International Journal of Hydrogen Energy*, 36, 4143–4147.

Grimes, C. A., Varghese, O. K. & Ranjan, S. 2008. Hydrogen generation by water splitting. *Light, Water, Hydrogen: The Solar Generation of Hydrogen by Water Photoelectrolysis*, New York: Springer, 35–113.

Grubb, W. 1959. Batteries with solid ion exchange electrolytes: I. Secondary cells employing metal electrodes. *Journal of the Electrochemical Society*, 106, 275.

Grubb, W. & Niedrach, L. 1960. Batteries with solid ion-exchange membrane electrolytes: II. Low-temperature hydrogen-oxygen fuel cells. *Journal of the Electrochemical Society*, 107, 131.

Hauch, A., Ebbesen, S. D., Jensen, S. H. & Mogensen, M. 2008. Highly efficient high temperature electrolysis. *Journal of Materials Chemistry*, 18, 2331–2340.

He, H.-Z., Zhang, Y., Li, Y. & Wang, P. 2021. Recent innovations of silk-derived electrocatalysts for hydrogen evolution reaction, oxygen evolution reaction and oxygen reduction reaction. *International Journal of Hydrogen Energy*, 46, 7848–7865.

Henglein, A. & Kormann, C. 1985. Scavenging of OH radicals produced in the sonolysis of water. *International Journal of Radiation Biology and Related Studies in Physics, Chemistry and Medicine*, 48, 251–258.

Hinnemann, B., Moses, P. G., Bonde, J., Jørgensen, K. P., Nielsen, J. H., Horch, S., Chorkendorff, I. & Nørskov, J. K. 2005. Biomimetic hydrogen evolution: MoS2 nanoparticles as catalyst for hydrogen evolution. *Journal of the American Chemical Society*, 127, 5308–5309.

Hino, R., Haga, K., Aita, H. & Sekita, K. 2004. 38. R&D on hydrogen production by high-temperature electrolysis of steam. *Nuclear Engineering and Design*, 233, 363–375.

Hirano, M., Inagaki, M., Mizutani, Y., Nomura, K., Kawai, M. & Nakamura, Y. 2000. Mechanical and electrical properties of Sc2O3-doped zirconia ceramics improved by postsintering with HIP. *Solid State Ionics*, 133, 1–9.

Holladay, J. D., Hu, J., King, D. L. & Wang, Y. 2009. An overview of hydrogen production technologies. *Catalysis Today*, 139, 244–260.

Hong, H. S., Chae, U.-S. & Choo, S.-T. 2008. The effect of ball milling parameters and Ni concentration on a YSZ-coated Ni composite for a high temperature electrolysis cathode. *Journal of Alloys and Compounds*, 449, 331–334.

Hong, H. S., Chae, U.-S., Choo, S.-T. & Lee, K. S. 2005. Microstructure and electrical conductivity of Ni/YSZ and NiO/YSZ composites for high-temperature electrolysis prepared by mechanical alloying. *Journal of Power Sources*, 149, 84–89.

Huang, C. 2013. Solar hydrogen production via pulse electrolysis of aqueous ammonium sulfite solution. *Solar energy*, 91, 394–401.

Hwang, G.-J., Lim, S.-G., Bong, S.-Y., Ryu, C.-H. & Choi, H.-S. 2015. Preparation of anion exchange membrane using polyvinyl chloride (PVC) for alkaline water electrolysis. *Korean Journal of Chemical Engineering*, 32, 1896–1901.

Islam, M. H., Lamb, J. J., Burheim, O. S. & Pollet, B. G. 2020. Ultrasound-assisted electrolytic hydrogen production. *Micro-Optics and Energy: Sensors for Energy Devices*, 73–84. https://doi.org/10.1007/978-3-030-43676-6_7

Jaramillo, T. F., Jørgensen, K. P., Bonde, J., Nielsen, J. H., Horch, S. & Chorkendorff, I. 2007. Identification of active edge sites for electrochemical H2 evolution from MoS2 nanocatalysts. *Science*, 317, 100–102.

Jia, J., Seitz, L. C., Benck, J. D., Huo, Y., Chen, Y., Ng, J. W. D., Bilir, T., Harris, J. S. & Jaramillo, T. F. 2016. Solar water splitting by photovoltaic-electrolysis with a solar-to-hydrogen efficiency over 30%. *Nature Communications*, 7, 13237.

Jiang, Y., Gao, J., Liu, M., Wang, Y. & Meng, G. 2007. Fabrication and characterization of Y2O3 stabilized ZrO2 films deposited with aerosol-assisted MOCVD. *Solid State Ionics*, 177, 3405–3410.

Kadier, A., Kalil, M. S., Abdeshahian, P., Chandrasekhar, K., Mohamed, A., Azman, N. F., Logroño, W., Simayi, Y. & Hamid, A. A. 2016a. Recent advances and emerging challenges in microbial electrolysis cells (MECs) for microbial production of hydrogen and value-added chemicals. *Renewable and Sustainable Energy Reviews*, 61, 501–525.

Kadier, A., Simayi, Y., Abdeshahian, P., Azman, N. F., Chandrasekhar, K. & Kalil, M. S. 2016b. A comprehensive review of microbial electrolysis cells (MEC) reactor designs and configurations for sustainable hydrogen gas production. *Alexandria Engineering Journal*, 55, 427–443.

Kaninski, M. P. M., Maksić, A. D., Stojić, D. L. & Miljanić, Š. S. 2004. Ionic activators in the electrolytic production of hydrogen-cost reduction-analysis of the cathode. *Journal of Power Sources*, 131, 107–111.

Kaninski, M. P. M., Nikolić, V. M., Potkonjak, T. N., Simonović, B. R. & Potkonjak, N. I. 2007. Catalytic activity of Pt-based intermetallics for the hydrogen production-Influence of ionic activator. *Applied Catalysis A: General*, 321, 93–99.

Kaninski, M. P. M., Saponjic, D. P., Nikolic, V. M., Zugic, D. L. & Tasic, G. S. 2011a. Energy consumption and stability of the Ni-Mo electrodes for the alkaline hydrogen production at industrial conditions. *International Journal of Hydrogen Energy*, 36, 8864–8868.

Kaninski, M. P. M., Saponjic, D. P., Perovic, I. M., Maksic, A. D. & Nikolic, V. M. 2011b. Electrochemical characterization of the Ni-W catalyst formed in situ during alkaline electrolytic hydrogen production-Part II. *Applied Catalysis A: General*, 405, 29–35.

Karimi, A., Kazemi, N., Tavakoli, O. & Pirbazari, A. E. 2022. Catalytic supercritical water gasification of black liquor along with lignocellulosic biomass. *International Journal of Hydrogen Energy*, 47, 16729–16740.

Kaya, M. F., Demir, N., Albawabiji, M. S. & Taş, M. 2017. Investigation of alkaline water electrolysis performance for different cost effective electrodes under magnetic field. *International Journal of Hydrogen Energy*, 42, 17583–17592.

Kerboua, K. & Hamdaoui, O. 2018a. Numerical estimation of ultrasonic production of hydrogen: Effect of ideal and real gas based models. *Ultrasonics Sonochemistry*, 40, 194–200.

Kerboua, K. & Hamdaoui, O. 2018b. Ultrasonic waveform upshot on mass variation within single cavitation bubble: Investigation of physical and chemical transformations. *Ultrasonics Sonochemistry*, 42, 508–516.

Kerboua, K., Hamdaoui, O. & Al-Zahrani, S. 2021a. Sonochemical production of hydrogen: A numerical model applied to the recovery of aqueous methanol waste under o xygen-argon atmosphere. *Environmental Progress & Sustainable Energy*, 40, e13511.

Kerboua, K., Hamdaoui, O., Islam, M. H., Alghyamah, A., Hansen, H. E. & Pollet, B. G. 2021b. Low carbon ultrasonic production of alternate fuel: Operational and mechanistic concerns of the sonochemical process of hydrogen generation under various scenarios. *International Journal of Hydrogen Energy*, 46, 26770–26787.

Kothari, R., Buddhi, D. & Sawhney, R. 2008. Comparison of environmental and economic aspects of various hydrogen production methods. *Renewable and Sustainable Energy Reviews*, 12, 553–563.

Kovač, A., Paranos, M. & Marciuš, D. 2021. Hydrogen in energy transition: A review. *International Journal of Hydrogen Energy*, 46, 10016–10035.

Koza, J. A., Mühlenhoff, S., Uhlemann, M., Eckert, K., Gebert, A. & Schultz, L. 2009. Desorption of hydrogen from an electrode surface under influence of an external magnetic field-In-situ microscopic observations. *Electrochemistry Communications*, 11, 425–429.

Koza, J. A., Mühlenhoff, S., Żabiński, P., Nikrityuk, P. A., Eckert, K., Uhlemann, M., Gebert, A., Weier, T., Schultz, L. & Odenbach, S. 2011. Hydrogen evolution under the influence of a magnetic field. *Electrochimica Acta*, 56, 2665–2675.

Kumar, S. S. & Himabindu, V. 2019. Hydrogen production by PEM water electrolysis-A review. *Materials Science for Energy Technologies*, 2, 442–454.

Laguna-Bercero, M. A. 2012. Recent advances in high temperature electrolysis using solid oxide fuel cells: A review. *Journal of Power sources*, 203, 4–16.

Lam, B. T. X., Chiku, M., Higuchi, E. & Inoue, H. 2015. Preparation of PdAg and PdAu nanoparticle-loaded carbon black catalysts and their electrocatalytic activity for the glycerol oxidation reaction in alkaline medium. *Journal of Power Sources*, 297, 149–157.

Liang, M., Yu, B., Wen, M., Chen, J., Xu, J. & Zhai, Y. 2009. Preparation of LSM-YSZ composite powder for anode of solid oxide electrolysis cell and its activation mechanism. *Journal of Power Sources*, 190, 341–345.

Lin, M. Y. & Hourng, L. W. 2014. Effects of magnetic field and pulse potential on hydrogen production via water electrolysis. *International Journal of Energy Research*, 38, 106–116.

Lin, M.-Y., Hourng, L.-W. & Hsu, J.-S. 2017a. The effects of magnetic field on the hydrogen production by multielectrode water electrolysis. *Energy Sources, Part A: Recovery, Utilization, and Environmental Effects*, 39, 352–357.

Lin, M.-Y., Hourng, L.-W. & Kuo, C.-W. 2012. The effect of magnetic force on hydrogen production efficiency in water electrolysis. *International Journal of Hydrogen Energy*, 37, 1311–1320.

Lin, M.-Y., Hourng, L.-W. & Wu, C.-H. 2017b. The effectiveness of a magnetic field in increasing hydrogen production by water electrolysis. *Energy Sources, Part A: Recovery, Utilization, and Environmental Effects*, 39, 140–147.

Liu, B., Wang, C. & Chen, Y. 2018. Surface determination and electrochemical behavior of IrO2-RuO2-SiO2 ternary oxide coatings in oxygen evolution reaction application. *Electrochimica Acta*, 264, 350–357.

Liu, H., Grot, S. & Logan, B. E. 2005. Electrochemically assisted microbial production of hydrogen from acetate. *Environmental Science & Technology*, 39, 4317–4320.

Liu, H.-B., Hu, Q., Pan, L.-M., Wu, R., Liu, Y. & Zhong, D. 2019a. Electrode-normal magnetic field facilitating neighbouring electrochemical bubble release from hydrophobic islets. *Electrochimica Acta*, 306, 350–359.

Liu, Y., Pan, L.-M., Liu, H., Chen, T., Yin, S. & Liu, M. 2019b. Effects of magnetic field on water electrolysis using foam electrodes. *International Journal of Hydrogen Energy*, 44, 1352–1358.

Liu, Z., Shi, Q., Zhang, R., Wang, Q., Kang, G. & Peng, F. 2014. Phosphorus-doped carbon nanotubes supported low Pt loading catalyst for the oxygen reduction reaction in acidic fuel cells. *Journal of Power Sources*, 268, 171–175.

Liya, E. Y., Hildemann, L. M. & Niksa, S. 1999. Characteristics of nitrogen-containing aromatic compounds in coal tars during secondary pyrolysis. *Fuel*, 78, 377–385.

Makarem, M., Kiani, M., Abbaspour, M., Farsi, M. & Rahimpour, M. 2022. Nanoparticle-enhanced hydrogen separation from CO2 in cylindrical and cubic microchannels: A 3D computational fluid dynamics simulation. *International Journal of Hydrogen Energy*, 48(32), 12045-12055.

Maksic, A. D., Miulovic, S. M., Nikolic, V. M., Perovic, I. M. & Kaninski, M. P. M. 2011. Energy consumption of the electrolytic hydrogen production using Ni-W based activators-Part I. *Applied Catalysis A: General*, 405, 25–28.

Marina, O. A., Bagger, C., Primdahl, S. & Mogensen, M. 1999. A solid oxide fuel cell with a gadolinia-doped ceria anode: Preparation and performance. *Solid State Ionics*, 123, 199–208.

Marina, O. A., Pederson, L. R., Williams, M. C., Coffey, G. W., Meinhardt, K. D., Nguyen, C. D. & Thomsen, E. C. 2007. Electrode performance in reversible solid oxide fuel cells. *Journal of the Electrochemical Society*, 154, B452.

Mason, T. J. 2003. Sonochemistry and sonoprocessing: The link, the trends and (probably) the future. *Ultrasonics Sonochemistry*, 10, 175–179.

Matsushima, H., Iida, T. & Fukunaka, Y. 2013. Gas bubble evolution on transparent electrode during water electrolysis in a magnetic field. *Electrochimica Acta*, 100, 261–264.

Matz, O. & Calatayud, M. 2017. Periodic DFT study of rutile IrO2: Surface reactivity and catechol adsorption. *The Journal of Physical Chemistry C*, 121, 13135–13143.

Merabet, N. & Kerboua, K. 2022. Sonolytic and ultrasound-assisted techniques for hydrogen production: A review based on the role of ultrasound. *International Journal of Hydrogen Energy*, 47, 17879–17893.

Merouani, S. & Hamdaoui, O. 2020. The sonochemical approach for hydrogen production. *Sustainable Green Chemical Processes and Their Allied Applications*, Springer Nature Switzerland AG, 1–29.

Merouani, S., Hamdaoui, O., Rezgui, Y. & Guemini, M. 2015a. Mechanism of the sonochemical production of hydrogen. *International Journal of Hydrogen Energy*, 40, 4056–4064.

Merouani, S., Hamdaoui, O., Rezgui, Y. & Guemini, M. 2015b. Sensitivity of free radicals production in acoustically driven bubble to the ultrasonic frequency and nature of dissolved gases. *Ultrasonics Sonochemistry*, 22, 41–50.

Merouani, S., Hamdaoui, O., Rezgui, Y. & Guemini, M. 2016. Computational engineering study of hydrogen production via ultrasonic cavitation in water. *International Journal of Hydrogen Energy*, 41, 832–844.

Millet, P., Dragoe, D., Grigoriev, S., Fateev, V. & Etievant, C. 2009. GenHyPEM: A research program on PEM water electrolysis supported by the European Commission. *International Journal of Hydrogen Energy*, 34, 4974–4982.

Moravvej, Z., Soroush, E., Makarem, M. A. & Rahimpour, M. R. 2021. Thermochemical routes for hydrogen production from biomass. In *Advances in Bioenergy and Microfluidic Applications*. (pp. 193–208). Elsevier.

Mortensen, P. M., Grunwaldt, J.-D., Jensen, P. A., Knudsen, K. & Jensen, A. D. 2011. A review of catalytic upgrading of bio-oil to engine fuels. *Applied Catalysis A: General*, 407, 1–19.

Ngoh, S. K. & Njomo, D. 2012. An overview of hydrogen gas production from solar energy. *Renewable and Sustainable Energy Reviews*, 16, 6782–6792.

Ni, M., Leung, M. K. & Leung, D. Y. 2008. Technological development of hydrogen production by solid oxide electrolyzer cell (SOEC). *International Journal of Hydrogen Energy*, 33, 2337–2354.

Nikolaidis, P. & Poullikkas, A. 2017. A comparative overview of hydrogen production processes. *Renewable and Sustainable Energy Reviews*, 67, 597–611.

Nikolic, V. M., Tasic, G. S., Maksic, A. D., Saponjic, D. P., Miulovic, S. M. & Kaninski, M. P. M. 2010. Raising efficiency of hydrogen generation from alkaline water electrolysis-Energy saving. *International Journal of Hydrogen Energy*, 35, 12369–12373.

O'Brien, J., Stoots, C., Herring, J., Lessing, P., Hartvigsen, J. & Elangovan, S. 2005. Performance measurements of solid-oxide electrolysis cells for hydrogen production. *Journal of Fuel Cell Science and Technology*, 2(3), 156–163. https://doi.org/10.1115/1.1895946

Ohmori, T., Tachikawa, K., Tsuji, K. & Anzai, K. 2007. Nickel oxide water electrolysis diaphragm fabricated by a novel method. *International Journal of Hydrogen Energy*, 32, 5094–5097.

Osada, N., Uchida, H. & Watanabe, M. 2006. Polarization behavior of SDC cathode with highly dispersed Ni catalysts for solid oxide electrolysis cells. *Journal of the Electrochemical Society*, 153, A816.

Pan, Y., Hu, W., Liu, D., Liu, Y. & Liu, C. 2015. Carbon nanotubes decorated with nickel phosphide nanoparticles as efficient nanohybrid electrocatalysts for the hydrogen evolution reaction. *Journal of Materials Chemistry A*, 3, 13087–13094.

Pinsky, R., Sabharwall, P., Hartvigsen, J. & O'Brien, J. 2020. Comparative review of hydrogen production technologies for nuclear hybrid energy systems. *Progress in Nuclear Energy*, 123, 103317.

Pollet, B. 2012. *Power Ultrasound in Electrochemistry: From Versatile Laboratory Tool to Engineering Solution*. John Wiley & Sons.

Pool, D. H., Stewart, M. P., O'Hagan, M., Shaw, W. J., Roberts, J. A., Bullock, R. M. & Dubois, D. L. 2012. Acidic ionic liquid/water solution as both medium and proton source for electrocatalytic H2 evolution by [Ni (P2N2) 2] 2+ complexes. *Proceedings of the National Academy of Sciences*, 109, 15634–15639.

Popczun, E. J., Read, C. G., Roske, C. W., Lewis, N. S. & Schaak, R. E. 2014. Highly active electrocatalysis of the hydrogen evolution reaction by cobalt phosphide nanoparticles. *Angewandte Chemie International Edition*, 53, 5427–5430.

Rahim, A. A., Tijani, A. S., Fadhlullah, M., Hanapi, S. & Sainan, K. 2015. Optimization of direct coupling solar PV panel and advanced alkaline electrolyzer system. *Energy Procedia*, 79, 204–211.

Rashid, M., AL Mesfer, M. K., Naseem, H. & Danish, M. 2015. Hydrogen production by water electrolysis: A review of alkaline water electrolysis, PEM water electrolysis and high temperature water electrolysis. *International Journal of Engineering and Advanced Technology*, 4(3), 80–93.

Rathinam, N. K., Sani, R. K. & Salem, D. 2018. Rewiring extremophilic electrocatalytic processes for production of biofuels and value-added compounds from lignocellulosic biomass. *Extremophilic Microbial Processing of Lignocellulosic Feedstocks to Biofuels, Value-Added Products, and Usable Power*, Springer International Publishing AG, part of Springer Nature Switzerland, 229–245.

Rosen, J., Hutchings, G. S. & Jiao, F. 2013. Ordered mesoporous cobalt oxide as highly efficient oxygen evolution catalyst. *Journal of the American Chemical Society*, 135, 4516–4521.

Rozain, C., Mayousse, E., Guillet, N. & Millet, P. 2016a. Influence of iridium oxide loadings on the performance of PEM water electrolysis cells: Part I-Pure IrO2-based anodes. *Applied Catalysis B: Environmental*, 182, 153–160.

Rozain, C., Mayousse, E., Guillet, N. & Millet, P. 2016b. Influence of iridium oxide loadings on the performance of PEM water electrolysis cells: Part II-Advanced oxygen electrodes. *Applied Catalysis B: Environmental*, 182, 123–131.

Sancheti, S. V. & Gogate, P. R. 2017. A review of engineering aspects of intensification of chemical synthesis using ultrasound. *Ultrasonics Sonochemistry*, 36, 527–543.

Sandeep, K., Kamath, S., Mistry, K., Kumar, A., Bhattacharya, S., Bhanja, K. & Mohan, S. 2017. Experimental studies and modeling of advanced alkaline water electrolyser with porous nickel electrodes for hydrogen production. *International Journal of Hydrogen Energy*, 42, 12094–12103.

Sapountzi, F. M., Gracia, J. M., Fredriksson, H. O. & Niemantsverdriet, J. H. 2017. Electrocatalysts for the generation of hydrogen, oxygen and synthesis gas. *Progress in Energy and Combustion Science*, 58, 1–35.

Sarkar, S. & Peter, S. C. 2018. An overview on Pd-based electrocatalysts for the hydrogen evolution reaction. *Inorganic Chemistry Frontiers*, 5, 2060–2080.

Sarno, M. & Ponticorvo, E. 2019. High hydrogen production rate on RuS2@ MoS2 hybrid nanocatalyst by PEM electrolysis. *International Journal of Hydrogen Energy*, 44, 4398–4405.

Sasikumar, G., Muthumeenal, A., Pethaiah, S. S., Nachiappan, N. & Balaji, R. 2008. Aqueous methanol eletrolysis using proton conducting membrane for hydrogen production. *International Journal of Hydrogen Energy*, 33, 5905–5910.

Seehra, M. & Bollineni, S. 2009. Nanocarbon boosts energy-efficient hydrogen production in carbon-assisted water electrolysis. *International Journal of Hydrogen Energy*, 34, 6078–6084.

Seetharaman, S., Balaji, R., Ramya, K., Dhathathreyan, K. & Velan, M. 2013. Graphene oxide modified non-noble metal electrode for alkaline anion exchange membrane water electrolyzers. *International Journal of Hydrogen Energy*, 38, 14934–14942.

Seh, Z. W., Kibsgaard, J., Dickens, C. F., Chorkendorff, I., Nørskov, J. K. & Jaramillo, T. F. 2017. Combining theory and experiment in electrocatalysis: Insights into materials design. *Science*, 355, eaad4998.

Selamet, Ö. F., Acar, M. C., Mat, M. D. & Kaplan, Y. 2013. Effects of operating parameters on the performance of a high-pressure proton exchange membrane electrolyzer. *International Journal of Energy Research*, 37, 457–467.

Sharma, S. & Ghoshal, S. K. 2015. Hydrogen the future transportation fuel: From production to applications. *Renewable and Sustainable Energy Reviews*, 43, 1151–1158.

Shimizu, N., Hotta, S., Sekiya, T. & Oda, O. 2006. A novel method of hydrogen generation by water electrolysis using an ultra-short-pulse power supply. *Journal of Applied Electrochemistry*, 36, 419–423.

Siracusano, S., Trocino, S., Briguglio, N., Baglio, V. & Aricò, A. S. 2018. Electrochemical impedance spectroscopy as a diagnostic tool in polymer electrolyte membrane electrolysis. *Materials*, 11, 1368.

Stojić, D. L., Marčeta, M. P., Sovilj, S. P. & Miljanić, Š. S. 2003. Hydrogen generation from water electrolysis-possibilities of energy saving. *Journal of Power Sources*, 118, 315–319.

Strmcnik, D., Uchimura, M., Wang, C., Subbaraman, R., Danilovic, N., van der Vliet, D., Paulikas, A. P., Stamenkovic, V. R. & Markovic, N. M. 2013. Improving the hydrogen oxidation reaction rate by promotion of hydroxyl adsorption. *Nature Chemistry*, 5, 300–306.

Suntivich, J., May, K. J., Gasteiger, H. A., Goodenough, J. B. & Shao-Horn, Y. 2011. A perovskite oxide optimized for oxygen evolution catalysis from molecular orbital principles. *Science*, 334, 1383–1385.

Symes, D. 2011. *Sonoelectrochemical (20 kHz) Production of Hydrogen from Aqueous Solutions*. University of Birmingham.

Tasic, G. S., Maslovara, S. P., Zugic, D. L., Maksic, A. D. & Kaninski, M. P. M. 2011. Characterization of the Ni-Mo catalyst formed in situ during hydrogen generation from alkaline water electrolysis. *International Journal of Hydrogen Energy*, 36, 11588–11595.

Terezo, A. & Pereira, E. 1999. Preparation and characterization of Ti/RuO2-Nb2O5 electrodes obtained by polymeric precursor method. *Electrochimica Acta*, 44, 4507–4513.

Terrones, M., Ajayan, P., Banhart, F., Blase, X., Carroll, D., Charlier, J.-C., Czerw, R., Foley, B., Grobert, N. & Kamalakaran, R. 2002. N-doping and coalescence of carbon nanotubes: Synthesis and electronic properties. *Applied Physics A*, 74, 355–361.

Tian, J., Liu, Q., Asiri, A. M. & Sun, X. 2014. Self-supported nanoporous cobalt phosphide nanowire arrays: An efficient 3D hydrogen-evolving cathode over the wide range of pH 0-14. *Journal of the American Chemical Society*, 136, 7587–7590.

Torres-Palma, R. & Serna-Galvis, E. 2018. *Sonolysis, Advanced Oxidation Processes for Waste Water Treatment*. Elsevier.

Ursua, A., Gandia, L. M. & Sanchis, P. 2011. Hydrogen production from water electrolysis: Current status and future trends. *Proceedings of the IEEE*, 100, 410–426.

van Ruijven, B. & van Vuuren, D. P. 2009. Oil and natural gas prices and greenhouse gas emission mitigation. *Energy Policy*, 37, 4797–4808.

Vermeiren, P., Adriansens, W., Moreels, J. & Leysen, R. 1998. Evaluation of the Zirfon(r) separator for use in alkaline water electrolysis and Ni-H2 batteries. *International Journal of Hydrogen Energy*, 23, 321–324.

Vilasmongkolchai, T., Songprakorp, R. & Sudaprasert, K. The behaviour of gas bubble during rest period of pulse-activated electrolysis hydrogen production. *MATEC Web of Conferences*, 2016. EDP Sciences, Chongqing, China, 14001.

Vincent, I., Choi, B., Nakoji, M., Ishizuka, M., Tsutsumi, K. & Tsutsumi, A. 2018. Pulsed current water splitting electrochemical cycle for hydrogen production. *International Journal of Hydrogen Energy*, 43, 10240–10248.

Wang, C.-C. & Chen, C.-Y. 2009. Water electrolysis in the presence of an ultrasonic field. *Electrochimica Acta*, 54, 3877–3883.

Wang, D.-Y., Gong, M., Chou, H.-L., Pan, C.-J., Chen, H.-A., Wu, Y., Lin, M.-C., Guan, M., Yang, J. & Chen, C.-W. 2015. Highly active and stable hybrid catalyst of cobalt-doped FeS2 nanosheets-carbon nanotubes for hydrogen evolution reaction. *Journal of the American Chemical Society*, 137, 1587–1592.

Wang, J., Cui, W., Liu, Q., Xing, Z., Asiri, A. M. & Sun, X. 2016. Recent progress in cobalt-based heterogeneous catalysts for electrochemical water splitting. *Advanced materials*, 28, 215–230.

Wang, M., Wang, Z., Gong, X. & Guo, Z. 2014. The intensification technologies to water electrolysis for hydrogen production-A review. *Renewable and Sustainable Energy Reviews*, 29, 573–588.

Wang, S., Lu, A. & Zhong, C.-J. 2021. Hydrogen production from water electrolysis: Role of catalysts. *Nano Convergence*, 8, 1–23.

Wang, Y., Zhang, B., Gong, Z., Gao, K., Ou, Y. & Zhang, J. 2013. The effect of a static magnetic field on the hydrogen bonding in water using frictional experiments. *Journal of Molecular Structure*, 1052, 102–104.

Wei, Z., Yao, L.-P., Juan, L. & Zong, Z.-M. 2008. Electrolytic reduction of Nantong coal and model compounds with oxygenic functional groups in an aqueous NaCl solution. *Journal of China University of Mining and Technology*, 18, 112–115.

Wu, J., Yang, Z., Sun, Q., Li, X., Strasser, P. & Yang, R. 2014. Synthesis and electrocatalytic activity of phosphorus-doped carbon xerogel for oxygen reduction. *Electrochimica Acta*, 127, 53–60.

Wu, R., Zhang, J., Shi, Y., Liu, D. & Zhang, B. 2015. Metallic WO2-carbon mesoporous nanowires as highly efficient electrocatalysts for hydrogen evolution reaction. *Journal of the American Chemical Society*, 137, 6983–6986.

Wu, X., Tayal, J., Basu, S. & Scott, K. 2011. Nano-crystalline RuxSn1−xO2 powder catalysts for oxygen evolution reaction in proton exchange membrane water electrolysers. *International Journal of Hydrogen Energy*, 36, 14796–14804.

Xu, W. & Scott, K. 2010. The effects of ionomer content on PEM water electrolyser membrane electrode assembly performance. *International Journal of Hydrogen Energy*, 35, 12029–12037.

Yang, X. & Irvine, J. T. 2008. (La 0.75 Sr 0.25) 0.95 Mn 0.5 Cr 0.5 O 3 as the cathode of solid oxide electrolysis cells for high temperature hydrogen production from steam. *Journal of Materials Chemistry*, 18, 2349–2354.

Yang, Z., Zhang, G. & Lin, B. 2015. Performance evaluation and optimum analysis of a photovoltaic-driven electrolyzer system for hydrogen production. *International Journal of Hydrogen Energy*, 40, 3170–3179.

Yeh, M.-H., Leu, Y.-A., Chiang, W.-H., Li, Y.-S., Chen, G.-L., Li, T.-J., Chang, L.-Y., Lin, L.-Y., Lin, J.-J. & Ho, K.-C. 2018. Boron-doped carbon nanotubes as metal-free electrocatalyst for dye-sensitized solar cells: Heteroatom doping level effect on tri-iodide reduction reaction. *Journal of Power Sources*, 375, 29–36.

Yilanci, A., Dincer, I. & Ozturk, H. K. 2009. A review on solar-hydrogen/fuel cell hybrid energy systems for stationary applications. Progress *in Energy and Combustion Science*, 35, 231–244.

Yin, L., Yang, T., Ding, X., He, M., Wei, W., Yu, T. & Zhao, H. 2018. Synthesis of phosphorus-iridium nanocrystals and their superior electrocatalytic activity for oxygen evolution reaction. *Electrochemistry Communications*, 94, 59–63.

Younas, M., Shafique, S., Hafeez, A., Javed, F. & Rehman, F. 2022. An overview of hydrogen production: Current status, potential, and challenges. *Fuel*, 316, 123317.

Yu, B., Zhang, W., Xu, J., Chen, J., Luo, X. & Stephan, K. 2012. Preparation and electrochemical behavior of dense YSZ film for SOEC. *International Journal of Hydrogen Energy*, 37, 12074–12080.

Zadeh, S. H. 2014. Hydrogen production via ultrasound-aided alkaline water electrolysis. *Journal of Automation and Control Engineering*, 2, 103–109.

Zeng, K. & Zhang, D. 2010. Recent progress in alkaline water electrolysis for hydrogen production and applications. Progress *in Energy and Combustion Science*, 36, 307–326.

Zhang, W., Cui, L. & Liu, J. 2020. Recent advances in cobalt-based electrocatalysts for hydrogen and oxygen evolution reactions. *Journal of Alloys and Compounds*, 821, 153542.

Zhou, Y., Ma, R., Candelaria, S. L., Wang, J., Liu, Q., Uchaker, E., Li, P., Chen, Y. & Cao, G. 2016. Phosphorus/sulfur Co-doped porous carbon with enhanced specific capacitance for supercapacitor and improved catalytic activity for oxygen reduction reaction. *Journal of Power Sources*, 314, 39–48.

Zhu, B., Albinsson, I., Andersson, C., Borsand, K., Nilsson, M. & Mellander, B.-E. 2006. Electrolysis studies based on ceria-based composites. *Electrochemistry Communications*, 8, 495–498.

Zini, G. & Tartarini, P. 2009. Hybrid systems for solar hydrogen: A selection of case-studies. *Applied Thermal Engineering*, 29, 2585–2595.

Zore, U. K., Yedire, S. G., Pandi, N., Manickam, S. & Sonawane, S. H. 2021. A review on recent advances in hydrogen energy, fuel cell, biofuel and fuel refining via ultrasound process intensification. *Ultrasonics Sonochemistry*, 73, 105536.

2 Hydrogen Production from Water Thermochemical Splitting

Mohammad Hasan Khademi

2.1 INTRODUCTION

With the continuous growth of population and industrialization, the energy demand is escalating with each passing day. Despite the significant amount of research conducted on renewable energies, the majority of our energy supplies today are still heavily reliant on fossil fuels. The utilization of fossil fuels leads to various issues such as global warming, CO_2 emissions, and the depletion of fuel resources, which will have repercussions in the foreseeable future. It is crucial to utilize an alternative fuel that is both environmentally sustainable and harmless. Hydrogen emerges as a pivotal answer to address the growing worries of global warming and serves as an eco-friendly choice for future energy alternatives. Hydrogen, as an energy carrier, shows great promise as a carbon-free fuel that is easily accessible, reliable, affordable, and safe.

The reported net demand for hydrogen in 2019 was approximately 70 million tons (Acar and Dincer, 2019). By 2050, H_2 has the potential to fulfill approximately 18% of the global energy demand, cater to around 20%–25% of transportation requirements, and consequently mitigate CO_2 emissions by 6 Gt/year (Uyar and Besikci, 2017). Hydrogen can be stored in chemical form to serve as an energy storage material. Additionally, fuel cells provide an efficient means of converting hydrogen into electricity, while combustion can transform it into thermal energy. H_2 has the potential to be utilized in cleaner transportation methods through fuel cells and as a more environmentally friendly energy source for industrial processes such as drying, cooling, heating, and power generation (Farsi et al., 2020). Therefore, hydrogen offers economically promising and energy-efficient solutions for energy needs.

H_2 can be produced from various sources. In 2019, the International Energy Agency revealed that the main sources of H_2 production are carbon-based fuels such as oil (0.7%), coal (27.3%), and natural gas (71.3%). Conversely, renewable sources contribute to less than 1% of hydrogen production. Water is the most plentiful renewable source for generating hydrogen, as it can be separated into oxygen and hydrogen when sufficient energy is supplied. There are various methods available for water splitting, which can be classified into five main categories based on the energy needed for the process.

DOI: 10.1201/9781003382270-3

These categories include H_2 production through mechanical energy (e.g., sonochemical methods using ultrasound) (Merabet and Kerboua, 2022), electrical energy (e.g., electrolysis technology) (Hassan et al., 2023), biological energy (e.g., biological processes using microorganisms) (Ashtitha and George, 2021), photonic energy (e.g., photoelectrochemical, photocatalysis, or photolysis method) (Shankar et al., 2023), and thermal energy (e.g., thermochemical cycles and thermolysis methods) (Budama et al., 2023). Nevertheless, most of these methods for hydrogen production are still in their early stages of research and development, lacking maturity, and show low efficiency when implemented on a large scale. Among the different technologies for water splitting, electrolysis, and thermochemical cycles combined with renewable energy sources may be suitable for large-scale H_2 production due to their cost-effectiveness and ability to generate a significant quantity of hydrogen (Lee et al., 2022).

This chapter focuses on a review and categorization of thermochemical water-splitting cycles (TWCs) as potential methods for large-scale H_2 production based on the energy sources (pure and hybrid thermochemical cycles) and the number of steps (two-, three-, and four-steps). Furthermore, this chapter provides detailed elaboration on some of the most promising TWCs, along with discussions on their current state, limitations, and challenges. Finally, a comparative analysis and evaluation of the ZnO/Zn, S-I, HyS, Cu-Cl, Mg-Cl, Fe-Cl, and V-Cl cycles are carried out, considering their environmental impact, costs, and exergy and energy efficiencies.

2.2 THERMOCHEMICAL WATER-SPLITTING CYCLE

The TWCs rely on a series of chemical reactions that repetitively decompose water. These reactions involve intermediate substances, all of which are recycled throughout the process. As a result, the overall reaction is equivalent to breaking down the water molecule into oxygen and hydrogen (T-Raissi, 2012). In theory, the sole necessity for this process is thermal energy. TWCs are designed to generate hydrogen by harnessing thermal energy and recycling materials for subsequent use. TWCs are not heavily reliant on catalysts and the only substance consumed in the cycle is H_2O, which serves as the hydrogen source. Additionally, all other materials can be recycled in this process (Safari and Dincer, 2020a). Below are the advantages of TWCs (Safari and Dincer, 2020a):

i. In most cases, the temperature range is typically between 500°C and 1,800°C.
ii. Oxygen-hydrogen separation does not necessitate the presence of a membrane.
iii. In pure TWCs, there is no input electricity, while hybrid TWCs have a low requirement for electricity.

TWCs can be classified into two categories: pure thermochemical cycles, which rely solely on thermal energy (Figure 2.1a), and hybrid thermochemical cycles, which utilize both thermal energy and another type of energy such as photonic and electrical energy (Figure 2.1b) (Safari and Dincer, 2020a). Hybrid TWCs involve the utilization

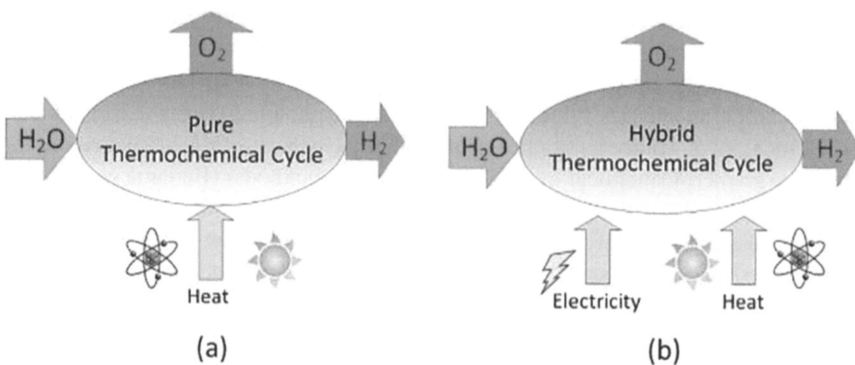

FIGURE 2.1 Overall diagram illustrating pure and hybrid TWCs (Safari and Dincer, 2020a).

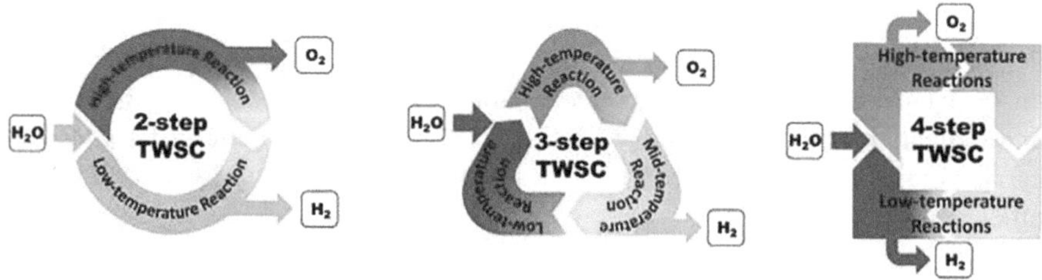

FIGURE 2.2 Categorizing the TWCs according to the number of reaction steps (Li et al., 2022).

of water, high-temperature heat generated by nuclear reactors or concentrated solar units, as well as photonic or electricity energy as inputs; while the outputs of these systems consist of oxygen and hydrogen. On the other hand, depending on the number of reactions involved in the cycle, TWCs can be categorized into cycles with two, three, four, or more steps (Figure 2.2) (Li et al., 2022). Typically, in two-step TWCs, a reaction temperature higher than 2,000 K is necessary. However, in certain three-step TWCs, this temperature requirement can be lowered to below 1,200 K, and in some four-step TWCs, it can be reduced even further to less than 800 K (Oruc and Dincer, 2021). Therefore, by increasing the number of reaction steps, it becomes possible to lower the reaction temperature. Additionally, electricity can be employed as a substitute for high-temperature reactions. However, incorporating electricity into the process often adds complexity to it.

The concept of TWC for hydrogen production was initially introduced by Funk and Reinstrom (1966) in the 1960s. Subsequently, in 1969, the European Community Joint Research Center compiled a set of 24 TWCs, often referred to as Mark Cycles (Beghi, 1985). Research on TWCs has seen a significant increase since 2003, primarily driven by the beginning of DOE's solar thermochemical hydrogen production research program (STCH) (Perret, 2011), which was subsequently followed by similar research projects in other countries (see Table 2.1). Despite the identification of over 300 cycles, the majority of them have encountered numerous problems and

TABLE 2.1

Overview of Global Research Projects on TWCs and Their Corresponding Countries and Institutions

Date	Country	Research Institute	Cycle	Reference
1970	Italy	Joint research center in Ispra	Mark version of cycles	Beghi (1985)
1972	United States	General Atomics (GA)	S-I	O'Keefe et al. (1982)
1975	United States	Westinghouse corporation	HyS	Brecher and Wu (1975)
1976	France	CEA France	CeO/Ce, S-I	Anzieu et al. (2006)
1978	Japan	Japan Atomic Energy Agency (JAEA)	S-I	Onuki et al. (1990), Kubo et al. (2004), Kasahara et al. (2014, 2017)
1980	Japan	University of Tokyo	UT-3 (Calcium Bromine)	Kameyama and Yoshida (1978)
1984	Germany	Julich RWTH Achen	S-I, V-Cl, Fe-Cl	Knoche et al. (1984a, 1984b)
1994	Japan	Tokyo Institute of Technology	CeO_2-MO_x (M=Cu, Ni, Fe)	Kaneko et al. (2007)
2002	Switzerland	Paul Scherrer Institute and ETH Zurich	Fe_3O_4/FeO, ZnO/Zn	Koepf et al. (2017)
2002	Germany	German Aerospace Center (DLR)	MO_x, HyS, S-I	Sattler (2019)
2003	United States	STCH project contributors	HyS, S-I	Perret (2011)
2003	United States	Savannah River National Laboratories (SRNL)	HyS	Gorensek and Summers (2009), Gorensek (2011), Gorensek et al. (2017, 2018)
2003	United States	Argonne National Laboratories (ANL)	S-I, Cu-Cl	Lewis et al. (2009a)
2004	Germany	HYTHEC project contributors	HyS, S-I	Le et al. (2007)
2005	China	Institute of Nuclear and Energy Technology (INET)	S-I	Zhang et al. (2010)
2006	South Korea	Korea Atomic Energy Research Institute (KAERI)	Ferrite cycles	Chang et al. (2007)
2007	Canada	Cu-Cl project contributors	Hybrid Cu-Cl	Naterer et al. (2009, 2013b, 2015)

(Continued)

TABLE 2.1 (*Continued*)

Overview of Global Research Projects on TWCs and Their Corresponding Countries and Institutions

Date	Country	Research Institute	Cycle	Reference
2007	France	CRNS France	MO_x (Perovskite material)	Abanades and Flamant (2006), Charvin et al. (2007), Chueh et al. (2010)
2007	Canada	University of Ontario Institute of Technology	Hybrid Cu-Cl, Mg-Cl	Dincer and Naterer (2014), Ozcan and Dincer (2016b, 2018)
2007	Canada	Atomic Energy of Canada Limited (AECL)	Hybrid Cu-Cl	Naterer et al. (2013a)
2008	Germany	HYCYCLES project contributors	HyS, S-I	Roeb et al. (2011)
2009	United States	Oak Ridge National Laboratory (ORNL)	UTC (Three-step Uranium)	Collins et al. (2010)
2012	Japan	Joint project of the University of Miyazaki, Niigata University, and Mitaka Kohki Co., LTD	Nonstoichiometric cerium oxide	Kodama et al. (2018)
2013	United States	Florida Solar Energy Center (FSEC)	HySA	Huang et al. (2014)
2013	Germany	SOL2H2 project contributors	Modified HyS	Odorizzi and Enginsoft (2016)
2015	Argentina	National Atomic Energy Commission (NAEC)	Co-Cl, Fe-Cl	Canavesio et al. (2015, 2020)
2016	Spain	Superior School of Experimental Science and Technology (ESCET)	Two-step cycles (Perovskite materials)	Angel and Marug (2016)
2017	United States	Sandia National Laboratories (SNL)- EETHP project	Simple binary oxide (ferrite) cycle	Ambrosini et al. (2017)

(Continued)

TABLE 2.1 (*Continued*)

Overview of Global Research Projects on TWCs and Their Corresponding Countries and Institutions

Date	Country	Research Institute	Cycle	Reference
2017	United States	National Renewable Energy Laboratory (NREL) and University of Colorado Boulder	Co doped-hercynite ($CoFe_2O_4$)	Hoskins et al. (2019)
2018	China	Chinese Academy of Science	UTC (Three-step Uranium)	Chen et al. (2019)
2019	Japan	Niigata University	Perovskite materials	Gokon et al. (2019)

limitations (Mehrpooya and Habibi, 2020). Only a handful of these cycles have been recognized as having potential, which is described in detail in Section 2.3.

2.2.1 Two-Step Thermochemical Cycle

In two-step TWCs, the production of hydrogen involves the utilization of metal, metal oxides, or reduced metal oxides, along with water. In the first stage, metal oxides are reduced to reduced metal oxide or pure metal, and oxygen is simultaneously released (reaction (2.1), known as activation or thermal reduction). In the second stage, the reduced metal oxides undergo a reaction with H_2O. This reaction releases hydrogen and converts the metal oxides back to their original form, allowing for recycling (reaction (2.2), known as hydrolysis). From a thermodynamic standpoint, the first stage is an endothermic process requiring temperatures above 1,500°C, while the second stage is exothermic and usually occurs at temperatures below 1,000°C (Mao et al., 2020).

$$MO_x(s) \rightarrow MO_{x-2n}(s) + nO_2(g) \qquad \Delta H > 0 \qquad (2.1)$$

$$MO_{x-2n}(s) + 2nH_2O(g) \rightarrow MO_x(s) + 2nH_2(g) \qquad \Delta H < 0 \qquad (2.2)$$

The two-step TWCs have the following characteristics (Mao et al., 2020):

i. Oxygen and hydrogen are generated through separate reaction processes and do not require separation.
ii. Each stage occurs at a lower temperature compared to the direct water thermolysis, although it still exceeds 1,500°C, depending on the specific redox material used.
iii. It enables the potential utilization of solar energy directly.

In general, while two-step cycles offer the advantage of simplicity with only two chemical reactors and inherent O_2-H_2 separation features, their relatively lower efficiency compared to electrolysis and the need for materials that can withstand high temperatures (1,700–3,000 K) are the main drawbacks of this type of TWCs (Jarret et al., 2016; Bulfin et al., 2017).

Several two-step TWCs such as ZnO/Zn, GeO_2/GeO, CdO/Cd, Co_3O_4/CoO, Mn_2O_3/MnO, CeO_2/Ce_2O_3, SnO_2/SnO, and Fe_3O_4/Fe have been proposed based on using redox pair reactions involving both volatile and non-volatile metal oxides (Xiao et al., 2012). Among the two-step TWCs, ZnO/Zn emerged as the most active cycle for hydrogen production. On the other hand, CeO_2/Ce_2O_3 exhibited notable advantages in terms of efficiency and cost (Lee et al., 2022).

2.2.2 Three-Step Thermochemical Cycle

The maximum temperature requirement for the two-step cycle can be reduced by replacing the highest temperature reaction with a two-step reaction process, allowing the construction of three-step TWCs (Safari and Dincer, 2020a). Tremendous efforts have been made to optimize the chemicals in three-step TWCs, effectively addressing the limitation of the two-step TWCs. As a result, the working temperature has been successfully reduced to below 1,500 K (Li et al., 2022). The series of reactions that repeat in a three-step TWC can be expressed by Equations (2.3)–(2.5) (Vitart et al., 2006):

$$H_2O + A \rightarrow 0.5H_2 \left(\text{or } 0.5O_2 \right) + AO \left(\text{or } AH_2 \right) \tag{2.3}$$

$$AO \left(\text{or } AH_2 \right) + B \rightarrow 0.5O_2 \left(\text{or } 0.5H_2 \right) + AB \tag{2.4}$$

$$AB \rightarrow A + B \tag{2.5}$$

The most widely recognized cycle in this category is the S-I cycle, also known as the General Electric cycle. In addition to the well-established S-I cycle, other three-step metal oxide cycles such as Co_3O_4/CoO, Mn_3O_4/MnO, Fe_3O_4/FeO, CeO_2/Ce_2O_3, and GeO_2/GeO have been explored by Charvin et al. (2007) and Bhosale (2020a, 2020b) to determine their compatibility with high-temperature solar reactors.

2.2.3 Four-Step Thermochemical Cycle

The four-step TWC would be a modified three-step TWC that compensates for its limitations. The fundamental stages of four-step cycles closely resemble three-step cycles for H_2 production, except with the distinct recombination step involving the reactants (Ishaq et al., 2018). The four-step TWCs involve the following reactions: (i) the reactant chemical compounds, like metal chlorides, undergo hydrolysis with water to produce oxides and hydride (reaction (2.6)), (ii) hydride derived from phase (i) decomposes, resulting in hydrogen production (reaction (2.7)), (iii) oxides derived from phase (i) undergo oxygen production (reaction (2.8)), and (iv) the reactants recombine to form the original compound (reaction (2.9)) (Lee et al., 2022).

$$MCl_{2n} + nH_2O \rightarrow MO_n + 2nHCl \tag{2.6}$$

$$2nHCl \rightarrow nCl_2 + nH_2 \tag{2.7}$$

$$MO_n \rightarrow M + 0.5nO_2 \tag{2.8}$$

$$M + nCl_2 \rightarrow MCl_{2n} \tag{2.9}$$

Among various types of metallic chlorides, V-Cl, Mg-Cl, Ce-Cl, Cu-Cl, and Fe-Cl were found to be effective for hydrogen production in the four-step TWCs. Among them, V-Cl and Cu-Cl have reported efficiencies of 46% and 43%, respectively, making them the most promising options. Fe-Cl has also been recommended for further studies to improve its efficiency (Lewis and Masin, 2009; Lewis et al., 2009b). The four-step TWC offers the significant benefit of lowering the maximum required temperature (e.g., 923 K for the Fe-Cl cycle) by introducing an additional reaction and facilitating the recombination of reactants. However, large-scale H_2 production faces challenges due to the formation of harmful chloride complexes or the decomposition of intermediates that can damage facilities (Lee et al., 2022).

2.3 PURE THERMOCHEMICAL CYCLE

Although many TWCs have been proposed in the literature, some of them, such as ZnO/Zn, S-I, Cu-Cl, Mg-Cl, Fe-Cl, and V-Cl, have the potential for large-scale H_2 production. In the following section, a brief explanation of each of these cycles will be given.

2.3.1 ZnO/Zn Cycle

The ZnO/Zn cycle is categorized as a volatile cycle due to the pyrolysis of ZnO into oxygen and zinc (in the gas phase) during the thermal reduction reaction. The reactions of this cycle can be represented by equations (2.10) and (2.11). The thermal reduction reaction is a high-temperature endothermic step. Figure 2.3 illustrates the various components of a solar ZnO/Zn TWC. These components include a water-splitting reactor, a cooler, and a solar reactor. In this process, zinc oxide is introduced into the solar reactor. By raising the temperature of the solar reactor, the thermal reduction of zinc oxide takes place. The gaseous products are then cooled using a cooler. The solid zinc is then transferred to the water-splitting reactor to produce hydrogen, while the oxygen is released from the cooler. Finally, the zinc oxide is recycled to the solar reactor (Mao et al., 2020).

Hydrolysis (850 K)

$$Zn(s) + H_2O(g) \rightarrow ZnO(s) + H_2(g) \tag{2.10}$$

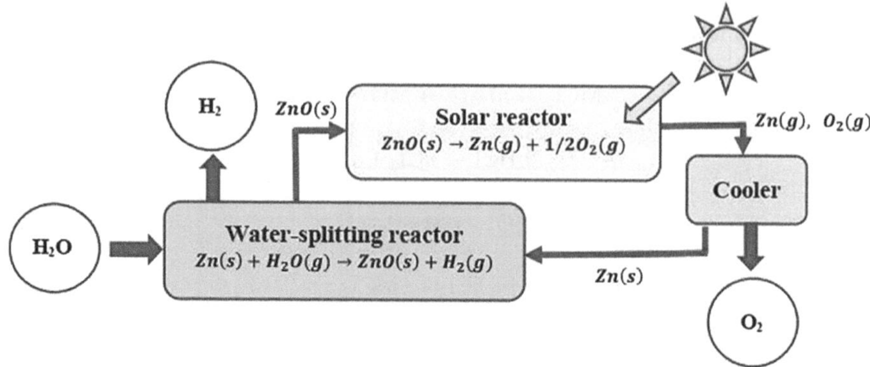

FIGURE 2.3   Schematic of the ZnO/Zn cycle.

Thermal reduction (2,350 K)

$$ZnO(s) \rightarrow Zn(g) + 1/2\,O_2(g) \tag{2.11}$$

Weidenkaff et al. (2000b) discovered two factors that influence this cycle:

i. The recombination of O_2 and zinc vapor produced during the thermal reduction reaction lowers the rate of ZnO decomposition.
ii. During the hydrolysis reaction, a thin film of ZnO forms on the surface of Zn, hindering further reaction.

The crucial aspect of this technology lies in the efficient separation of O_2 and Zn (g). Weidenkaff et al. (2000b) employed a tube furnace with a temperature gradient to rapidly cool down the zinc vapor, resulting in the formation of solid zinc. The study demonstrated that quenching with inert gases such as nitrogen or argon effectively achieved the separation of zinc vapor and oxygen. Fletcher (1999) suggested a direct electrolysis approach for liquid zinc oxide. In this approach, oxygen and zinc vapor are generated at separate electrodes, resulting in minimal electrical energy consumption and preventing recombination between oxygen and zinc. However, it is necessary to maintain a temperature above 2,248 K to melt the ZnO for this process.

Steinfeld (2002) introduced a method for analyzing the exergy efficiency of the ZnO/Zn cycle, which helps determine the irreversibility and optimal exergy efficiency. The results revealed that the cycle's maximum exergy efficiency reached 29%, with irreversibility primarily stemming from the quenching process of zinc vapor and re-radiation loss in the solar reactor. Economic assessment for this cycle indicated that hydrogen production cost was \$7.98/kg H_2 (reported by Charvin et al. (2008)) and \$5/kg H_2 (reported by Steinfeld (2002)).

An $H_2O/Zn/Fe_3O_4$ system was proposed by Tamaura et al. (2001) where the reduction temperature exceeded the boiling point of Zn. After quenching, Zn was deposited onto the surface of Fe_3O_4. The thermal reduction step predominantly yielded Fe_3O_4 and zinc oxide as products, as O_2 and Zn combined to form ZnO. Unoxidized zinc reduced Fe^{3+} to Fe^{2+} and infiltrated the Fe_3O_4 lattice, resulting in the formation of $ZnFe_2O_4$ through hydrolysis. In the absence of a quenching device, zinc oxide separated from $ZnFe_2O_4$ and accumulated on the walls of the reaction cell during the thermal reduction step (Tamaura et al., 2004). These reactions can be represented by equations (2.12) and (2.13).

Thermal reduction

$$3ZnFe_2O_4 \rightarrow 2Fe_3O_4 + 3ZnO + 0.5O_2 \tag{2.12}$$

Water-splitting reaction

$$3ZnO + 2Fe_3O_4 + H_2O \rightarrow 3ZnFe_2O_4 + H_2 \tag{2.13}$$

2.3.2 S-I Cycle

The S-I cycle was initially introduced by General Atomics in 1970 (Safari and Dincer, 2020a). The S-I cycle, identified as a highly promising cycle in current research, exhibits the capability for large-scale production, as it can be integrated with high-temperature nuclear (Zhang et al., 2010) and solar energies (Sattler et al., 2017; Yilmaz and Selbas, 2017; Cumpston et al., 2020; Mehrpooya et al., 2021). This cycle is characterized by three primary reactions (2.14)–(2.16) (Mehrpooya and Habibi, 2020). The exothermic reaction (2.14), commonly referred to as the Bunsen reaction, involves H_2O, SO_2, and I_2 as reactants, resulting in the formation of two immiscible acids when excess water and iodine are present. The acids generated in this phase are subsequently decomposed in the endothermic reactions (2.15) and (2.16), ultimately yielding hydrogen and oxygen as final products, along with other by-products. These by-products are then cycled back into the Bunsen reaction (Mohd and Nandong, 2017). Figure 2.4 illustrates a schematic diagram of the S-I cycle.

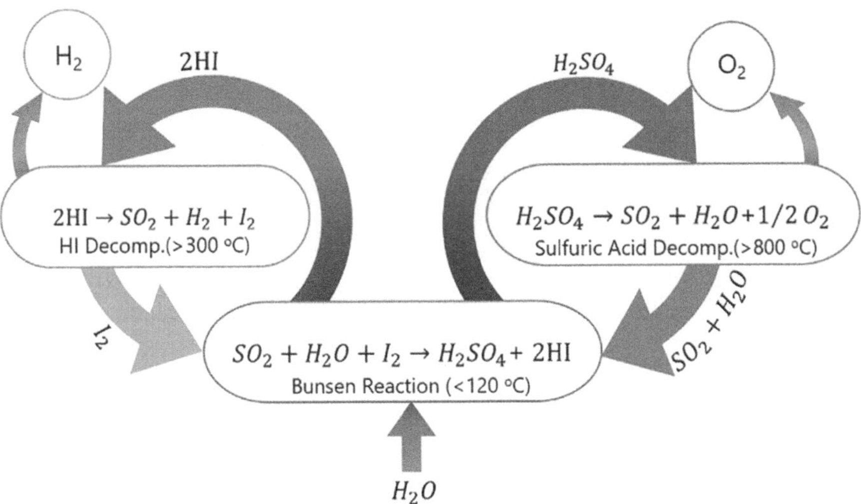

FIGURE 2.4 Schematic of the S-I cycle (Oruc and Dincer, 2021).

Bunsen reaction (293–393 K)

$$I_2(s) + SO_2(g) + H_2O(l) \rightarrow H_2SO_4(aq) + 2HI(aq) \tag{2.14}$$

$$\Delta H = -75 \pm 15 \text{ kJ/mol}$$

H_2SO_4 decomposition (1,073–1,273 K)

$$\left.\begin{array}{l} H_2SO_4 \rightarrow SO_3 + H_2O \\ SO_3 \rightarrow SO_2 + 0.5O_2 \end{array}\right\} \begin{array}{l} H_2SO_4(g) \rightarrow SO_2(g) + H_2O(g) \\ + 0.5O_2(g) \end{array} \tag{2.15}$$

$$\Delta H = 183 \pm 3 \text{ kJ/mol}$$

HI decomposition (573–773 K)

$$2HI(g) \rightarrow I_2(g) + H_2(g) \tag{2.16}$$

$$\Delta H = 12 \text{kJ/mol}$$

As depicted in Figure 2.5, reaction (2.14) necessitates a Bunsen reactor, two purification reactors, a liquid–liquid separator, and some supplementary equipment. Within the reactor, the Bunsen reaction is executed with an excess of water, a thermodynamically favorable condition, and an excess of iodine to forestall any undesirable side reactions. This surplus iodine induces the automatic separation of the H_2SO_4 into distinct heavy and light phases. The lighter phase comprises H_2O and H_2SO_4, while the denser phase contains H_2O, I_2, and HI. Most of the surplus iodine exits the reactor together with the dense phase (Giaconia et al., 2009). The liquid phases are subsequently separated using the liquid–liquid separator. Given the relatively low concentrations of HI and H_2SO_4 in the light and heavy phases, respectively, the Bunsen reverse reaction is carried out in the subsequent reactors to facilitate purification (Fu et al., 2016a).

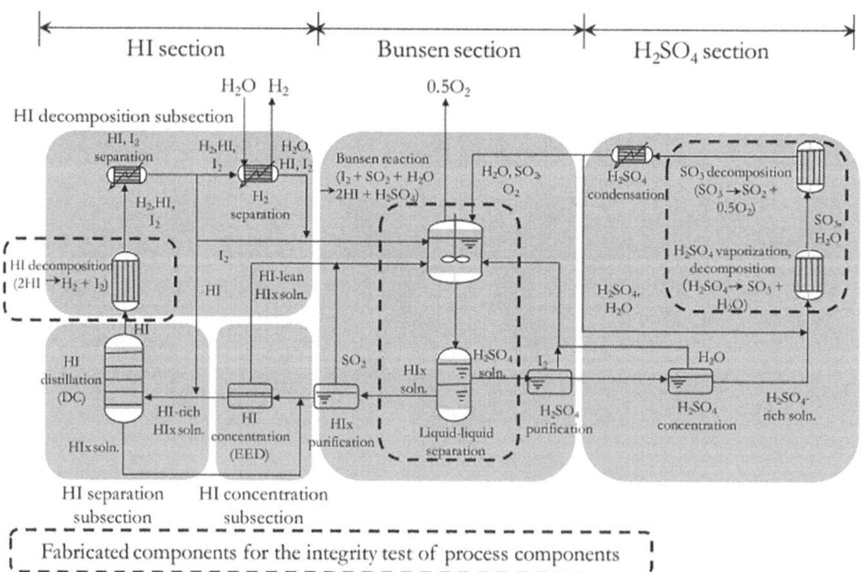

FIGURE 2.5 A flow diagram of the S-I cycle system (Kasahara et al., 2017).

Reaction (2.15) is carried out through a combination of a decomposition reactor and a flash tank. In the flash tank, the greater volatility of water compared to acid causes water to evaporate, resulting in the concentration of sulfuric acid. By heating this concentrated H_2SO_4 and allowing it to evaporate, both water vapor and SO_3 are generated. Subsequently, in the decomposition reactor, SO_3 is broken down into O_2 and SO_2. Following condensation, the H_2SO_4 formed from unreacted water and SO_3 cycles back to the decomposition unit. The remaining water, oxygen, and SO_2 then reenter the Bunsen reactor. Ultimately, O_2 is extracted as a by-product of this process (Lee et al., 2009).

In reaction (2.16), the key apparatus includes a distillation column, an electro-electrodialysis (EED) unit, and a decomposition reactor (Ping et al., 2018). Following the removal of sulfur dioxide from the HIx phase during the purification process, HIx is directed to the concentration stage. Due to the azeotropic constraint, conventional distillation alone cannot effectively concentrate the acid phase. Three approaches have been suggested for this stage: reactive distillation, extractive distillation with H_3PO_4, and the combination of conventional distillation with EED. The first approach poses challenges in terms of the harsh operating environment, while the second approach involves the energy required for recycling H_3PO_4 and the use of a catalyst (Xu et al., 2018a). Electrolysis (third approach) emerges as a promising option for several reasons. Despite the first approach, it doesn't involve surplus materials like H_3PO_4. Moreover, its operating pressure and temperature are relatively lower compared to reactive distillation, and it's readily implementable in manufacturing (Chen et al., 2013). The process involves the reduction of iodine ($I_2 + 2e \rightarrow 2I^-$) at the cathode and the oxidation of iodide anions ($2I^- \rightarrow I_2 + 2e$) at the anode. The H_2 cation then moves from the anode to the cathode via the electrolyte. Throughout this process, the concentration of HI in the HIx phase rises at the cathode (Wang et al., 2014). Subsequently, HI is separated from this solution through distillation, and in the ensuing unit, it undergoes decomposition into hydrogen and I. In the S-I process, a significant portion of the undecomposed I_2 and HI is separated through condensation. The final step of this process involves extracting hydrogen from iodine through contact with external H_2O (Kasahara et al., 2017).

The purification unit is critically important in this TWC because, without it, side reactions (2.17) and (2.18) which are known as the hydrogen sulfide formation and sulfur formation reaction, respectively, and their resulting products can lead to issues such as equipment pipe clogging (Mehrpooya and Habibi, 2020).

$$H_2SO_4 + 8HI \rightarrow 4H_2S + 4I_2 + 4H_2O \qquad (2.17)$$

$$H_2SO_4 + 6HI \rightarrow 4S + 3I_2 + 4H_2O \qquad (2.18)$$

The optimization of the Bunsen reaction is noted as the central solution in this cycle. The primary challenges reported in this regard are as follows (Safari and Dincer, 2020a):

   i. Operating at high temperatures (more than 800°C) is necessary.
  ii. The reactants used are highly corrosive and pose risks to industrial applications.
 iii. Separating hydrogen iodide from H_2SO_4 involves the presence of excessive iodine, resulting in a significant loss of iodine.

Numerous researchers have undertaken efforts to address these challenges and potentially develop S-I cycle-based systems that are more commercially viable. For instance, Zhu et al. (2017), Liberatore et al. (2019), Park et al. (2020), and Zhou et al. (2007) have suggested a novel open-loop SI TWC. This cycle aims to produce electric power, sulfuric acid, and hydrogen for emerging markets. The process eliminates the need for the H_2SO_4 decomposition step by utilizing SO_2 from flue gas. This approach reduces costs in the industrial sector and yields marketable products such as H_2 and H_2SO_4. As a result, the overall cost of the S-I system is reduced, facilitating the commercialization of this TWC.

Another challenge of the S-I cycle is that the equilibrium decomposition of HI is inherently low, even at relatively high temperatures. To enhance this process, factors like temperature elevation or hydrogen removal using a membrane need to be employed (Mehrpooya and Habibi, 2020). Moreover, to enhance the conversion of hydrogen iodide to hydrogen, an investigation was conducted on the decomposition reaction within electrochemical cells (Xu et al., 2018b).

A distinct S-I cycle has been proposed by Lanchi et al. (2009), involving the integration of nickel metal into the H_2 production stage within the S-I TWC. Subsequently, nickel recycling occurs during the NiI_2 decomposition phase. This approach aims to alleviate challenges encountered in the S-I process, such as the low reaction rate in the hydrogen iodide decomposition stage and the reliance on precious catalysts.

Due to the low conversion and reaction rates in the decomposition reactions (HI and SO_3 decomposition), the utilization of catalysts becomes imperative. Noteworthy catalysts exhibiting superior performance in HI decomposition include Ni(3%)-Co(1%)/AC (Singhania and Bhaskarwar, 2018), CeO_2-Ni-300 (Singhania, 2018), AC (Zhang et al., 2015), Pd-CeO_2_300 (Singhania et al., 2018), Ni(2.5%)-Pt(2.5%)/AC (Singhania et al., 2016), 12%Ni/AC (Fu et al., 2016b), AC-CS (Fu et al., 2016a), rGO-HH, and Pt/rGO-EG (Bo et al., 2014). Regarding SO_3 decomposition, $CuCr_2O_4$, $CuFe_2O_4$, and Pt/SiC catalysts were examined. $CuCr_2O_4$ demonstrated superior activity, while Pt/SiC exhibited higher stability (Zhang et al., 2012).

Park et al. (2019) highlighted an additional enhancement in the S-I process using a novel reactor-separator network to prevent oxygen contact during the Bunsen reaction, which can otherwise decrease the efficiency of H_2SO_4 production. Moreover, Mandal and Jana (2020) developed a reactive distillation column to address equilibrium limitations and the creation of azeotropes in the hydrogen iodide decomposition reaction. This innovative approach allows for simultaneous HI decomposition and H_2 separation, resulting in improved overall efficiency.

2.3.3 Cu-Cl Cycle

The Cu-Cl TWC, being studied by the National Chemical Laboratory for Industry for the first time in 1976, shows promising performance in H_2 production and efficiency (Dokiya and Kotera, 1976). In comparison to other TWCs, its advantages include a lower maximum temperature (lower than 550°C) and the ability to integrate with low-grade sources such as nuclear energy and with lower temperature waste heat from industries (Safari and Dincer, 2020a). The Cu-Cl process and its separation units are also easier to handle.

The Cu-Cl cycles come in various forms, ranging from three to five steps, depending on the number of reactions, with two types for each. In the first type, the electrolysis of CuCl yields Cu, while in the second type, the electrolysis of HCl and CuCl yields hydrogen. When comparing the various steps and types of the Cu-Cl TWC, it was found that a decrease in the number of steps reduces the production of solids and related issues, such as deposition, and low mass and heat transfer. However, it also introduces new challenges such as higher temperature requirements and complex separation as a result of combining several reactions. Due to the greater severity of these issues within the three-step TWC, it is not recommended for implementation in the industry. The second type also offers benefits such as fewer solid species and lower reaction temperatures, but it does encounter challenges like side reactions, a corrosive environment, and the adverse impact of increasing current and voltage on the concentration of species (Wang et al., 2009; Orhan et al., 2012).

A first-type five-step Cu-Cl TWC is displayed in Figure 2.6. During this TWC, $CuCl_2(s)$ and water vapor undergo a reaction in a fluidized bed reactor (S1) to yield

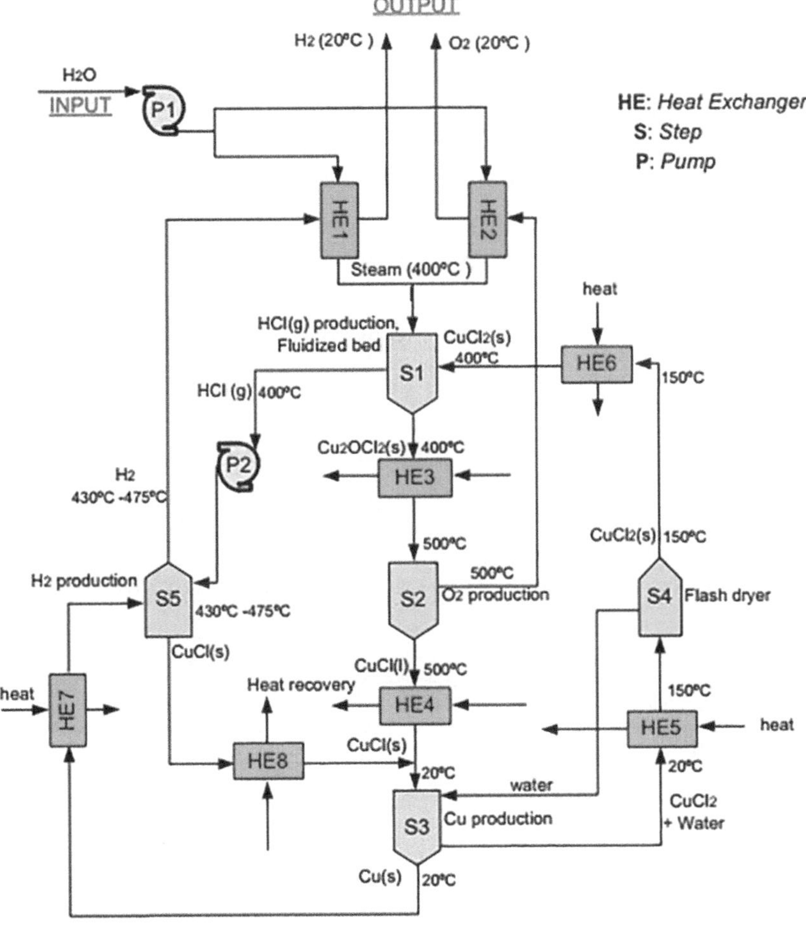

FIGURE 2.6   A flow diagram of the first-type five-step Cu-Cl process (Orhan et al., 2011).

HCl(g) and Cu_2OCl_2(s). HCl(g) is sent into the fifth reactor (S5) via a pump (P2), while the Cu_2OCl_2(s) in the second reactor (S2) undergoes decomposition, yielding CuCl(l) as a recycling agent and O_2 as a by-product. Within the electrolyzer, the reaction of CuCl from the S2 and S5, as well as water acting as the catalyst, results in the generation of $CuCl_2$(aq) and Cu(s). In the S5, the reaction between HCl(g) and Cu(s) yields H_2 as the ultimate product and CuCl, which is then cycled back to the S2. Meanwhile, $CuCl_2$(aq) transforms to a solid phase within a spray dryer and is reintroduced to the S1. The H_2O exiting from the dryer is utilized in the third reactor (S3). HE1 and HE2, the initial and subsequent heat exchangers, lower the temperature of oxygen and hydrogen to the surrounding environment, while their heat is harnessed to vaporize the incoming H_2O to cycle (Orhan et al., 2011).

Hydrolysis (400°C)

$$2CuCl_2(s) + H_2O(g) \rightarrow 2HCl(g) + CuO^*CuCl_2(s) \tag{2.19}$$

Thermolysis (O_2 production) (500°C)

$$CuO^*CuCl_2(s) \rightarrow 2CuCl(l) + 1/2O_2(g) \tag{2.20}$$

Electrolysis (25°C–80°C)

$$4CuCl(aq) \rightarrow 2CuCl_2(aq) + 2Cu(s) \tag{2.21}$$

Drying (> 100°C)

$$CuCl_2(aq) \rightarrow CuCl_2(s) \tag{2.22}$$

Chlorination (H_2 production) (450°C–475°C)

$$2Cu(s) + 2HCl(g) \rightarrow 2CuCl(l) + H_2(g) \tag{2.23}$$

Figure 2.7 indicates a schematic diagram of a first-type four-step Cu-Cl TWC. The process combines the drying and hydrolysis reactions of the five-step TWC into a

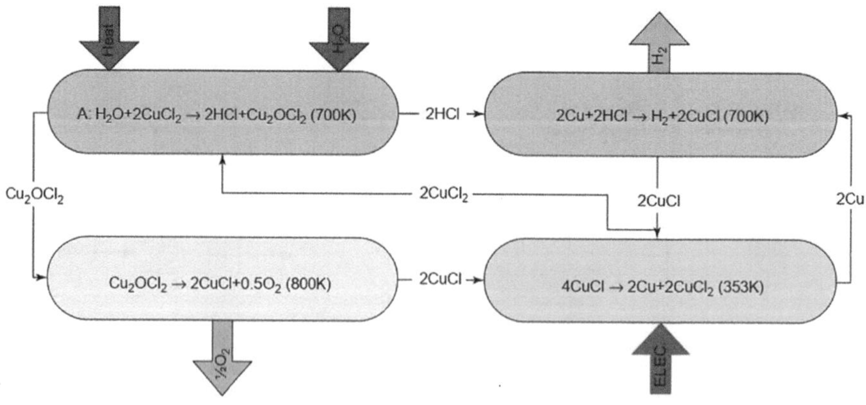

FIGURE 2.7 A schematic diagram of the first-type four-step Cu-Cl process (Dincer and Bicer, 2018).

single integrated step. The procedure comprises separators, dryers, reactors, and additional auxiliary apparatus. In each reactor, the chemical reaction takes place, and the products are separated by the separator. The spray dryer carries out the physical drying reaction and handles both drying and atomization (Dincer and Bicer, 2018). The reactions attributed to this cycle are as follows (Mehrpooya and Habibi, 2020):

Hydrolysis (375°C–400°C)

$$2CuCl_2(aq) + H_2O(g) \rightarrow 2HCl(g) + CuO^*CuCl_2(s) \tag{2.24}$$

Thermolysis (O_2 production) (500°C–530°C)

$$CuO^*CuCl_2(s) \rightarrow 2CuCl(l) + 1/2O_2(g) \tag{2.25}$$

Electrolysis (30°C–80°C)

$$4CuCl(aq) \rightarrow 2CuCl_2(aq) + 2Cu(s) \tag{2.26}$$

Chlorination (H_2 production) (450°C–475°C)

$$2Cu(s) + 2HCl(g) \rightarrow 2CuCl(l) + H_2(g) \tag{2.27}$$

The reactions that occurred in the first-type three-step Cu-Cl TWC can be represented by equations (2.28)–(2.30) (Mehrpooya and Habibi, 2020). A novel first-type three-step Cu-Cl cycle is illustrated in Figure 2.8. In this process, $CuCl_2$(s) is introduced as the feedstock in the O_2 production unit, and the resulting $CuCl_2$(aq) from the electrolysis unit is not utilized as a recycling agent. The $CuCl_2$(aq) exits mixer1 and is heated to the temperature of the first reactor (R1) in the O_2 production unit by a heater (H1). O_2, HCl, and CuCl are generated within the reactor and then separated using a liquid–gas separator. After reaching ambient temperature in the first cooler (C1), O_2 is extracted as a by-product, while HCl and CuCl proceed to the H_2 production and electrolysis units, respectively. The CuCl exiting from the third and first separators are combined in the mixer2, where their temperature is decreased to the

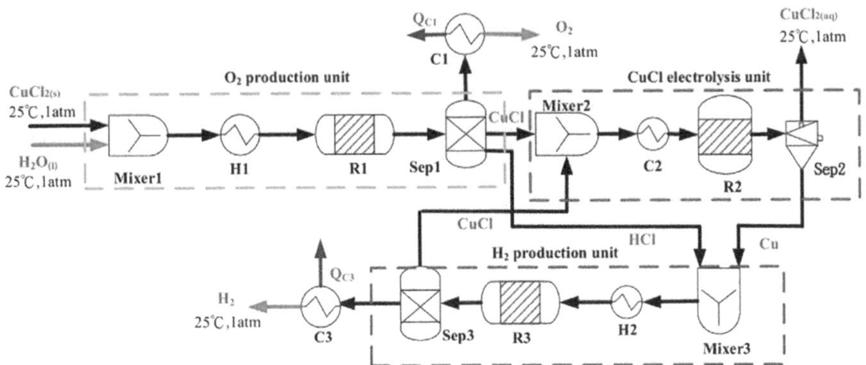

FIGURE 2.8 A flow diagram of the first-type three-step Cu-Cl system (Wu et al., 2016).

operational temperature of the electrolyzer. The electrolyzer generates $CuCl_2$ and Cu, which are subsequently separated using a solid–liquid separator. After reaching the reaction temperature, a blend of Cu from the second separator and HCl from the first separator is introduced into the third reactor. In this reactor, H_2 and liquid CuCl are generated. Ultimately, copper chloride is returned to the electrolysis unit as a recycling agent, and H_2 is extracted as the final product following the cooling process (Wu et al., 2016).

Hydrolysis (O_2 production) (400°C–600°C)

$$2CuCl_2\left(aq\right) + H_2O\left(g\right) \rightarrow 2HCl\left(g\right) + 2CuCl\left(1\right) + 1/2O_2\left(g\right) \qquad (2.28)$$

Electrolysis (20°C–80°C)

$$4CuCl\left(aq\right) \rightarrow 2CuCl_2\left(aq\right) + 2Cu\left(s\right) \qquad (2.29)$$

Chlorination (H_2 production) (430°C–475°C)

$$2Cu\left(s\right) + 2HCl\left(g\right) \rightarrow 2CuCl\left(1\right) + H_2\left(g\right) \qquad (2.30)$$

A second-type five-step Cu-Cl TWC is demonstrated in Figure 2.9. This process involves heating a mixture of $H_2O(l)$ and $CuCl_2(s)$ to 400°C using the first heater (H1) and then introducing it into a spray reactor to hydrolyze $CuCl_2(aq)$ and produce $HCl(g)$, $CuCl_2(s)$, and $CuO(s)$. The heat from the exit stream of the hydrolysis reactor is recovered by connecting it to the first cooler (C1), and hydrochloric acid is then obtained from a vapor–solid separator and directed into an isentropic compressor. Following the compressor, the HCl stream's temperature rises from 25°C to 619°C, while its pressure goes up from 1 to 23.7 atm. Subsequently, the third cooler (C3) is installed to lower the temperature of the hydrochloric acid stream from 619°C to 100°C. $CuCl_2(s)$ and $CuO(s)$ are conveyed via a solid conveyor and then subjected to heating by the second heater (H2). The solid reactants are introduced into a reactor to facilitate the O_2 and chlorine production reactions at 1 atm and 500°C. Assuming complete consumption of $CuO(s)$, $Cl_2(g)$, and $CuCl_2(s)$, the exit stream from the reactor is cooled to 25°C using the second cooler. Following this, $CuCl(s)$ and $O_2(g)$ can be completely separated using a solid-vapor separator. The purified O_2 is released at 1 atm and 25°C, while HCl is combined with $CuCl(s)$ at 100°C. The combination of $H_2O(g)$, $HCl(g)$, and $CuCl(s)$ is directed into an electrolyzer where $H_2(g)$ is generated at the cathode and $CuCl_2(aq)$ at the anode. The recycled $CuCl_2(aq)$ transforms into $CuCl_2(s)$ after the drying process and is subsequently transported by another solid conveyor and combined with the initial feed. The purified hydrogen stream from the electrochemical cell, operating at 24 bar and 100°C, is directed into a gas turbine for electricity generation and then released at 1 atm and 25°C (Wu et al., 2017). The reactions in this cycle are as follows (Mehrpooya and Habibi, 2020):

Hydrolysis (400°C)

$$2CuCl_2\left(s\right) + H_2O\left(g\right) \rightarrow 2HCl\left(g\right) + CuO\left(s\right) + CuCl_2\left(s\right) \qquad (2.31)$$

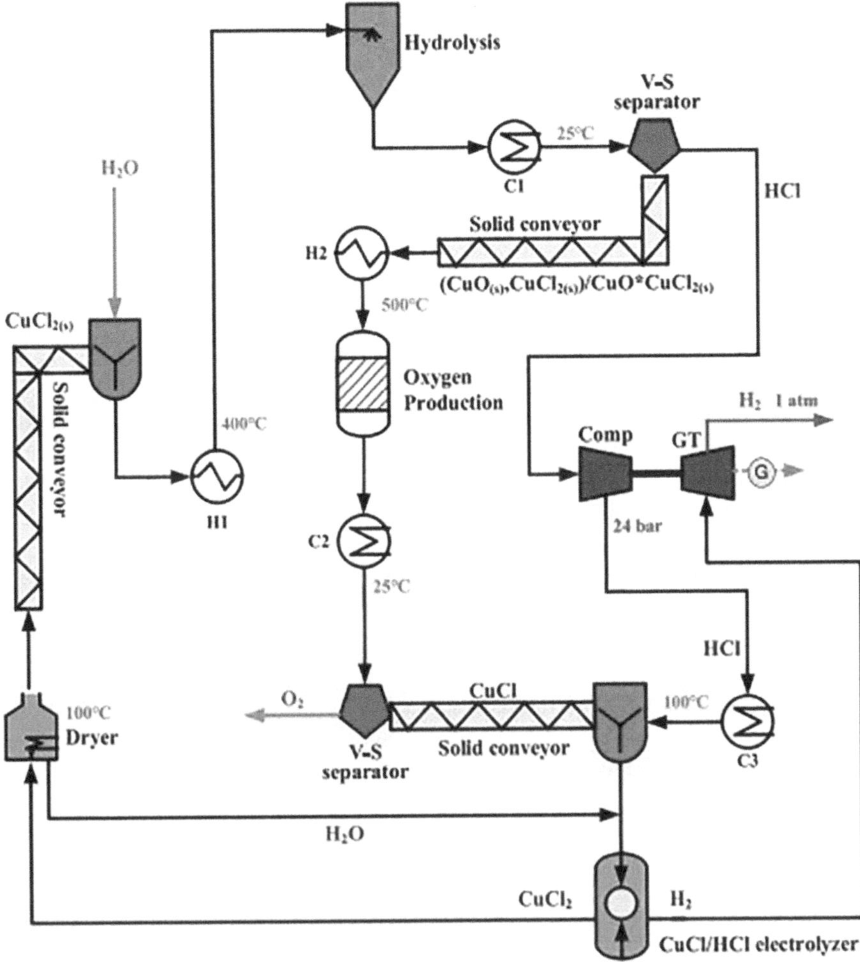

FIGURE 2.9 A flow diagram of the second-type five/four-step Cu-Cl process (Wu et al., 2017).

O$_2$ production (500°C)

$$CuCl_2(s) \rightarrow CuCl(l) + 1/2Cl_2(g) \tag{2.32}$$

$$CuO(s) + 1/2Cl_2(g) \rightarrow CuCl(l) + 1/2O_2(g) \tag{2.33}$$

Electrolysis (H$_2$ production) (100°C)

$$2CuCl(l) + 2HCl(l) \rightarrow 2CuCl_2(aq) + H_2(g) \tag{2.34}$$

Drying (100°C)

$$CuCl_2(aq) \rightarrow CuCl_2(s) \tag{2.35}$$

Importantly, if $CuCl_2(s)$ and $CuO(s)$ are replaced with CuO*CuCl2 as the intermediate species in the hydrolysis reaction, the second-type five-step Cu-Cl TWC transforms into the second-type four-step Cu-Cl TWC (see Figure 2.9). In this regard, reactions (2.31)–(2.33) are replaced with reactions (2.36) and (2.37) as follows (Mehrpooya and Habibi, 2020):

Hydrolysis (400°C)

$$2CuCl_2(s) + H_2O(g) \rightarrow 2HCl(g) + CuO^*CuCl_2(s) \tag{2.36}$$

O_2 production (500°C)

$$CuO^*CuCl_2(s) \rightarrow 2CuCl(1) + 1/2O_2(g) \tag{2.37}$$

By replacing $CuCl_2(s)$ in the hydrolysis unit (reaction (2.36)) with $CuCl_2(aq)$ and eliminating the drying reaction (equation (2.35)), the second-type four-step Cu-Cl TWC transforms into the second-type three-step Cu-Cl cycle (Mehrpooya and Habibi, 2020). Figure 2.10 presents a schematic diagram of this TWC.

Electrolysis constitutes a crucial stage in all Cu-Cl cycles, which may be conducted using either anionic or cationic membranes. Soltani et al. (2019) conducted an electrochemical analysis to model the electrolysis unit of a Cu-Cl TWC in a non-equilibrium state. They investigated the impact of temperature on factors like ohmic overpotentials, required voltage, and activation. Gong et al. (2009), Balashov et al. (2011), and Kim et al. (2011) have studied the impact of various membranes and different parameters on the performance of the electrolysis. Vaghasia et al. (2018) performed a study to suggest the best operating conditions for CuCl/HCl electrolysis.

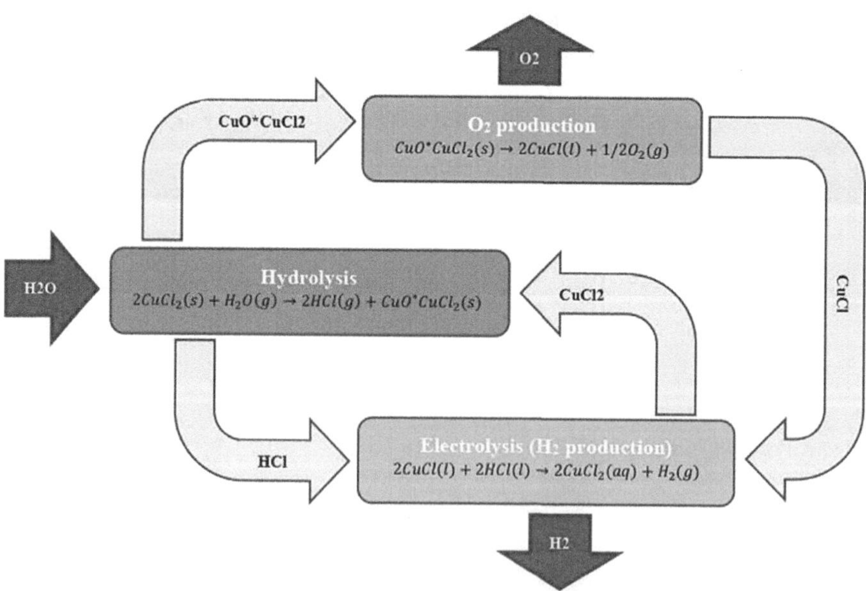

FIGURE 2.10　A schematic diagram of the second-type three-step Cu-Cl process.

Their results indicate that raising the molarity of the anode solution (H_2O, HCl, CuCl) enhances hydrogen production.

Wu et al. (2017) conducted an energy analysis on the first- and second-type of the three- to five-step Cu-Cl cycle, integrated with the Brayton cycle. The study revealed that, with 100% and 72% heat recovery, the first type of the three- and five-step cycles exhibited the highest energy efficiencies. However, from a commercial standpoint, the second type is favored over the first type due to reduced solid production and lower temperature of the exiting H_2. Additionally, the four-step TWC is preferred to the five-step in the second-type configuration as it eliminates the production of toxic Cl_2 gas and has lower energy losses and equipment.

Osuolale et al. (2018) conducted a comparison of the first-type three-, four-, and five-step cycles, evaluating their exergy and energy efficiencies as well as the factors influencing them. The exergy efficiencies for the five-, four-, and three-step TWCs were found to be 59.64%, 44.74%, and 78.21%, respectively, while the energy efficiencies were 50%, 49%, and 35%, respectively. In the case of the four-step TWC, reducing the reference temperature and enhancing the temperature of the hydrolysis unit resulted in improved energy efficiency. Meanwhile, for the five-step cycle, increasing the reference temperature led to an improvement in exergy efficiency.

Ishaq and Dincer (2019) utilized Aspen software to simulate the first-type three- and five-step, as well as the second-type four-step Cu-Cl TWCs and conducted exergy and energy analyses for each. Their findings revealed that in the three- and five-step TWCs, the highest heat consumption and exergy destruction were associated with O_2 and H_2 production units, respectively. In the case of the four-step TWC, both were related to the O_2 production unit. Furthermore, a comparison of exergy and energy efficiencies indicated that the four-step TWC exhibited higher efficiencies than the others.

Ratlamwala and Dincer (2012), Al-Zareer et al. (2017, 2018, 2020), and Sayyaadi (2017) designed some novel Cu-Cl systems by adding the Rankine cycle, Brayton cycle, and Kalina cycle. Their analysis was conducted in terms of exergy and energy efficiency.

2.3.4 MG-CL CYCLE

The Mg-Cl TWC is a promising alternative that offers favorable exergy and energy efficiency. It has the potential to rival other low-temperature cycles, including the Cu-Cl TWC. The cycle's maximum low temperature opens up possibilities for integration with various energy sources, including solar, nuclear, and waste heat from other power plants (Balta et al., 2014). The cycle is based on the two-step Hallett air products cycle, incorporating the aqueous HCl electrolysis and reverse Deacon reaction (Mehrpooya and Habibi, 2020). Challenges such as the potential reaction between products in the first step, the high potential required due to the aqueous state of HCl in the second step, the corrosive properties of the species, and high-temperature conditions (800°C) have hindered the success of this cycle (Simpson et al., 2006). By addressing these challenges and reducing the TWC temperature to approximately 450°C–550°C, the Mg-Cl TWC was introduced as a viable low-temperature alternative. The initial Mg-Cl cycle is a three-step process comprising electrolysis, chlorination, and hydrolysis. The combined effects of the first two reactions result in the reverse Deacon reaction (Simpson et al., 2006).

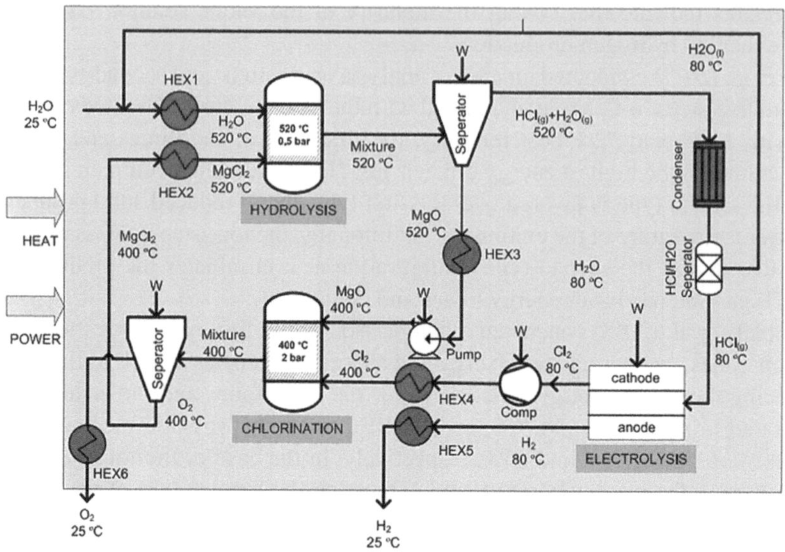

FIGURE 2.11　A flow diagram of the first-type three-step Mg-Cl process (Ozcan and Dincer, 2014).

Figure 2.11 illustrates a schematic of the first-type three-step Mg-Cl cycle. In the first step, which is endothermic, $MgCl_2$ produced in the chlorination unit undergoes a reaction with water vapor in the hydrolysis reactor (reaction (2.38)). Following product separation, the magnesium oxide is cooled to 400°C and introduced into the chlorination reactor. The H_2O is separated from the HCl-H_2O mixture and recycled back to the hydrolysis reactor, while the remaining hydrochloric acid is transferred to the electrolytic cell. At this stage, Cl_2 produced at the cathode reacts with MgO once the chlorination reactor reaches its operating temperature. Hydrogen is generated as the final product at the anode and subsequently cooled to the surrounding temperature using a heat exchanger (reaction (2.40)). The resulting exothermic reaction yields $MgCl_2$ and O_2 (reaction (2.39)), which are then separated in a separator. The by-product O_2 is cooled to the surrounding temperature, similar to H_2, while the $MgCl_2$ is reused as the reactant in the hydrolysis unit (Ozcan and Dincer, 2014). The reactions in this cycle are as follows (Ozcan and Dincer, 2014):

Hydrolysis (450°C–550°C)

$$MgCl_2(s) + H_2O(g) \rightarrow MgO(s) + 2HCl(aq) \quad \Delta H = 92.65 \text{ kJ/mol } H_2 \quad (2.38)$$

Chlorination (450°C–500°C)

$$MgO(s) + Cl_2(g) \rightarrow MgCl_2(s) + 1/2O_2(g) \quad \Delta H = -35.33 \text{ kJ/mol } H_2 \quad (2.39)$$

Electrolysis (70°C–90°C)

$$2HCl(aq) \rightarrow H_2(g) + Cl_2(g) \quad 1.8 \text{ V} \quad (2.40)$$

In the second-type three-step Mg-Cl TWC, the exothermic hydrolysis reaction occurs at a relatively low-temperature range of 240°C–300°C. The reaction between H_2O and $MgCl_2$ leads to the formation of HCl and MgOHCl (reaction (2.41)). In the second reactor, the Cl_2 reacts with MgOHCl, resulting in the generation of steam, oxygen, and $MgCl_2$ (reaction (2.42)). The $MgCl_2$ is then directed to the hydrolysis reactor, while O_2 is separated from the H_2O. In the final stage, HCl undergoes electrolysis to produce H_2 and chlorine (reaction (2.43)). The chlorine gas is directed to the chlorination reactor, while this system continues to generate O_2 and H_2 (Mehrpooya and Habibi, 2020). Ozcan and Dincer (2016b) reported that one challenge of this process is the generation of O_2 and steam during the chlorination reaction, necessitating a relatively difficult separation step as a result of the solubility of O_2.

Hydrolysis (240°C–300°C)

$$2MgCl_2(s) + 2H_2O(g) \rightarrow 2MgOHCl + 2HCl(aq) \tag{2.41}$$

Chlorination (450°C)

$$2MgOHCl + Cl_2 \rightarrow 2MgCl_2 + H_2O + 0.5O_2 \quad \Delta H = 116.3 \text{ kJ/mol } H_2 \tag{2.42}$$

Electrolysis (90°C)

$$2HCl(aq) \rightarrow H_2(g) + Cl_2(g) \quad 1.8 \text{ V} \tag{2.43}$$

In the first- and second-type of the three-step Mg-Cl TWC, as a result of the high H_2O-to-HCl ratio in the products of the hydrolysis reaction, hydrochloric acid is present in an aqueous phase, necessitating increased electrical voltage and work in the electrolysis unit (Mehrpooya and Habibi, 2020). Therefore, a four-step Mg-Cl TWC (Figure 2.12) is suggested to tackle this issue. The magnesium oxide generated in this innovative TWC exhibits greater reactivity compared to that of the first-type three-step Mg-Cl TWC.

The hydrolysis reaction is conducted at a reduced temperature, similar to that in the second-type three-step cycle (reaction (2.44)). The gaseous HCl and MgO are obtained through the endothermic decomposition of MgHOCl (reaction (2.45)). In this process, the introduction of an inert gas serves to remove the hydrochloric acid from the surface of MgO, preventing their reaction and lowering the temperature while enhancing the decomposition reaction rate. The MgO like the chlorination reaction in the first-type three-step TWC undergoes a reaction with Cl_2 to yield O_2 and $MgCl_2$ (reaction (2.46)). The HCl generated in reactions (2.45) and (2.44) is subsequently directed to the dry and aqueous electrolysis steps, respectively (reactions (2.47) and (2.48)) (Ozcan and Dincer, 2016a).

Hydrolysis (280°C)

$$MgCl_2(s) + H_2O(g) \rightarrow MgOHCl + HCl(aq) \quad \Delta H = -17.82 \text{ kJ/mol } H_2 \tag{2.44}$$

Decomposition (450°C)

$$MgOHCl \rightarrow MgO + HCl(g) \quad \Delta H = 118.1 \text{ kJ/mol } H_2 \tag{2.45}$$

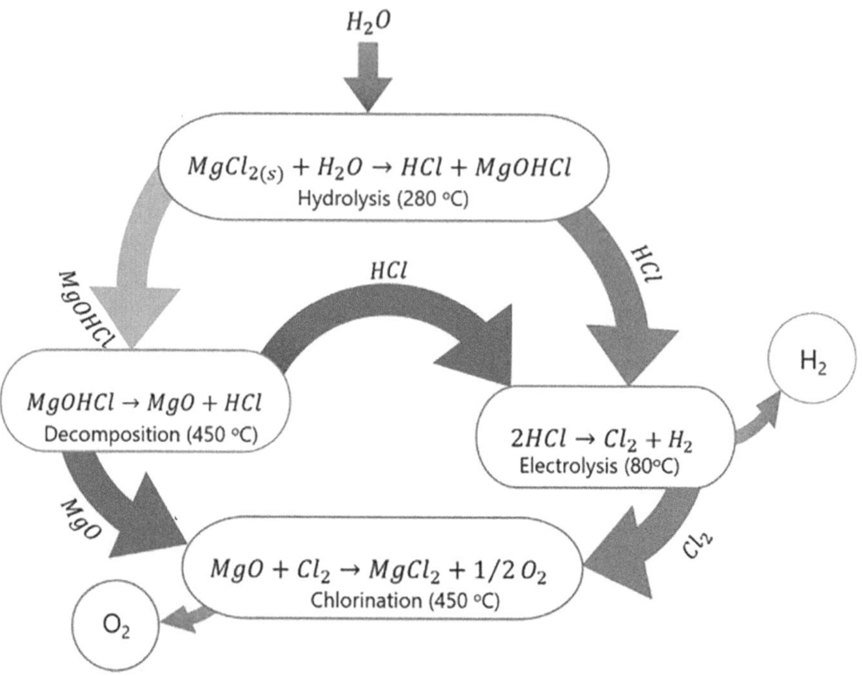

FIGURE 2.12   A schematic diagram of the four-step Mg-Cl TWC (Oruc and Dincer, 2021).

Chlorination (450°C–500°C)

$$MgO + Cl_2 \rightarrow MgCl_2 + 0.5O_2 \qquad \Delta H = -31.51 \text{ kJ/mol } H_2 \qquad (2.46)$$

Dry electrolysis (70°C)

$$2HCl(g) \rightarrow H_2(g) + Cl_2(g) \qquad 1.4V \qquad (2.47)$$

Aqueous electrolysis (70°C)

$$2HCl(aq) \rightarrow H_2(g) + Cl_2(g) \qquad 1.8V \qquad (2.48)$$

To minimize electrical work and enhance efficiency during the electrolysis step, an additional step was incorporated into the existing four-step TWC to capture dry hydrochloric acid from the H₂O-HCl mixture during the hydrolysis step. This involved utilizing MgO as an absorbent, as it reacts with HCl. Reactions (2.49)–(2.51) pertain to this specific step. The reported exergy and energy efficiencies of the cycle were 53.4% and 43.4%, respectively, when achieving 100% hydrochloric acid capture (Ozcan and Dincer, 2016b, 2018).

$$MgO + 2HCl \rightarrow MgCl_2 + H_2O \qquad (2.49)$$

$$MgO + H_2O \rightarrow Mg(OH)_2 \tag{2.50}$$

$$MgO + HCl \rightarrow MgOHCl \tag{2.51}$$

Ozcan and Dincer (2017) conducted an exergoeconomic and thermodynamic analysis, including multi-objective optimization, for the four-step Mg-Cl TWC with 30% hydrochloric acid capture. H_2 production and annual costs as well as exergy and energy efficiencies were 3.67 \$/kg H_2, \$458.5 million, 53%, and 44.3%, respectively.

2.3.5 Fe-Cl Cycle

Funk (1976) conducted the initial investigation into four-step Fe-Cl TWC. A schematic diagram of this Fe-Cl TWC is shown in Figure 2.13. This Fe-Cl TWC consists of four reactions, including one at low temperature (chlorination), one at moderate temperature (decomposition), and two at high temperature (hydrolysis and O_2 production) as follows (Safari and Dincer, 2020b):

Hydrolysis (650°C)

$$6FeCl_2 + 8H_2O \rightarrow 2Fe_3O_4 + 12HCl + 2H_2 \tag{2.52}$$

Chlorination (125°C)

$$2Fe_3O_4 + 16HCl \rightarrow 4FeCl_3 + 2FeCl_2 + 8H_2O \tag{2.53}$$

Decomposition (425°C)

$$4FeCl_3 \rightarrow 4FeCl_2 + 2Cl_2 \tag{2.54}$$

O_2 production (650°C)

$$2Cl_2 + 2H_2O \rightarrow 4HCl + O_2 \tag{2.55}$$

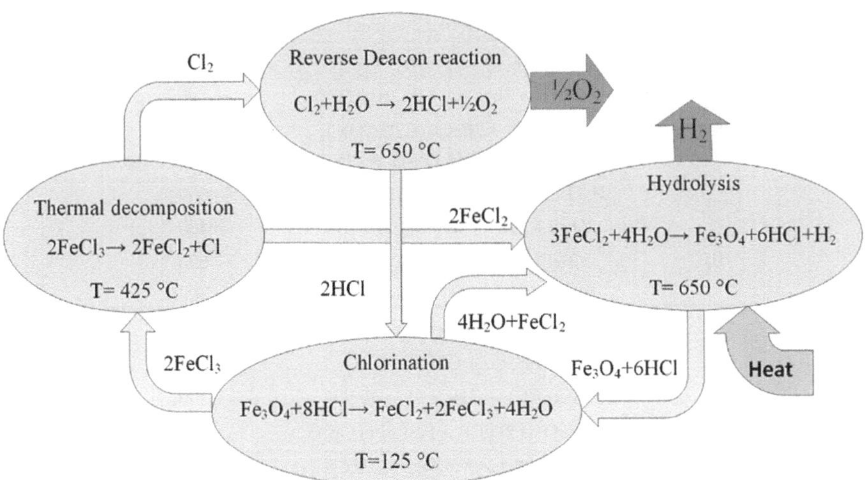

FIGURE 2.13 A schematic diagram of the Fe-Cl TWC (Safari and Dincer, 2020b).

Van Velzen and Langenkamp (1978) investigated this Fe-Cl TWC for the production of one mole of H_2 (referred to as Mark 15) and revealed a relatively lower energy efficiency of 20.9%. This lower efficiency renders it less economically viable compared to other H_2 production routes. On the other hand, the cost-effective nature of the chemicals and the in-depth understanding of the chemical properties of chlorides and iron oxides have garnered interest in this cycle (Canavesio et al., 2015). However, the investigations have not yielded satisfactory results thus far. The Fe-Cl TWC has encountered some significant issues: (i) the hydrolysis of $FeCl_2$, (ii) the heat required, and (iii) the limited reactivity of thermal decomposition of $FeCl_3$, where Fe_2Cl_6 is formed instead of Cl_2 and $FeCl_2$ (Tsutsumi, 2009).

Knoche et al. (1976) proposed eighteen different reactions within the Fe-Cl TWC. These reactions combine to form 35 multistep cycles. Table 2.2 showcases four cycles that are thermodynamically feasible for Fe-Cl TWC. The Fe-Cl TWC enables the heat necessary for H_2 production to be sourced from solar or nuclear power plants.

The literature contains several reports that analyze the performance of the Fe-Cl TWC from various perspectives. Canavesio et al. (2015) conducted experimental studies on the four-step Fe-Cl TWC, focusing on investigating the validity of thermochemical reactions. Their study involved studying the reaction kinetics and the parameters that influence them, with the aim of enhancing the hydrogen production rate and cycle efficiency. Utgikar and Ward (2006) performed a life cycle assessment on the Fe-Cl TWC, considering it a viable option for hydrogen production based on nuclear power. The research examined the environmental effects of an H_2 production

---

TABLE 2.2

Thermodynamically Feasible Fe-Cl TWCs (Andress and Martin, 2011)

Cycle	Reaction	Temperature (K)
(I)	$H_2 + Fe_3O_4 \rightarrow H_2O + 3FeO$	1,100
	$3H_2O + 3FeCl_2 \rightarrow 6HCl + 3FeO$	1,100
	$Fe + 2Fe(OH)_3 \rightarrow 3H_2 + O_2 + Fe_3O_4$	300
	$6HCl + 6FeO \rightarrow 3FeCl_2 + Fe + 2Fe(OH)_3$	1,200
(II)	$6HCl + 2Fe_2O_3 \rightarrow 2FeCl_3 + 2Fe(OH)_3$	300
	$H_2 + 2Fe(OH)_3 \rightarrow 4H_2O + O_2 + 2Fe$	1,100
	$3H_2O + 2FeCl_3 \rightarrow 6HCl + Fe_2O_3$	600
	$3H_2O + 2Fe \rightarrow 3H_2 + Fe_2O_3$	400
(III)	$6FeCl_3 \rightarrow 3Cl_2 + 6FeCl_2$	600
	$3Cl_2 + 2FeO \rightarrow O_2 + 2FeCl_3$	300
	$12HCl + 4FeO \rightarrow 4H_2O + 2H_2 + 4FeCl_3$	300
	$6H_2O + 6FeCl_2 \rightarrow 12HCl + 6FeO$	1,100
(IV)	$Fe + 3Fe(OH)_2 \rightarrow 3H_2 + O_2 + 4FeO$	1,200
	$6HCl + 6FeO \rightarrow 2FeCl_3 + Fe + 3Fe(OH)_2$	300
	$3H_2O + 2FeCl_3 \rightarrow 6HCl + Fe_2O_3$	600
	$H_2 + Fe_2O_3 \rightarrow H_2O + 2FeO$	1,100

facility that utilized a Fe-Cl TWC integrated with a very high-temperature reactor. The system's life cycle assessment revealed an acidification potential and a GWP equivalent to $11.252\,g\ SO_2$-eq/kg H_2 and $2,515\,g\ CO_2$-eq/kg H_2, respectively. Safari and Dincer (2020b) used the ASPEN Plus software package to simulate critical processes like reverse Deacon reaction and hydrolysis and conducted parametric studies to assess the impact of various operating parameters on the performance of each step.

2.3.6 V-Cl Cycle

McRea and his team initially investigated the V-Cl TWC, which has been documented as the most efficient among the chlorine family cycles (Safari and Dincer, 2022). The V-Cl cycle as a pure TWC relies solely on heat to generate H_2 from H_2O, with a maximum process temperature of 925°C (Lewis and Masin, 2009). This cycle includes two high-temperature reactions (equations (2.57) and (2.59)), one moderate-temperature (equation (2.58)), and one low-temperature reaction (equation (2.56)) (Balta et al., 2016):

$$2VCl_2\,(s) + 2HCl\,(aq) \rightarrow 2VCl_3\,(s) + H_2\,(g)\ 25 - 120°C \qquad (2.56)$$

$$4VCl_3\,(s) \rightarrow 2VCl_4\,(g) + 2VCl_2\,(s)\ 750 - 766°C \qquad (2.57)$$

$$2VCl_4\,(l) \rightarrow 2VCl_3\,(s) + Cl_2\,(g)\ 200°C \qquad (2.58)$$

$$Cl_2\,(g) + H_2O\,(g) \rightarrow 2HCl\,(g) + 1/2O_2\,(g)\ 850 - 925°C \qquad (2.59)$$

In the existing literature, various chemical reactions are outlined for sequential execution in order to achieve greater efficiency within this cycle. For instance, Knoche et al. (1984a, b) have suggested the following two reactions instead of reaction (2.59), which is commonly referred to as the reverse Deacon reaction:

$$Cl_2\,(g) + 1/3V_2O_5\,(l) \rightarrow 2/3VOCl_3\,(g) + 1/2O_2\,(g)\ 875°C \qquad (2.60)$$

$$H_2O\,(g) + 2/3VOCl_3\,(g) \rightarrow 1/3V_2O_5\,(l) + 2HCl\,(g)\ 125°C \qquad (2.61)$$

Lewis and Masin (2009) highlighted that the H_2 production phase of this cycle (reaction (2.56)) can be replaced by the following reactions:

$$2VCl_2\,(s) + (2nH_2O + 2HCl)(l) \rightarrow 2(VCl_3.nH_2O)(s) + H_2\,(g)\ 125°C \quad (2.62)$$

$$2(VCl_3.nH_2O)(s) \rightarrow 2VCl_3\,(s) + H_2O\,(g)\ 160°C \qquad (2.63)$$

Moreover, Amendola (2005) described two types of TWC in his patent application. In the first type, VCl_3 is decomposed to Cl_2 and VCl_2 at approximately 525°C (reaction (2.64)). In the second step, the Cl_2 from the previous reaction is combined with H_2O to yield HCl and O_2 (reaction (2.65)). Lastly, HCl is utilized in a reaction with VCl_2 to regenerate VCl_3 and H_2. The first type constitutes a dry process involving the following reactions:

$$2VCl_3(s) \rightarrow 2VCl_2(s) + Cl_2(g)\ 525°C \tag{2.64}$$

$$Cl_2(g) + H_2O(g) \rightarrow 2HCl(g) + 1/2O_2(g)\ 100°C \tag{2.65}$$

$$2VCl_2(s) + 2HCl(g) \rightarrow 2VCl_3(s) + H_2(g)\ 300°C \tag{2.66}$$

The second type involves a wet process, where H_2 is generated from an HCl aqueous solution, as outlined below.

$$2VCl_3(s) \rightarrow 2VCl_2(s) + Cl_2(g)\ 525°C \tag{2.67}$$

$$Cl_2(g) + H_2O(g) \rightarrow 2HCl(g) + 1/2O_2(g)\ 100°C \tag{2.68}$$

$$2VCl_2(s) + 2HCl(aq) \rightarrow 2VCl_3(aq) + H_2(g)\ 30°C \tag{2.69}$$

$$2VCl_3(aq) \rightarrow 2VCl_3(s) \tag{2.70}$$

The wet process offers the advantage of not requiring the removal of HCl from oxygen, as the products of the reaction (2.68) are condensed to facilitate the decomposition of hydrochloric acid at 30°C. However, the wet process necessitates an additional reaction to dry VCl_3 (reaction (2.69)). Figure 2.14 demonstrates a schematic diagram of this TWC.

Knoche et al. (1984a, b) conducted an experimental investigation of the V-Cl TWC and carried out exergy and energy evaluation using the experimental values for reaction steps. A new integration of V-Cl TWC with algal biomass gasification

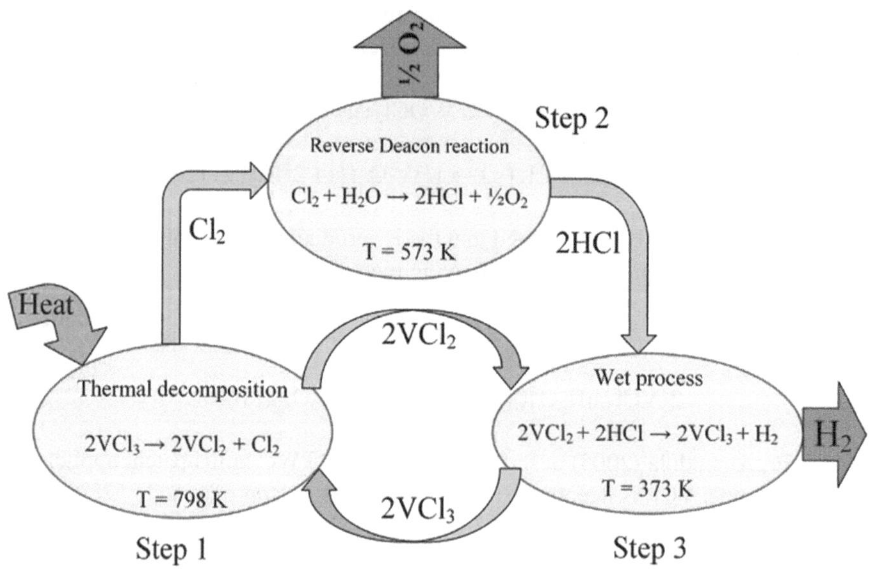

FIGURE 2.14 A schematic diagram of the V-Cl TWC (Safari and Dincer, 2022).

was modeled and optimized by Safari and Dincer (2022). Optimal values for environmental impacts, total cost per unit of exergy, and overall exergy efficiency were 14.46 g CO_2-eq/kWh, 6.36 \$/GJ, and 60.45%, respectively. Balta et al. (2016) performed a comparative evaluation of different chlorine TWCs. They found that the three-step V-Cl TWC demonstrated the highest exergy and energy efficiencies, leading them to recommend further investigation of this cycle.

2.4 HYBRID THERMOCHEMICAL CYCLE

To enhance efficiency and decrease GWP and costs, it is possible to implement thermochemical water splitting within hybrid cycles (Allen et al., 2014), which can be integrated with renewable energy sources like electricity, nuclear heat waste, wind energy, or solar energy. By combining these hybrid thermochemical cycles with electricity, it is possible to achieve greater energy efficiency of up to 50% through the recycling of industrial waste heat and the reduction of required maximum temperatures (Lee et al., 2022).

The two-step HyS TWC called a Westinghouse cycle, is a widely recognized process for producing hydrogen on a large scale. It is referred to as a combined electrochemical-thermochemical cycle. The hybrid sulfur cycle follows a two-step process. As depicted in Figure 2.15, this cycle involves a thermal decomposition step of H_2SO_4 to H_2O and SO_2, which consumes heat and produces O_2 at high temperatures (more than 800°C). Additionally, there is an electrochemical unit for sulfur oxide electrolysis, where electricity is consumed and H_2 is produced at T = 100°C (Sattler et al., 2017). The voltage needed for the electrolysis of SO_2 is approximately 0.17 V, significantly less than the voltage required for water electrolysis (1.23 V) (Allen et al., 2014). The analysis from an electrochemical perspective led to the conclusion that achieving high overall efficiency is possible by operating the electrolyzer at low voltage while producing concentrated sulfuric acid.

Bilgen (1988) conducted the initial assessment of this cycle, revealing a total energy efficiency of 40%. In the 2000s, Commonwealth Scientific and Industrial

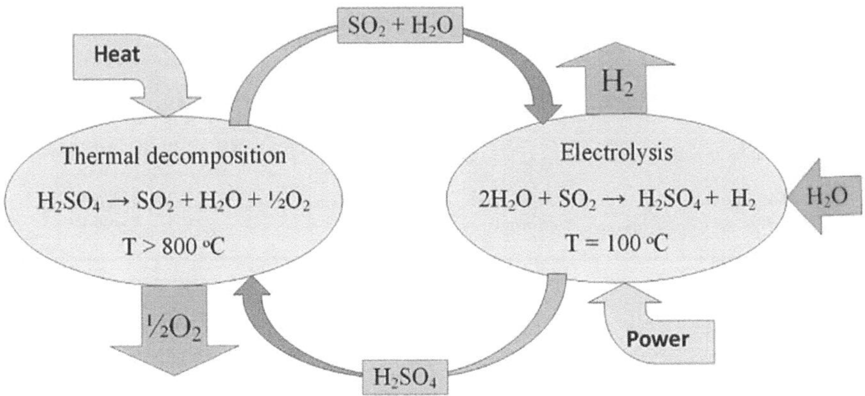

FIGURE 2.15 Schematic diagram of a HyS TWC (Safari and Dincer, 2020a).

Research Organization (CSIRO) and Savannah River National Laboratory (SRNL) have primarily examined the combination of this TWC with concentrated solar energy. They reported a plant efficiency of 27% and a H_2 production cost ranging from 3.19 to 5.57 \$/kg (Corgnale and Summers, 2011; Allen et al., 2014). The main difficulties encountered in the HyS cycle are the reduction of sulfur oxide at high temperatures and the presence of corrosive materials (O'Brien et al., 2010; Roeb et al., 2013). To address the issue of high-temperature sulfur oxide reduction, iron oxide catalysts are frequently employed to enhance the reaction rate during this step. Additionally, silicon carbide (SiC) has been identified as a suitable material for withstanding the corrosive substances generated during the process (Corgnale and Summers, 2011).

Some hybrid cycles use other forms of energy, such as photonic energy, in addition to thermal energy. Huang et al. (2006) introduced a new family of hybrid sulfur-ammonia (HySA) photo-thermochemical cycles. In this innovative approach, they incorporated ammonia as the working reagent. Figure 2.16 demonstrates a schematic diagram of the HySA TWC. In this hybrid TWC, the inputs include H_2O, heat, and photonic energy; and similar to other TWCs, the outputs consist solely of oxygen and hydrogen. The following list outlines the five main reactions of the HySA cycle (Safari and Dincer, 2020a):

Chemical absorption (25°C)

$$SO_2(g) + 2NH_3(g) + H_2O(l) \rightarrow (NH_4)_2 SO_3(aq) \tag{2.71}$$

Electrolytic oxidation (80°C–150°C)

$$(NH_4)_2 SO_3(aq) + H_2O(l) \rightarrow (NH_4)_2 SO_4(aq) + H_2(g) \tag{2.72}$$

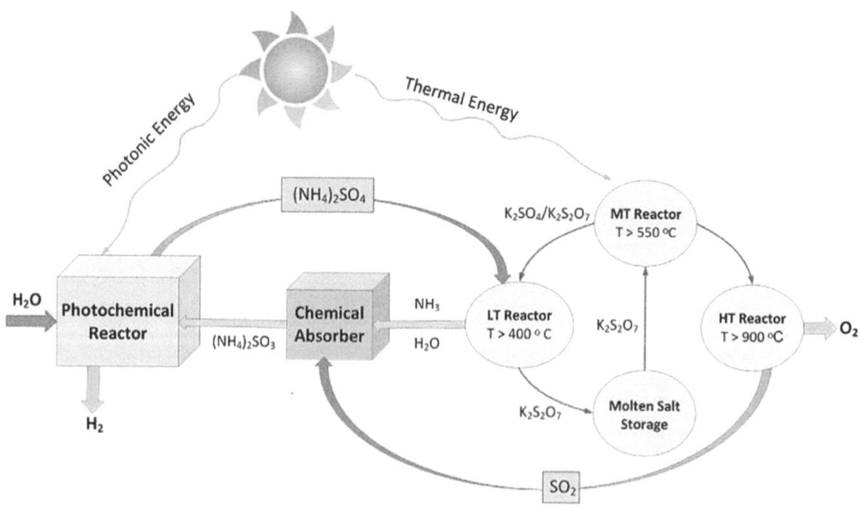

FIGURE 2.16　Schematic diagram of a HySA TWC (Safari and Dincer, 2020a).

Solar thermal conversion (400°C)

$$(NH_4)_2 SO_3 (aq) + K_2SO_4 (l) \rightarrow K_2S_2O_7 (l) + 2NH_3 (g) + H_2O(g) \quad (2.73)$$

Solar thermal conversion (550°C)

$$K_2S_2O_7 (l) \rightarrow K_2SO_4 (l) + SO_3 (g) \quad (2.74)$$

Solar thermal conversion (850°C)

$$SO_3 (g) \rightarrow SO_2 (g) + 1/2O_2 (g) \quad (2.75)$$

In reaction (2.72), ammonium sulfite undergoes an oxidation reaction at a temperature close to ambient, resulting in the production of hydrogen. In the low-temperature reactor, potassium sulfate reacts with ammonium sulfate to form potassium pyrosulfate (reaction (2.73)), creating a sub-cycle among the other thermochemical reactions. The potassium pyrosulfate formed in the previous step is then fed into the lower-temperature reactor, where it undergoes decomposition to produce K_2SO_4 and SO_3, effectively closing the sub-cycle. The combination of $K_2S_2O_7$ and K_2SO_4 forms a liquid melt, enabling the separation and transfer of chemicals in reactions (2.73) and (2.74) to take place. In the presence of a catalyst, the O_2 production unit takes place at a high temperature. The separation of O_2 from sulfur dioxide occurs in the presence of H_2O during the reaction (2.71).

Ammonium sulfate can be utilized as a substitute for H_2SO_4 to minimize the corrosion of facilities caused by sulfur oxide intermediates. However, in hybrid cycles, the use of ammonium sulfate salt requires reactors for chemical adsorption and additional steps, as well as a separate decomposition cycle for ammonium sulfate. The primary benefit of the HySA TWC is the integrated utilization of different parts of solar irradiance. The photonic portion (UV, visible) is utilized for the hydrogen production unit, while the thermal portion (infrared) is employed for the oxygen production step (Littlefield et al., 2012). However, due to the absence of a decision between dual solar fields or beam-splitting mirrors, photolytic sulfur ammonia has not reached a level of maturity that allows for a definitive conceptual design to be established (Safari and Dincer, 2020a).

The combination of a metal oxide and metal sulfate can be employed for hydrogen production through TWCs, in conjunction with solar energy. This is advantageous due to the lower cost and thermal reduction temperature associated with this approach. The process of thermally reducing $MgSO_4$ to MgO is extensively utilized in the two-step TWCs (Bhosale, 2020) as follows:

Solar thermal reduction

$$MgSO_4 \rightarrow MgO + SO_2 + 1/2O_2 \quad (2.76)$$

Thermochemical water reduction

$$MgO + H_2O + SO_2 \rightarrow MgSO_4 + H_2 \quad (2.77)$$

2.5   COMPARISON BETWEEN VARIOUS TWCS

This section presents a comparative assessment of TWCs belonging to the sulfur and chlorine families. S-I and HyS from the sulfur family group, V-Cl, Cu-Cl, Mg-Cl, and Fe-Cl from the chlorine family group, and ZnO/Zn from the metal oxide cycle group are compared from economical, environmental, exergetic, and energetic viewpoints. Table 2.3 outlines the chosen TWCs, highlighting the unique characteristics of each cycle and addressing the primary obstacles faced during their development.

The TWCs mentioned in this study were selected based on priority assessments conducted by Balta et al. (2016). Various comparative evaluations of TWCs have been performed in the literature, including comparisons with other hydrogen production methods. For instance, Acar and Dincer (2015) reviewed and evaluated various hydrogen production technologies, considering factors such as efficiency, cost, and sustainability. Balta et al. (2016) conducted a comparative evaluation of TWCs within the chlorine family, identifying Mg-Cl and Cu-Cl as more sustainable in terms of operating temperature, while V-Cl exhibited higher exergy efficiency. Further investigations were recommended for the V-Cl cycle in order to address the existing technical challenges.

TABLE 2.3

Distinct Features and Major Challenges of Various TWCs (Safari and Dincer, 2020a)

TWC	Distinct Features	Major Challenges
ZnO/Zn	• No side reaction • H_2 generation in two stages	• Expensive materials for high-temperature solar reactor
S-I	• The process does not generate any harmful by-products or emissions • High thermal efficiency	• High-temperature solar for integration • Corrosive materials
Cu-Cl	• Capability to utilize low-grade waste heat for enhancing energy efficiency and potentially reducing the cost of materials • Lower operating temperatures	• Corrosive operating fluids • Solids handling between process
Mg-Cl	• Promising to be integrated with nuclear and solar • Low temperature	• Fast formation of $MgCl_2$ hydrates • MgO chlorination • Development of electrochemical reaction • Optimization of Hydrolysis reaction
Fe-Cl	• Abundance of materials and low cost • Known thermodynamics	• Dimerization of $FeCl_3$ to Fe_2Cl_6 • High-temperature $FeCl_3$ decomposition • Low efficiency
V-Cl	• Capable of integration with high-temperature solar energy • High efficiency	• Unknown thermodynamics • Slow kinetics of chlorination reaction • High-temperature and separation requirements for the reverse Deacon reaction
HyS	• SO_2 as the only intermediate • Cheap materials	• Corrosive materials • High temperature H_2SO_4 decomposition

---

El-Emam and Khamis (2018) presented a review of research projects on hydrogen production by the International Atomic Energy Agency. The study also involved a comparison of the exergy and energy efficiencies of various TWCs within the chlorine family. In another study, El-Emam and Ozcan (2019) assessed the options for large-scale H_2 production based on nuclear energy. They selected HyS, S-I, Ca-Br, Mg-Cl, and Cu-Cl for evaluation, focusing on their performance when integrated with different nuclear or solar technologies.

Figure 2.17 illustrates the comparative assessment of the selected TWCs in terms of exergy and energy efficiency based on the lower heating value (El-Emam and Ozcan, 2019; Safari and Dincer, 2020a). Among the studied TWCs, it can be observed that V-Cl exhibits the highest exergy and energy efficiency, whereas Fe-Cl demonstrates the lowest.

Additionally, based on the data reported by Safari and Dincer (2020a), Figure 2.18 provides a comparative analysis of various TWCs focused on GWP and hydrogen

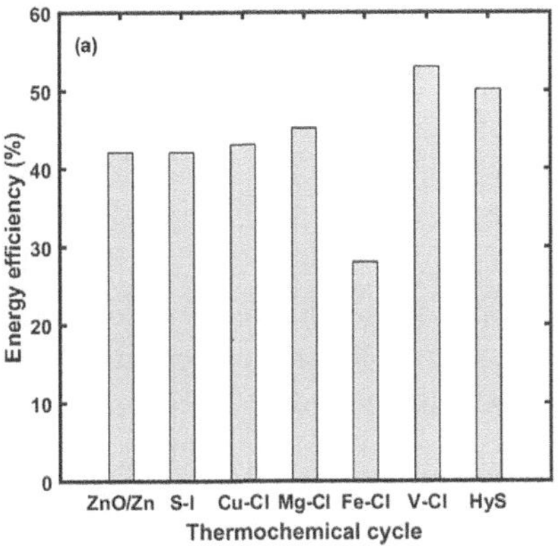
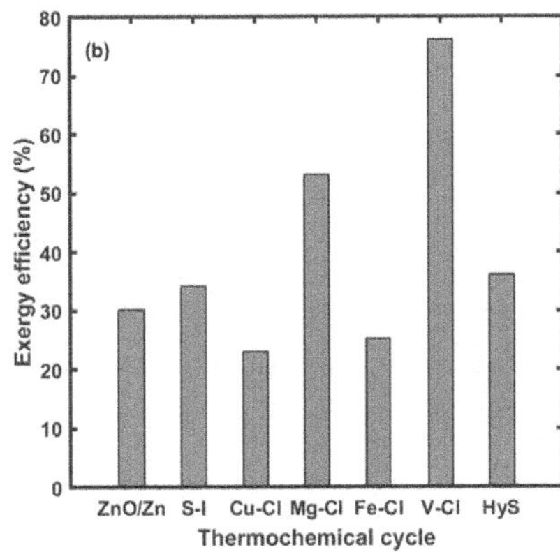

FIGURE 2.17 Comparison between various TWCs in terms of (a) energy efficiency, and (b) exergy efficiency.

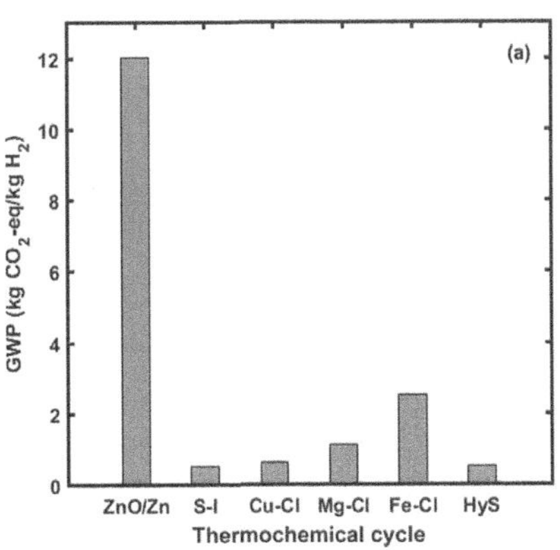
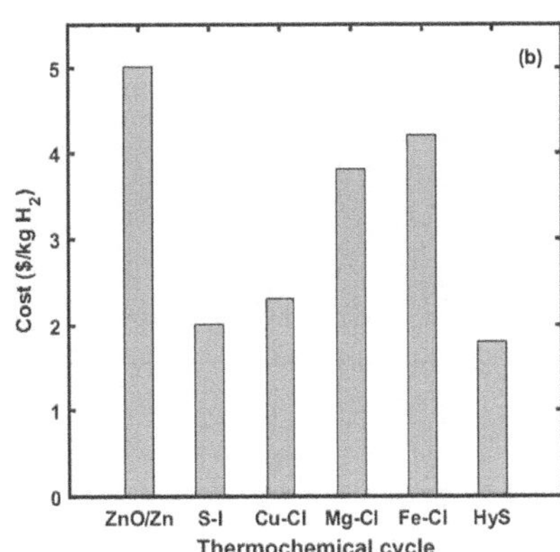

FIGURE 2.18 Comparison between various TWCs in terms of (a) GWP, and (b) cost.

production cost. As illustrated in Figure 2.18, the cost of producing 1 kg of hydrogen using ZnO/Zn is high because of the expensive materials involved in solar integration. However, by improving solar heat production, we can greatly reduce the GWP and cost. Among the selected TWCs, the sulfurine family cycles have the lowest cost of less than \$2 per kg of hydrogen. Additionally, environmental analysis indicates a low GWP of 0.5 and 0.48 for the S-I and HyS TWCs, respectively. However, the durability of sulfur-based cycles remains a challenge due to the high-temperature SO_3 decomposition unit and corrosion. The Cu-Cl process offers a cost of \$2.24 per kg of hydrogen and a low GWP of 0.55 kg CO_2-eq per kg of hydrogen. A comparison between the S-I and Cu-Cl cycles shows that both cycles have similar energy efficiency and are highly reliant on heat recovery. They also share a similar cost, which is lower than that of direct electrolysis. However, in the case of Cu-Cl, the cost of electrolysis hinges on the price of electricity, which tends to be more expensive than thermal energy (Wang et al., 2010). Considering the lower maximum temperature requirement of the Cu-Cl TWC (550°C) and its strong compatibility with fourth-generation nuclear reactors, it is regarded as a dependable and promising cycle for H_2 production on a large scale.

2.6 CONCLUSION

To obtain hydrogen from sustainable and renewable water, extremely high temperatures are necessary for the thermolysis method. Hence, extensive research has been conducted on thermochemical water-splitting processes that utilize the redox potentials of chemical reagents, like metal oxides, in order to lower the necessary maximum temperature. To further reduce system temperature, numerous two-, three-, and four-step TWCs are suggested for H_2 production. Nevertheless, only a limited number of these methods have been successfully implemented in conjunction with sustainable energy sources for the production of hydrogen. Two-step cycles are primarily under development for scenarios where a high-temperature solar reactor is accessible. In addition, thermochemical cycles based on sulfur and chlorine families are considered highly promising options for integration with nuclear reactors due to their exceptional efficiency and compatibility. This chapter conducted a comprehensive comparison of representative TWCs from the perspectives of the environment, economics, exergy, and energy. The concluding remarks can be summarized as follows:

i. Thermochemical H_2 production, in combination with high-temperature solar reactors, shows promise with two-step TWCs that utilize metal oxide redox pairs such as ZnO/Zn.

ii. The sulfur family thermochemical cycles, specifically HyS and S-I, are favorable options for integration with high-temperature nuclear reactors and exhibit low GWP in the H_2 production process. Nevertheless, corrosion remains a significant challenge that hinders the development of these cycles.

iii. The Mg-Cl and Cu-Cl cycles are promising due to their potential for low-cost hydrogen production and compatibility with relatively low-temperature industrial waste heat. Additionally, other chlorine TWCs like Fe-Cl show potential for further research.

iv. The V-Cl cycle exhibits high efficiency, but there are still technical limitations to its integration with an energy source. Further investigation is required to address these challenges and optimize the cycle.

More investigation is required to tackle the issues related to cycles approaching commercial viability, such as lower costs and higher efficiencies. This is essential for enabling large-scale thermochemical H_2 production in the future. Because of the extensive process equipment required for thermochemical cycles, the next studies can focus on streamlining the cycles to reduce the number of reactions, equipment, and overall cost.

REFERENCES

Abanades, S., Flamant, G. 2006. Thermochemical hydrogen production from a two-step solar-driven water-splitting cycle based on cerium oxides. *Solar Energy* 80, 1611–23.

Acar, C., Dincer, I. 2015. Review and evaluation of hydrogen production methods for better sustainability. *Int. J. Hydrogen Energy* 40, 11094–111.

Acar, C., Dincer, I. 2019. Review and evaluation of hydrogen production options for better environment. *J. Clean. Prod.* 218, 835–49.

Allen, J.A., Rowe, G., Hinkley, J.T., Donne, S.W. 2014. Electrochemical aspects of the hybrid sulfur cycle for large scale hydrogen production. *Int. J. Hydrogen Energy* 39, 11376–89.

Al-Zareer, M., Dincer, I., Rosen, M.A. 2017. Performance analysis of a supercritical water-cooled nuclear reactor integrated with a combined cycle, a Cu-Cl thermochemical cycle and a hydrogen compression system. *Appl. Energy* 195, 646–58.

Al-Zareer, M., Dincer, I., Rosen, M.A. 2018. Assessment and analysis of hydrogen and electricity production from a Generation IV lead-cooled nuclear reactor integrated with a copper-chlorine thermochemical cycle. *Int. J. Energy Res.* 42, 91–103.

Al-Zareer, M., Dincer, I., Rosen, M.A. 2020. Analysis and assessment of the integrated generation IV gas-cooled fast nuclear reactor and copper-chlorine cycle for hydrogen and electricity production. *Energy Convers. Manag.* 205, 112387.

Ambrosini, A., Babiniec, M., Miller, J.M. 2017. *Renewable Hydrogen Production via Thermochemical/Electrochemical Coupling.* Laboratory Directed Research & Development (LDRD) project. Albuquerque, NM: Sandia National Laboratory.

Amendola, S. 2005. Thermochemical hydrogen produced from a vanadium decomposition cycle. *U.S. Patent Application* 20050013771.

Andress, R.J., Martin, L.L. 2011. A systematic hierarchical thermodynamic analysis of hydrogen producing iron chlorine reaction clusters. *Ind. Eng. Chem. Res.* 50, 1278–93.

Angel, J., Marug, J. 2016. Perovskite materials for hydrogen production by thermochemical water splitting. *Int. J. Hydrogen Energy* 41, 19329–38.

Anzieu, P., Carles, P., Le Duigou, A., Vitart, X., Lemort, F. 2006. The sulphur-iodine and other thermochemical process studies at CEA. *Int. J. Nucl. Hydrogen Product Appl.* 2, 144–53.

Ashtitha, S., George, S.C. 2021. Chapter 23- Splitting of water: Biological and non-biological approaches. In S. Sahay (ed.), *Handbook of Biofuels*, Academic Press (Elsevier): Netherlands, 453–69.

Balashov, V.N., Schatz, R.S., Chalkova, E., Akinfiev, N.N., Fedkin, M.V., Lvov, S.N. 2011. CuCl electrolysis for hydrogen production in the Cu-Cl thermochemical cycle. *J. Electrochem. Soc.* 158, B266–75.

Balta, M.T., Dincer, I., Hepbasli, A. 2014. Performance assessment of solar-driven integrated Mg-Cl cycle for hydrogen production. *Int. J. Hydrogen Energy* 39, 20652–61.

Balta, M.T., Dincer, I., Hepbasli, A. 2016. Comparative assessment of various chlorine family thermochemical cycles for hydrogen production. *Int. J. Hydrogen Energy* 41, 7802–13.

Beghi, G.E. 1985. A decade of research on thermochemical hydrogen at the joint research center - Ispra. Hydrogen systems. *Presented at the International Symposium on Hydrogen Systems,* 7–11 May 1985, Beijing, China; 153–71.

Bhosale, R.R. 2020a. A novel three-step GeO2/GeO thermochemical water splitting cycle for solar hydrogen production. *Int. J. Hydrogen. Energy* 45(10), 5816–28.

Bhosale, R.R. 2020b. Solar hydrogen production via thermochemical magnesium oxide - magnesium sulfate water splitting cycle. *Fuel* 275, 117892.

Bilgen, E. 1988. Solar hydrogen production by hybrid thermochemical processes. *Solar Energy* 41(2), 199–206.

Bo, Z., Zhang, X., Yu, K., Wang, Z., Lin, X., Yan, J., Cen, K. 2014. Metal-free and Pt-decorated graphene-based catalysts for hydrogen production in a sulfur-iodine thermochemical cycle. *Ind. Eng. Chem. Res.* 53, 11920–8.

Brecher, L.E., Wu, C.K. 1975. Electrolytic decomposition of water. Westinghouse Electric Corp. US Patent No. 3888750.

Budama, V.K., Duarte, J.P.R., Roeb, M., Sattler, C. 2023. Potential of solar thermochemical water-splitting cycles: A review. *Solar Energy,* 249, 353–66.

Bulfin, B., Vieten, J., Agrafiotis, C., Roeb, M., Sattler, C. 2017. Applications and limitations of two step metal oxide thermochemical redox cycles: A review. *J. Mater. Chem. A* 5, 18951–66.

Canavesio, C., Nassini, D., Nassini. H.E., Bohé, A.E. 2020. Study on an original cobalt-chlorine thermochemical cycle for nuclear hydrogen production. *Int. J. Hydrogen Energy* 45(49), 26090–103.

Canavesio, C., Nassini, H.E., Bohé, A.E. 2015. Evaluation of an iron-chlorine thermochemical cycle for hydrogen production. *Int. J. Hydrogen Energy* 40, 8620–32.

Chang, J., Kim, Y., Lee, K., Lee, Y., Lee, W.J., Noh, J., Kim, M., Lim, H., Shin, Y., Bae, K., Jung, K., 2007. A study of a nuclear hydrogen production demonstration plant. *Nucl. Eng. Technol.* 39, 111–22.

Charvin, P., Abanades, S., Florent, L., Gilles, F. 2008. Analysis of solar chemical processes for hydrogen production from water splitting thermochemical cycles. *Energy Convers. Manag.* 49, 1547–56.

Charvin, P., Abanades, S., Lemort, F., Flamant, G. 2007. Hydrogen production by three-step solar thermochemical cycles using hydroxides and metal oxide systems. *Energy Fuels* 21, 2919–28.

Chen, A., Liu, C., Liu, Y., Zhang, L. 2019. Uranium thermochemical cycle used for hydrogen production. *Nucl. Eng. Technol.* 51, 214–20.

Chen, S., Wang, R., Zhang, P., Wang, L., Xu, J., Ke, Y. 2013. HIx concentration by electro-electrodialysis using stacked cells for thermochemical water-splitting IS process. *Int. J. Hydrogen Energy* 38, 3146–3153.

Chueh, W.C., Falter, C., Abbott, M., Scipio, D., Furler, P., Haile, S.M., Steinfeld, A., 2010. High-flux solar-driven thermochemical dissociation of CO2 and H2O using nonstoichiometric ceria. *Science* 330, 1797–801.

Collins, J.L., Dole, L.R., Ferrada, J.J., Forsberg C.W., Haire, M.J., Hunt, R.D., Lewis, B.E., Wymer, R.G., 2010. Carbonate thermochemical cycle for the production of hydrogen. US Patent No. 7,666,387 B2.

Corgnale, C., Summers, W.A. 2011. Solar hydrogen production by the hybrid sulfur process. *Int. J. Hydrogen Energy* 36, 11604–19.

Cumpston, J., Herding, R., Lechtenberg, F., Offermanns, C., Thebelt, A., Roh, K. 2020. Design of 24/7 continuous hydrogen production system employing the solar-powered thermochemical S-I cycle. *Int J Hydrogen Energy* 45, 24383–96.

Dincer, I., Bicer, Y. 2018. Solar thermochemical energy conversion. *Comprehensive Energy Systems*, Elsevier: Netherlands, 895–946.

Dincer, I., Naterer, G.F. 2014. Overview of hydrogen production research in the Clean Energy Research Laboratory (CERL) at UOIT. *Int. J. Hydrogen Energy* 39 (35), 20592–613.

Dokiya, M., Kotera, Y. 1976. Hybrid cycle with electrolysis using Cu-Cl system. *Int. J. Hydrogen Energy* 1, 117–21.

El-Emam, R.S., Khamis, I. 2018. Advances in nuclear hydrogen production: Results from an IAEA international collaborative research project. *Int. J. Hydrogen Energy* 44, 19080–8.

El-Emam, R.S., Ozcan, H. 2019. Comprehensive review on the techno-economics of sustainable large-scale clean hydrogen production. *J. Cleaner Prod.* 220, 593–609.

Farsi, A., Dincer, I., Naterer, G.F. 2020. Review and evaluation of clean hydrogen production by the copperechlorine thermochemical cycle. *J. Cleaner Prod.* 276, 123833.

Fletcher, E.A. 1999. Solarthermal and solar quasi-electrolytic processing and separations: Zinc from Zinc Oxide as an example. *Ind. Eng. Chem. Res.* 38, 2275–82.

Fu, G., He, Y., Zhang, Y., Zhu, Y., Wang, Z., Cen, K. 2016b. Catalytic performance and durability of Ni/AC for HI decomposition in sulfur-iodine thermochemical cycle for hydrogen production. *Energy Convers. Manag.* 117, 520–7.

Fu, G., Wang, Z., Zhang, Y., Huang, Z., Liu, J., Zhou, J., Cen, K. 2016a. Effect of raw material sources on activated carbon catalytic activity for HI decomposition in the sulfur-iodine thermochemical cycle for hydrogen production. *Int. J. Hydrogen Energy* 41, 7854–60.

Funk, J.E. 1976. Thermochemical production of hydrogen via multistage water splitting processes. *Int. J. Hydrog Energy* 1, 33–43.

Funk, J.E., Reinstrom, R.M. 1966. Energy requirements in production of hydrogen from water. *Ind. Eng. Chem. Proc. Des. Dev.* 5, 336–42.

Giaconia, A., Caputo, G., Sau, S., Prosini, P.P., Pozio, A., Francesco, M.D., Tarquini, P., Nardi, L., 2009. Survey of Bunsen reaction routes to improve the sulfur-iodine thermochemical water-splitting cycle. *Int. J. Hydrogen Energy* 34, 4041–8.

Gokon, N., Hara, K., Sugiyama, Y., Bellan, S., Kodama, T., Hyun-seok, C., 2019. Thermochemical two-step water splitting cycle using perovskite oxides based on LaSrMnO3 redox system for solar H2 production. *Thermochim Acta* 680, 178374.

Gong, Y., Chalkova, E., Akinfiev, N.N., Balashov, V., Fedkin, M., Lvov, S.N. 2009. CuCl-HCl electrolyzer for hydrogen production via Cu-Cl thermochemical cycle. *ECS Transactions* 19 (10), 21–32.

Gorensek, MB. 2011. Hybrid sulfur cycle flowsheets for hydrogen production using high-temperature gas-cooled reactors. *Int. J. Hydrogen Energy* 36, 12725–41.

Gorensek, M.B., Corgnale, C., Staser, J.A., Weidner, J.W. 2018. Solar thermochemical hydrogen (STCH) processes. *Electrochem. Soc. Interface* 27, 53–6.

Gorensek, M.B., Corgnale, C., Summers, W.A. 2017. Development of the hybrid sulfur cycle for use with concentrated solar heat. I. Conceptual design. *Int. J. Hydrogen Energy* 42, 20939–54.

Gorensek, M.B., Summers, W.A. 2009. Hybrid sulfur flowsheets using PEM electrolysis and a bayonet decomposition reactor. *Int. J. Hydrogen Energy* 34, 4097–114.

Hassan, N.S., Jalil, A.A., Rajendran, S., Khusnun, N.F., Bahari, M.B., Johari, A., Kamaruddin, M.J., Ismail, M., 2023. Recent review and evaluation of green hydrogen production via water electrolysis for a sustainable and clean energy society, *Int. J. Hydrogen Energy*, 52, 420–441.

Hoskins, A.L., Millican, S.L., Czernik, C.E., Alshankiti, I., Netter, J.C., Wendelin, T.J., Musgrave, C.B., Weimer, A.W., 2019. Continuous on-sun solar thermochemical hydrogen production via an isothermal redox cycle. *Appl. Energy* 249, 368–76.

Huang, C., Adebiyi, O., Muradov, N., T-Raissi, A. 2006. Light photolysis of ammonium sulfite aqueous solution for the production of hydrogen. Proc.16th World Hydrogen Energy Conf, Lyon, France; (June 13–16).

Huang, C., T-Raissi, A., Muradov, N. 2014. Solar metal sulfate-ammonia based thermochemical water splitting cycle for hydrogen production. US Patent No. 8691068B1.

Ishaq, H., Dincer, I. 2019. A comparative evaluation of three CuCl cycles for hydrogen production. *Int. J. Hydrogen Energy* 44, 7958–68.

Ishaq, H., Dincer, I., Naterer, G.F. 2018. Industrial heat recovery from a steel furnace for the cogeneration of electricity and hydrogen with the copper-chlorine cycle. *Energy Convers. Manag.* 171, 384–97.

Jarret, C., Chueh, W., Yuan, C., Kawajiri, Y., Sandhage, K.H., Henry, A. 2016. Critical limitations on the efficiency of two-step thermochemical cycles. *Solar Energy* 123, 57–73.

Kameyama, H., Yoshida, K. 1978. Br-Ca-Fe Water-decomposition cycles for hydrogen production. *Proc. 2nd World Hydrogen Energy Conference,* Zurich, Switzerland; 829–50.

Kaneko, H., Miura, T., Ishihara, H., Taku, S., Yokoyama, T., Nakajima, H., Tamaura, Y., 2007. Reactive ceramics of CeO2-MOx (M=Mn, Fe, Ni, Cu) for H2 generation by two-step water splitting using concentrated solar thermal energy. *Energy* 32(656), 663.

Kasahara, S., Iwatsuki, J., Takegami, H., Tanaka, N., Noguchi, H., Kamiji, Y., Onuki, K., Kubo, S., 2017. Current R&D status of thermochemical water splitting iodine-sulfur process in Japan Atomic Energy Agency. *Int. J. Hydrogen Energy* 42, 13477–85.

Kasahara, S., Tanaka, N., Noguchi, H., Iwatsuki, J., Takegami, H., Yan, X.L., Kubo, S., 2014. JAEA's R & D on the thermochemical hydrogen production IS process. *Proceedings of the HTR,* Weihai, China; October 27–31.

Kim, S., Schatz, R., Khurana, S., Fedkin, M., Wang, C., Lvov, S. 2011. Advanced CuCl electrolyzer for hydrogen production via the Cu-Cl thermochemical Cycle. *ECS Transactions* 35, 257.

Knoche, K.F., Cremer, H., Steinborn, G., 1976. A thermochemical process for hydrogen production. *Int. J. Hydrogen Energy* 1, 23–32.

Knoche, K.F., Schuster, P., Ritterbex, T. 1984a. Thermochemical production of hydrogen by a vanadium/chlorine cycle. Part 1: An energy and exergy analysis of the process. *Int. J. Hydrogen Energy* 9, 457–72.

Knoche, K.F., Schuster, P., Ritterbex, T. 1984b. Thermochemical production of hydrogen by a vanadium/ chlorine cycle. Part 2: Experimental investigation of the individual reactions. *Int. J. Hydrogen Energy* 9, 7792–801.

Kodama, T., Gokon, N., Cho, H.S., Bellan, S., Matsubara, K., Inoue, K. 2018. Particle fluidized bed receiver / reactor with a beam-down solar concentrating optics: Performance test of two-step water splitting with ceria particles using 30-kWth sun- simulator. *AIP Conference Proceedings 2033,* 130009. https://doi.org/10.1063/1.5067143.

Koepf, A., Alxneit, I., Wieckert, C., Meier, A. 2017. A review of high temperature solar driven reactor technology: 25 years of experience in research and development at the Paul Scherrer Institute. *Appl. Energy* 188, 620–51.

Kubo, S., Kasahara, S., Okuda, H., Terada, A., Tanaka, N., Inaba, Y., Ohashi, H., Inagaki, Y., Onuki, K., Hino, R., 2004. A pilot test plan of the thermochemical iodine sulfur process. *Nucl. Eng. Des.* 233, 355–62.

Lanchi, M., Caputo, G., Liberatore, R., Marrelli, L., Sau, S., Spadoni, A., Tarquini, P. 2009. Use of metallic Ni for H2 production in S-I thermochemical cycle: Experimental and theoretical analysis. *Int. J. Hydrogen Energy* 34, 1200–7.

Le, A., Borgard, J., Larousse, B., Doizi, D., Allen, R., Ewan, B.C., Priestman, G.H., Elder, R., Devonshire, R., Ramos, V., Cerri, G., Salvini, C., Giovannelli, A., Maria, G.D., Corgnale, C., Brutti, S., Roeb, M., Noglik, A., Rietbrock, P., Mohr, S., Oliveira, L.D., Monnerie, N., Schmitz, M., Sattler, C., Martinez, A.O., Manzano, D.L., Rojas, J.C., Dechelotte, S., Baudouin, O., 2007. HYTHEC: An EC funded search for a long term massive hydrogen production route using solar and nuclear technologies. *Int. J. Hydrogen Energy* 32, 1516–29.

Lee, B.J., No, H.C., Yoon, H.J., Jin, H.G., Kim, Y.S., Lee, J.I. 2009. Development of a flowsheet for iodine-sulfur thermo-chemical cycle based on optimized Bunsen reaction. *Int. J. Hydrogen Energy* 34, 2133–43.

Lee, J.E., Shafiq, I., Hussain, M., Lam, S.S., Rhee, G.H., Park, Y-K. 2022. A review on integrated thermochemical hydrogen production from water, *Int. J. Hydrogen Energy*, 47, 4346–56.

Lewis, M.A., Ferrandon, M., Tatterson, D. 2009a. R&D status for the Cu-Cl thermochemical cycle; DOE Hydrogen Program FY 2009 Annual Progress Report. IL, US: Argonne National Laboratory.

Lewis, M.A., Ferrandon, M.S., Tatterson, D.F., Mathias, P. 2009b. Evaluation of alternative thermochemical cycles - Part III further development of the Cu-Cl cycle. *Int. J. Hydrogen Energy* 34, 4136–45.

Lewis, M.A., Masin, J.G. 2009. The evaluation of alternative thermochemical cycles Part II: The down-selection process. *Int. J. Hydrogen Energy*. 34, 4125–35.

Li, X., Sun, X., Song, Q., Yang, Z., Wang, H., Duan, Y. 2022. A critical review on integrated system design of solar thermochemical water-splitting cycle for hydrogen production, *Int. J. Hydrogen Energy*, 47, 33619–42.

Liberatore, R., Lanchi, M., Caputo, G., Felici, C., Giaconia, A., Sau, S., Tarquini, P., 2019. Hydrogen production by flue gas through sulfur-iodine thermochemical process: Economic and energy evaluation. *Int. J. Hydrogen Energy* 37, 8939–53.

Littlefield, J., Wang, M., Brown, L.C., Herz, R.K., Talbot, J.B. 2012. Process modeling and thermochemical experimental analysis of a solar sulfur ammonia hydrogen production cycle. *Energy Procedia* 29, 616–23.

Mandal, S., Jana, A.K. 2020. Simulating reactive distillation of HIx (HIeH2OeI2) system in sulphur-iodine cycle for hydrogen production. *Nucl. Eng. Technol.* 52, 279–86.

Mao, Y., Gao, Y., Dong, W., Wu, H., Song, Z., Zhao, X., Sun, J., Wang, W., 2020. Hydrogen production via a two-step water splitting thermochemical cycle based on metal oxide— A review. *Appl. Energy* 267, 114860.

Mehrpooya, M., Ghorbani, B., Ekrataleshian, A., Mousavi, S.A. 2021. Investigation of hydrogen production by sulfur-iodine thermochemical water splitting cycle using renewable energy source. *Int. J. Energy Res.* 45, 14845–69.

Mehrpooya, M., Habibi, R. 2020. A review on hydrogen production thermochemical water-splitting cycles. *J. Cleaner Prod.* 275, 123836.

Merabet, N.H., Kerboua, K. 2022. Chapter 19- The sonochemical and ultrasound-assisted production of hydrogen: Energy efficiency for the generation of an energy carrier. In O. Hamdaoui, K., Kerboua (eds), *Energy Aspects of Acoustic Cavitation and Sonochemistry*, Elsevier: Netherlands, 313–29.

Mohd, N., Nandong, J. 2017. Multi-scale control of bunsen section in iodine-sulphur thermochemical cycle process. *Chem. Prod. Process Model* 12, 1–15.

Naterer, G.F., Suppiah, S., Lewis, M., Gabriel, K., Dincer, I., Rosen, M.A., Fowler, M., Rizvi, G., Easton, E.B., Ikeda, B.M., Kaye, M.H., Lu, L., Pioro, I., Spekkens, P., Tremaine, P., Mostaghimi, J., Avsec, J., Jiang, J., 2009. Recent Canadian advances in nuclear-based hydrogen production and the thermochemical Cu-Cl cycle. *Int. J. Hydrogen Energy* 34, 2901–17.

Naterer, G.F., Suppiah, S., Stolberg, L., Lewis, M., Wang, Z., Daggupati, V., Gabriel, K., Dincer, I., Rosen, M.A., Spekkens, P., Lvov, S.N., Fowler, M., Tremaine, P., Mostaghimi, J., Easton, E.B., Trevani, L., Rizvi, G., Ikeda, B.M., Kaye, M.H., Lu, L., Pioro, I., Smith, W.R., Secnik, E., Jiang, J., Avsec, J., 2013a. Canada's program on nuclear hydrogen production and the thermochemical Cu-Cl cycle. *Int. J. Hydrogen Energy* 35, 10905–26.

Naterer, G.F., Suppiah, S., Stolberg, L., Lewis, M., Wang, Z., Dincer, I., Rosen, M.A., Gabriel, K., Secnik, E., Easton, E.B., Pioro, I., Lvov, S., Jiang, J., Mostaghimi, J., Ikeda, B.M., Rizvi, G., Lu, L., Odukoya, A., Spekkens, P., Fowler, M., Avsec, J., 2013b. Progress of international hydrogen production network for the thermochemical Cu-Cl cycle. *Int. J. Hydrogen Energy* 38, 740–59.

Naterer, G.F., Suppiah, S., Stolberg, L., Lewis, M., Wang, Z., Rosen, M.A., Dincer, I., Gabriel, K., Odukoya, A., Secnik, E., Easton, E.B., Papangelakis, V., 2015. Progress in thermochemical hydrogen production with the copper e chlorine cycle. *Int. J. Hydrogen Energy* 40, 6283–95.

O'Brien, J.A., Hinkley, J.T., Donne, S.W., Lindquist, S.E. 2010. The electrochemical oxidation of aqueous sulfur dioxide: A critical review of work with respect to the hybrid sulfur cycle. *Electrochem. Acta* 55, 573–91.

Odorizzi, S., Enginsoft, S. 2016. SOL2HY2-Solar to Hydrogen Hybrid Cycles Project. Publishable Summary; Available online: https://cordis.europa.eu/docs/ results/325/325320/ final1-sol2hy2-publishable-summary-final-report-24-2.pdf

O'Keefe, D.R., Allen, C.L., Besenbruch, G.E., Brown, L., Norman, J., Sharp, R., McCorkle, K., 1982. Preliminary results from Bench-scale testing of a sulfur-iodine Thermochemical water-splitting cycle. *Int. J. Hydrogen Energy* 7, 381–92.

Onuki, K.S., Shimizu, N.H., Fujita, S., Ikezoe, Y., Sato, S., Machi, S. 1990. Studies on an iodine sulfur process for thermochemical hydrogen production. *8th World Hydrogen Energy Conf,* Honolulu and Waikoloa, U.S; 547.

Orhan, M.F., Dincer, I., Rosen, M.A. 2011. Design of systems for hydrogen production based on the Cu-Cl thermochemical water decomposition cycle: Configurations and performance. *Int. J. Hydrogen Energy* 36, 11309–20.

Orhan, M.F., Dinçer, I., Rosen, M.A. 2012. Efficiency comparison of various design schemes for copper-chlorine (Cu-Cl) hydrogen production processes using Aspen Plus software. *Energy Convers. Manag.* 63, 70–86.

Oruc, O., Dincer, I. 2021. Assessing the potential of thermo-chemical water splitting cycles: A bridge towards clean and sustainable hydrogen generation. *Fuel* 286, 119325.

Osuolale, F., Ogunleye, O., Fakunle, M., Busari, A., Abolanle, Y. 2018. Comparative studies of Cu-Cl thermochemical water decomposition cycles for hydrogen production. *International Conference on Renewable Energy,* E3S Web of Conferences. EDP Sciences, 61, 00009.

Ozcan, H., Dincer, I. 2014. Performance investigation of magnesium-chloride hybrid thermochemical cycle for hydrogen production. *Int. J. Hydrogen Energy* 39, 76–85.

Ozcan, H., Dincer, I. 2016a. Comparative performance assessment of three configurations of magnesium-chlorine cycle. *Int. J. Hydrogen Energy* 41, 845–56.

Ozcan, H., Dincer, I. 2016b. Modeling of a new four-step magnesium-chlorine cycle with dry HCl capture for more efficient hydrogen production. *Int. J. Hydrogen Energy* 41, 7792–801.

Ozcan, H., Dincer, I. 2017. Exergoeconomic optimization of a new four-step magnesium-chlorine cycle. *Int. J. Hydrogen Energy* 42, 2435–45.

Ozcan, H., Dincer, I. 2018. Experimental investigation of an improved version of the four-step magnesium-chlorine cycle. *Int. J. Hydrogen Energy* 43, 5808–19.

Park, J., Ifaei, P., Ba-Alawi, A.H., Safder, U., Yoo, C. 2020. Hydrogen production through the sulfureiodine cycle using a steam boiler heat source for risk and techno-socio-economic cost (RSTEC) reduction. *Int. J. Hydrogen Energy* 45(28), 14578–93.

Park, J., Lee, S., Lee, I. 2019. Study of alternative reactor—Separator network in Bunsen process of sulfur-iodine cycle for hydrogen production. *J. Chem. Eng. Jpn.* 52, 638–49.

Perret, R. 2011. Solar Thermochemical Hydrogen Production Research (STCH); Thermochemical Cycle Selection and Investment Priority. Prepared by Sandia National Laboratories. Albuquerque, NM 87185 and Livermore, CA 94550.

Ping, Z., Laijun, W., Songzhe, C., Jingming, X. 2018. Progress of nuclear hydrogen production through the iodine-sulfur process in China. *Renew. Sustain. Energy Rev.* 81, 1802–12.

Ratlamwala, T., Dincer, I. 2012. Energy and exergy analyses of a Cu-Cl cycle based integrated system for hydrogen production. *Chem. Eng. Sci.* 84, 564–73.

Roeb, M., Thomey, D., De Oliveira, L., Sattler, C., Fleury, G., Pra, F., Tochon, P., Brevet, A., Roux, G., Gruet, N., Mansilla, C., LeNaour, F., Poitou, S., Allen, R.W.K., Elder, R., Kargiannakis, G., Agrafiotis, C., Zygogianni, A., Pagkoura, C., Konstandopoulos, A.G., Giaconia, A., Sau, S., Tarquini, P., Haussener, S., Steinfeld, A., Canadas, I., Orden, A., Ferrato, M., 2013. Sulphur based thermochemical cycles: Development and assessment of key components of the process. *Int. J. Hydrogen Energy* 38, 6197–204.

Roeb, M., Thomey, D., Graf, D., Sattler, C. 2011. HycycleS: A project on nuclear and solar hydrogen production by sulphur-based thermochemical cycles. *Int. J. Nucl. Hydrogen Product Appl.* 2, 202–26.

Safari, F., Dincer, I. 2020a. A review and comparative evaluation of thermochemical water splitting cycles for hydrogen production. *Energy Convers. Manag.* 205, 112182.

Safari, F., Dincer, I. 2020b. A study on the Fe-Cl thermochemical water splitting cycle for hydrogen production. *Int. J. Hydrogen Energy* 45(38), 18867–75.

Safari, F., Dincer, I. 2022. Assessment and multi-objective optimization of a vanadium-chlorine thermochemical cycle integrated with algal biomass gasification for hydrogen and power production. *Energy Convers. Manag.* 253, 115132.

Sattler, C. 2019. Solar thermochemical cycles for fuel production in Germany. *J. Solar Energy Eng.* 141(2), 020304.

Sattler, C., Roeb, M., Agrafiotis, C., Thomey, D. 2017. Solar hydrogen production via Sulphur based thermochemical water-splitting. *Solar Energy* 156, 30–47.

Sayyaadi, H. 2017. A conceptual design of a dual hydrogen-power generation plant based on the integration of the gas-turbine cycle and copper chlorine thermochemical plant. *Int. J. Hydrogen Energy* 42, 28690–709.

Shankar, V., Dharani, S., Ravi, A., SaravanaVadivu, A. 2023. A concise review: MXene-based photo catalytic and photo electrochemical water splitting reactions for the production of hydrogen, *Int. J. Hydrogen Energy* 48(57), 21654–73.

Simpson, M.F., Herrmann, S.D., Boyle, B.D. 2006. A hybrid thermochemical electrolytic process for hydrogen production based on the reverse Deacon reaction. *Int. J. Hydrogen Energy* 31, 1241–6.

Singhania, A. 2018. Catalytic decomposition of hydrogen-iodide over nanocrystalline ceria promoted by transition metal oxides for hydrogen production in sulfur-iodine thermo-chemical cycle. *Catal. Lett.* 148, 1416–22.

Singhania, A., Bhaskarwar, A.N. 2018. Performance of activated-carbon-supported Ni, Co, and NieCo catalysts for hydrogen iodide decomposition in a thermochemical water-splitting sulfur-iodine cycle. *Energy Technol.* 6, 1104–11.

Singhania, A., Krishnan, V.V., Bhaskarwar, A.N., Bhargava, B., Parvatalu, D. 2018. Hydrogen-iodide decomposition over PdCeO2 nanocatalyst for hydrogen production in sulfur-iodine thermochemical cycle. *Int. J. Hydrogen Energy* 43, 3886–91.

Singhania, A., Krishnan, V.V., Bhaskarwar, A.N., Bhargava, B., Parvatalu, D., Banerjee, S. 2016. Catalytic performance of bimetallic Ni-Pt nanoparticles supported on activated carbon, gamma-alumina, zirconia, and ceria for hydrogen production in sulfur-iodine thermochemical cycle. *Int. J. Hydrogen Energy* 41, 10538–46.

Soltani, R., Dincer, I., Rosen, M.A. 2019. Kinetic and electrochemical analyses of a CuCl/HCl electrolyzer. *Int. J. Energy Res.* 43, 6890–906.

Steinfeld, A. 2002. Solar hydrogen production via a two-step water-splitting thermochemical cycle based on Zn/ZnO redox reactions. *Int. J. Hydrogen Energy* 27, 611–9.

Tamaura, Y., Kojima, N., Hasegawa, N., Inoue, M., Uehara, R., Gokon, N., Kaneko, H. 2001. Stoichiometric studies of H2 generation reaction for H2O/Zn/Fe3O4 system. *Int. J. Hydrogen Energy* 26, 917–22.

Tamaura, Y., Uehara, R., Hasegawa, N., Kaneko, H., Aoki, H. 2004. Study on solid-state chemistry of the ZnO/Fe3O4/H2O system for H2 production at 973-1073K. *Solid State Ionics* 172, 121–4.

Tsutsumi, A. 2009. *Energy Carriers and Conversion Systems - Vol. I - Thermochemical Cycles.* Oxford, UK: Eolss Publishing Group Co, Ltd.

T-Raissi, A. 2012. Water splitting: Thermochemical. *Encyclopedia of Inorganic and Bioinorganic Chemistry*, John Wiley & Sons, Ltd: United States.

Utgikar, V., Ward, B. 2006. Life cycle assessment of ISPRA Mark 9 thermochemical cycle for nuclear hydrogen production. *J. Chem. Technol. Biotechnol.* 81, 1753–9.

Uyar, T.S., Besikci, D. 2017. Integration of hydrogen energy systems into renewable energy systems for better design of 100% renewable energy communities. *Int. J. Hydrogen Energy* 42(4), 2453–6.

Vaghasia, R., Jianu, O.A., Rosen, M.A. 2018. Experimental study of effect of anolyte concentration and electrical potential on electrolyzer performance in thermochemical hydrogen production using the Cu-Cl cycle. *Int. J. Hydrogen Energy* 43, 4160–6.

Van Velzen, D., Langenkamp, H., 1978. Problems around Fe-Cl cycles. Int. J. Hydrogen Energy 3, 419–429.

Vitart, X., Le Duigou, A., Carles, P. 2006. Hydrogen production using the sulfur-iodine cycle coupled to a VHTR: An overview. *Energy Convers. Manag.* 47, 2740–7.

Wang, Z., Chen, S., Zhang, P., Wang, L., Xu, J., Wang, S. 2014. Evaluation on the electro-electrodialysis stacks for hydrogen iodide concentrating in iodine-sulphur cycle. *Int. J. Hydrogen Energy* 39, 13505–11.

Wang, Z., Naterer, G., Gabriel, K., Gravelsins, R., Daggupati, V. 2009. Comparison of different copper-chlorine thermochemical cycles for hydrogen production. *Int. J. Hydrogen Energy* 34, 3267–76.

Wang, Z., Naterer, G., Gabriel, K., Gravelsins, R., Daggupati, V. 2010. Comparison of sulfur-iodine and copper-chlorine thermochemical hydrogen production cycles. *Int. J. Hydrogen Energy* 35, 4820–30.

Weidenkaff, A., Reller, A., Wokaun, A., Steinfeld, A. 2000b. Thermogravimetric analysis of the ZnO/Zn water splitting cycle. *Thermochim Acta* 359, 69–75.

Wu, W., Chen, H.Y., Hwang, J-J. 2017. Energy analysis of a class of copper-chlorine (Cu-Cl) thermochemical cycles. *Int. J. Hydrogen Energy* 42(25), 15990–6002.

Wu, W., Chen, H.Y., Wijayanti, F. 2016. Economic evaluation of a kinetic-based copperchlorine (CuCl) thermochemical cycle plant. *Int. J. Hydrogen Energy* 41, 16604–12.

Xiao, L., Wu, S.Y., Li, Y.R. 2012. Advances in solar hydrogen production via two-step water-splitting thermochemical cycles based on metal redox reactions. *Renew Energy* 41, 1–12.

Xu, S., He, Y., Fu, G., Dai, F., Zhang, Y., Wang, Z. 2018a. Effect of iodine precipitation on HI separation subsection in sulfur-iodine cycle for hydrogen production. *Int. J. Hydrogen Energy* 43, 10896–904.

Xu, S., He, Y., Huang, B., Zhang, Y., Wang, Z., Cen, K. 2018b. Decomposition of hydriodic acid by electrolysis in the thermochemical water sulfur-iodine splitting cycle. *Int. J. Hydrogen Energy* 43, 3597–604.

Yilmaz, F., Selbas, R. 2017. Thermodynamic performance assessment of solar based Sulfur-Iodine thermochemical cycle for hydrogen generation. *Energy* 140, 520–9.

Zhang, P., Chen, S.Z., Wang, L.J., Xu, J.M. 2010. Overview of nuclear hydrogen production research through iodine sulfur process at INET. *Int. J. Hydrogen Energy* 35, 2883–7.

Zhang, P., Su, T., Chen, Q., Wang, L., Chen, S., Xu, J., 2012. Catalytic decomposition of sulfuric acid on composite oxides and Pt/SiC. *Int. J. Hydrogen Energy* 37, 760–764.

Zhang, Y., Wang, R., Lin, X., Wang, Z., Liu, J., Zhou, J., Cen, K. 2015. Catalytic performance of different carbon materials for hydrogen production in sulfur-iodine thermochemical cycle. *Appl. Catal. B Environ.* 166, 413–22.

Zhou, J., Zhang, Y., Wang, Z., Yang, W., Zhou, Z., Liu, J., Cen, K., 2007. Thermal efficiency evaluation of open-loop SI thermochemical cycle for the production of hydrogen, sulfuric acid and electric power. *Int. J. Hydrogen Energy* 32, 567–75.

Zhu, Z., Ma, Y., Qu, Z., Fang, L., Zhang, W., Yan, N. 2017. Study on a new wet flue gas desulfurization method based on the Bunsen reaction of sulfur-iodine thermochemical cycle. *Fuel* 195, 33–7.

Section II

Biomass and Wastes Thermochemical Conversion Processes

3 Pyrolysis Process for Conversion of Lignin to Hydrogen

Sina Mosallanezhad, Parvin Kiani,
and Mohammad Reza Rahimpour

3.1 INTRODUCTION

Producing bio-oil from renewable biomass is a feasible method for using bioenergy due to its low emissions and, crucially, its sustainability (Mosallanezhad et al., 2023). Nevertheless, using bio-oil directly as a fuel for engines poses difficulties due to its high water content, oxygen-containing chemical compounds, and heavy organic compounds with high molecular weights (Lehto et al., 2014). Hence, hydrotreatment is essential to boost the biofuel's heating value and thermal stability by removing excess oxygen from bio-oil (Oh et al., 2020). Hydrotreatment of bio-oil usually necessitates elevated pressure and temperature to optimize the catalytic effectiveness for hydrodeoxygenation (Zhang et al., 2020). Nevertheless, the process is hindered by the quick deactivation of the catalyst caused by coke generation at high temperatures (Kadarwati, 2016).

The creation of coke results from the breaking and repolymerization of the abundant oxygen-containing organic compounds present in bio-oil (Schmitt et al., 2018). Within this group of organic substances, several complex molecules, including pyrolytic lignin (PL), have a significant impact on coke formation. Electron-rich benzene rings in PL enable electrophilic substitution, forming more significant organic compounds (Hu et al., 2013). In addition, PL's duration in a catalyst bed is much greater than that of light organics. This extended residence period leads to the creation of carbonaceous deposits on the catalyst surface during the processes of polymerization or breaking. Highly complex organic compounds, such as PL, may readily generate reactive radicals that need to be stabilized by active hydrogen radicals (Gholizadeh et al., 2016). Insufficient hydrogen radicals during the hydrotreatment process might cause heavy organic radicals to recondense, producing coke. Hence, the creation of coke poses a significant obstacle to the commercialization of bio-oil hydrotreatment. To tackle the challenges posed by the high propensity of bio-oil to generate coke, it is imperative first to comprehend the features of PL.

Furthermore, apart from its use in biofuels, PL in bio-oil may be utilized as a raw material for synthesizing aromatic hydrocarbons like benzene, toluene, and xylene. This is due to the high concentration of benzene rings found in PL.

DOI: 10.1201/9781003382270-5

PL may also be used to synthesize resins via the condensation reaction with aldehydes (Sukhbaatar et al., 2009).

Multiple studies have examined the compositions of lignin (Wang et al., 2017), the process of converting lignin into biofuel by hydrotreatment (Hu et al., 2013), and the process of converting lignin into chemicals. In particular, Wang et al. (2017) conducted a thorough examination of the composition and thermal decomposition characteristics of lignin. Brebu et al. provided a concise summary of the thermal breakdown characteristics of lignin. A comprehensive analysis of Kraft lignin (KL) chemistry, specifically emphasizing lignin structure, was made possible by a thorough examination conducted by Chakar and Ragauskas (2004). Azadi et al. (2013) comprehensively analyzed the advancements made in transforming lignin into liquid fuels, chemicals, and hydrogen. Bu et al. (2012) examined the process of hydrotreatment of lignin utilizing different catalysts and methods. Mu et al. (2013) examined the hydrotreatment process on a model component formed from lignin. Amen-Chen et al. (2001) concisely summarized the creation of individual phenol molecules during biomass and lignin pyrolysis. In a review by Effendi et al. (2008), the feasibility of manufacturing phenolic resin from lignin and biomass was investigated. These review articles provide valuable information on the current advancements in converting bio-oil or lignin-derived substances into biofuels or chemicals.

Nevertheless, limited research has examined the characteristics and reaction patterns of PL as it transforms into fuels, chemicals, radical scavengers, antioxidants, resins, carbon materials, binders, and other systems. This research performed techno-economic studies on several approaches to broaden the application range of PL. The findings serve as valuable reference information for assessing the viability of using PL as a raw material for diverse products.

3.2 LIGNIN RESOURCES

Lignin, a plentiful and versatile natural polymer that accounts for 15%–35% of the dry weight of wood, is a crucial component of the cell walls of vascular plants. In lesser concentrations, it may also be found in the tissues of non-woody plants, including grasses and agricultural waste. Lignin is made up of complex aromatic molecules, which make it difficult to degrade. In the paper and pulp business, lignin is regarded as a waste product and is often burnt to provide energy. However, there is considerable interest in using lignin as a feedstock to create biofuels and other compounds as a possible renewable energy source (Azuma and Tetsuo, 1988). It is possible to get lignin from various lignocellulosic biomass sources, including wood, agricultural wastes like wheat straw and maize stover, and energy crops like switchgrass and Miscanthus. The lignin content and type of different lignocellulosic biomass might impact the production and quality of the resultant biofuel or chemical products. In conclusion, more study is required to maximize the conversion and exploitation of lignin, which represents a potentially significant and underutilized resource for generating renewable energy and chemicals (Dutta et al., 2018). Figure 3.1 shows different sources of lignin.

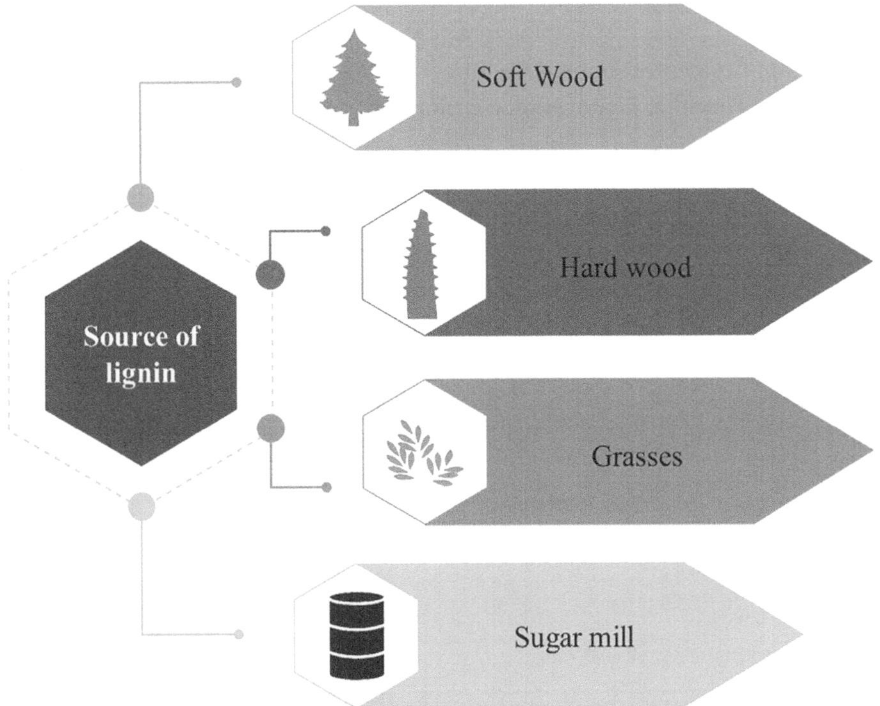

FIGURE 3.1 Different lignin resources.

3.3 LIGNIN INTERUNIT CONNECTIONS AND STRUCTURE

Coniferyl, sinapyl, and coumaryl alcohols—which have zero, one, and two methoxyl groups, respectively—are the three main phenylpropane monomers randomly polymerized to form lignin, an amorphous copolymer that is cross-linked. As seen in Table 3.1, various plants and species have varying amounts of lignin and variable ratios between these core monomer units. Softwoods are the highest in lignin concentration, followed by hardwoods and grasses. Hardwood lignins comprise about equal percentages of sinapyl alcohol (syringyl structure) and coniferyl alcohol (guaiacol structure), whereas coniferyl alcohol is the main component of softwood lignin (Azadi et al., 2013).

Most of the connections in lignin consist of ether bonds, often exceeding two-thirds of the total. Hardwood lignin has about 1.5 times the number of β-O-4-linkages compared to softwood lignin. Lignin has many functional groups, such as methoxyl, phenolic hydroxyl, aliphatic hydroxyl, benzyl alcohol, noncyclic benzyl ether, and carbonyl groups. These groups contribute to the reactivity of lignin in different chemical processes. The β-O-4-aryl ether bonds are the most often occurring coupling connections in polymerization. The other significant connections consist of 4-O-5-diaryl ether, 5-5-biphenyl, β-5-phenylcoumaran, β-O-4-aryl ether,

TABLE 3.1
The Lignin Concentration and Chemical Structures of the Three Main Monomers Found in Lignocellulosic Biomass

	Lignin (wt%)	Phenylpropane unit (%)		
Structure		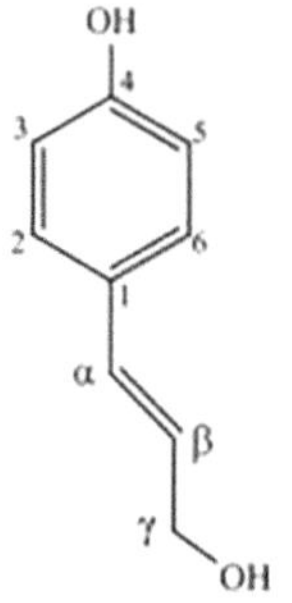	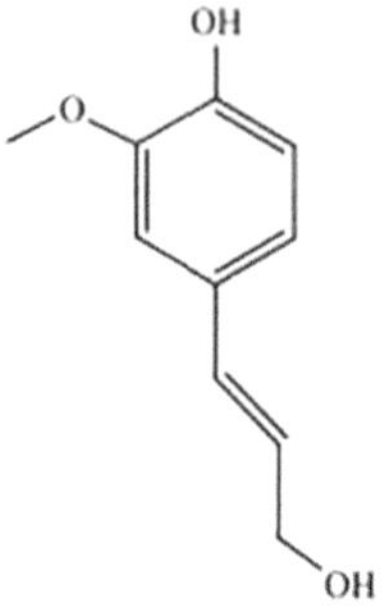	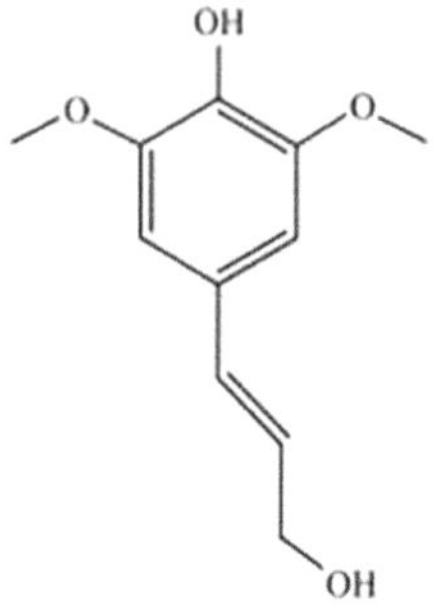
		Coumaryl	Coniferyl	Sinapyl
Softwood	27–33	–	90–95	5–10
Hardwood	18–25	–	50	50
Grasses	17–24	5	75	25

Source: Adopted from Azadi et al. (2013).

β-1-(1,2-diaryl propane), and β–β-retinol connections, as seen in Table 3.2 (Zakzeski et al., 2010). The ratios of these connections exhibit substantial variation across different types of wood and plants. Table 3.2 also provides the estimated proportions of primary connections found in hardwood and softwood lignin.

3.4 PYROLYSIS OF LIGNIN

Lignin, cellulose, and hemicellulose constitute one of the principal elements in lignocellulosic biomass (Raveendran et al., 1996). The substantial quantities of lignin generated through industrial and agricultural practices have led to the exploration of pyrolysis technology as an efficient means of conversion (Lu et al., 2020). Unfortunately, the intricate interplay between cellulose, hemicellulose, and lignin influences the pyrolytic behavior of lignin, introducing complexities that render the pyrolysis process more challenging. Following delignification, various types of lignin can be extracted from untreated biomass, exhibiting high purity and uniformity. These removed lignin types can be subsequently subjected to pyrolysis, yielding low-molecular-weight compounds referred to as pyrolytic oils. Predominantly composed of phenolic compounds, aldehydes, acids, and other aromatic hydrocarbons, these pyrolytic oils are the outcome of the pyrolysis of lignin (Liu et al., 2016). The radical reaction during lignin pyrolysis is conventionally categorized into initial, primary, and charring phases. This discussion commences with an overview of the general processes involved in lignin pyrolysis before delving into specific details regarding its pyrolytic characteristics and the distribution of resultant products.

TABLE 3.2
Proportions of Connections between Different Units in Hardwood and Softwood Lignins

C_9–O–C_9	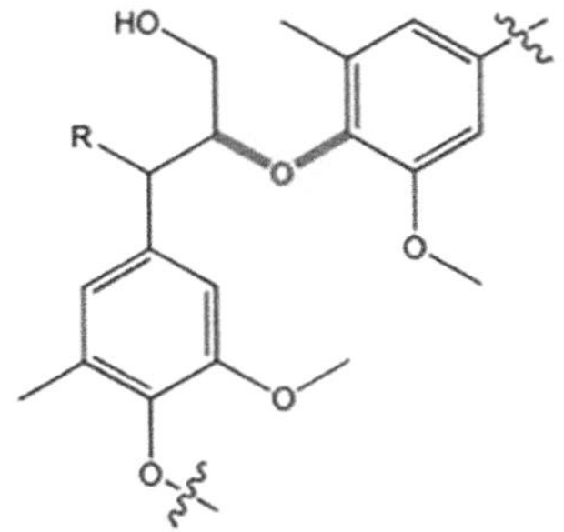	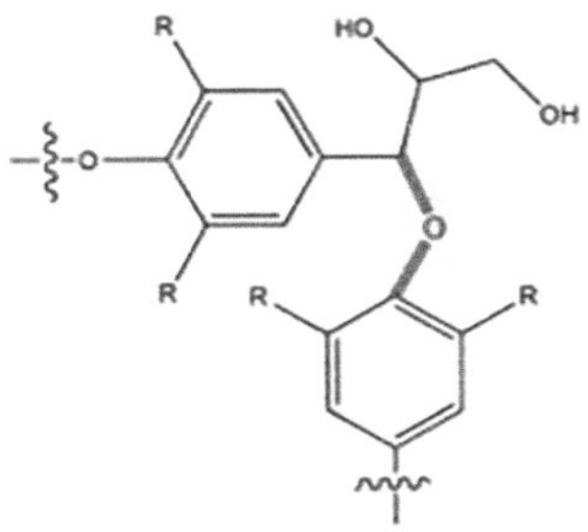	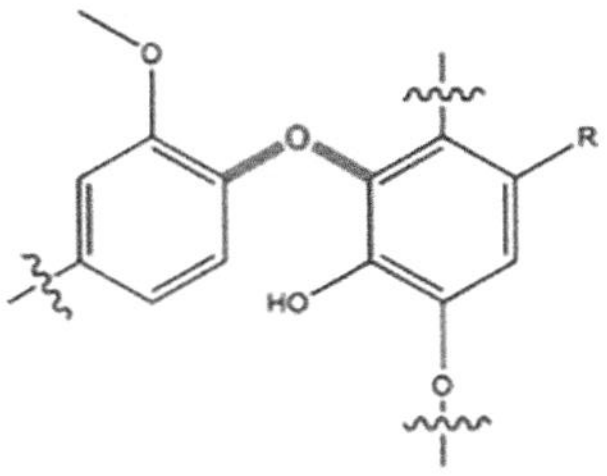
Linkages	β-O-4	α-O-4	4-O-5
Softwood (%)	46	6–8	3.5–4
Hardwood (%)	60	6–8	6.5

(*Continued*)

TABLE 3.2 (*Continued*)

Proportions of Connections between Different Units in Hardwood and Softwood Lignins

C_9–C_9

Linkages	β–5	β–1	β–β	5–5
Softwood (%)	9–12	7	2	9.5–11
Hardwood (%)	6	7	3	4.5

Source: Adopted from Azadi et al. (2013).

3.4.1 GENERAL PYROLYSIS

For investigating how lignin pyrolysis works, scientists can use techniques like gas chromatography/mass spectrometry (GC/MS), thermogravimetric analysis (TG), and Fourier transforms infrared spectroscopy (FTIR) to track mass loss, volatile evolution, and decomposition product distribution, respectively. For instance, Yang et al. (2020) used a TG-FTIR-GC/MS coupling research to examine the pyrolysis of lignin-rich residue obtained from Arundo donax. Data on bio-oil composition, carbon number, and atomic ratio (C/H, C/O) were collected from pyrolytic investigations to propose a pyrolysis process and possible reaction pathway (Davda et al., 2005). According to Doherty et al. (2011), TG analysis may be used to detect the three main phases of lignin pyrolysis: water evaporation at 200°C, primary breakdown at 250°C–500°C, and carbonization at temperatures more than 500°C. The benzene ring would release several light vapors as functional groups such as carbonyl, carboxyl, and methoxy escaped. Lastly, according to Wünning (Wünning, 2001), lignin biochar containing several functional groups, such as -OH and -OCH3, would be produced via carbonization. The functional groups in biochar would break more easily and release lighter gasses as the temperature rose. Repolymerization, in contrast, would transform bio-oil active aromatic molecules into biochar.

3.4.2 PYROLYSIS OF LIGNIN (INITIAL STEP)

The lignin pyrolysis process comprises three distinct phases: the initial, primary, and charring stages, each characterized by unique pyrolytic features and mechanisms (Figure 3.2). Li et al. (2020) conducted a comprehensive examination of the initial stage's reaction pathway, revealing that the intricate initial reactions of lignin pyrolysis predominantly occur within the temperature range of 160°C–330°C. Below 160°C, removing external and internal water leads to a slight mass decrease. Within this temperature range, a mass loss of approximately 20% occurs, forming substantial

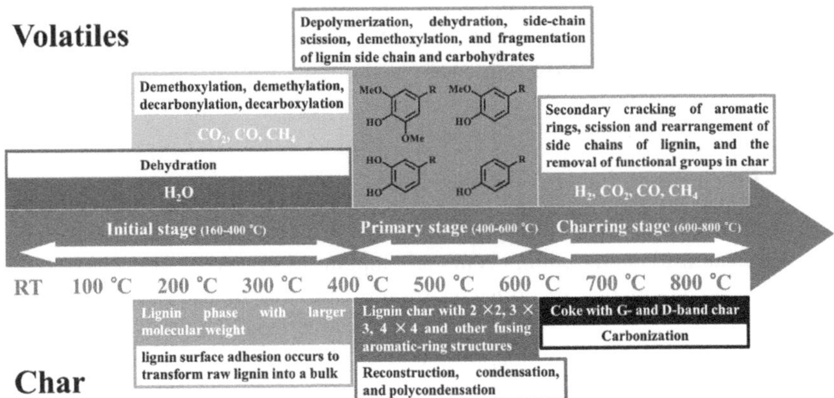

FIGURE 3.2 Distinctive pyrolytic activities are seen throughout the initial, primary, and charring phases. Adopted from Lu and Gu (2022).

quantities of volatile products. Notably, the initial stage witnesses the release of light gaseous products, such as H_2O, between 160°C and 330°C and 330°C and 400°C due to dehydroxylation and hydroxyl cracking within the lignin hydrogen bond network. Additionally, CO_2 is produced between 160°C–270°C and 330°C–400°C through fatty ether bond cleavage, oxidation, carboxylic acid groups, and cleavage of carbonyl groups. The diverse activation energies of these light gases stem from their generation from various sources with differing bond energies (Faravelli et al., 2010). At temperatures exceeding 150°C, the polymerization process forms a condensed lignin phase characterized by higher molecular weight. Further temperature elevation to 330°C results in the orderly production of oil-based components, including lignin monomers, dimers, and trimers (Li et al., 2020). The evolution of functional groups indicates that hydroxyl and hydrocarbon groups in bulk lignin are temperature-induced, with their content decreasing as the temperature rises, leading to the generation of more light gases. This phenomenon occurs due to the transformation of raw lignin into a bulk state facilitated by the surface adhesion of lignin.

When the pyrolytic temperature rises, side chains release from lignin's non-conjugated, conjugated guaiacyl, aromatic, and syringyl groups to generate light gases. All of these groups may be transformed into aromatic compounds. Finally, biochar with fused aromatic rings is produced by the structural polymerization of condensed lignin or the repolymerization of aromatic molecules.

Chua et al. (2019) looked at the structural changes made to the char after the first stage. The resulting lignin char's concentration of the THF-insoluble fraction has a higher molecular weight as the temperature rises (beyond the lignin's softening temperature of around 140°C (Binod et al., 2017), suggesting the presence of a repolymerization activity. When the temperature rises to 300°C, the THF-soluble fraction nearly completely transforms into volatile products. The remaining portion repolymerizes into the THF-insoluble fraction, which has an aromatic ring structure of mono, two to three fused, and three to five fused rings (Wang et al., 2012). The production of light gases and oil component chemicals (such as phenolic monomers, dimers, or trimers) causes the aromatic structure of char to be decreased between RT and 250°C. In the THF-soluble fraction, the H/C and O/C ratios decreased as the temperature rose from 100°C to 300°C, demonstrating that certain mono-aromatics would repolymerize into condensed phases with fused-ring structures in the THF-insoluble fraction.

3.4.3 Pyrolysis of Lignin (Primary Step)

After the first stage, the primary stage of lignin pyrolysis, which involves more intensive structural degradation, occurs at temperatures between 400°C and 600°C. Phenolic molecules, which are both temperature- and time-induced, play a substantial role at this stage (Klemetsrud et al., 2017). The production of products is significantly influenced by the pyrolytic temperature and residence duration (Zhao et al., 2018). The most excellent yield of phenolic products may be reached at 450°C, according to Lin et al.'s investigation (Liu et al., 2016) into the impact of temperature and time on the production of phenolic products. Due to the increased lignin conversion rate with rising temperature, the generation of phenolic monomers is dramatically increased.

However, more phenolic monomers would be converted into light gases due to the secondary reaction that would take place with the lengthening of the reaction period. The reaction equilibrium, on the other hand, would not necessarily result in a noticeable increase in the production of products at 600°C when the residence time was increased. Lignin structure impacts the pyrolytic behavior in the primary stage in addition to reaction time and temperature (Ansari et al., 2019).

3.4.4 Pyrolysis of Lignin (Charring Stage)

It is well known from the primary research on lignin pyrolysis that the structure of lignin char changes noticeably with increasing temperature. Zheng et al. investigated the development of char structure during rapid lignin pyrolysis at temperatures ranging from 200°C to 800°C (Zheng et al., 2022).

3.5 PL APPLICATIONS

3.5.1 Hydrogen Production

Hydrogen stands as a vital raw material within the fine-chemical industry, and its production is often facilitated through the steam-reforming process of small organic compounds such as acetic acid, alcohol, and methane (Kiani et al., 2023). In addition to these more minor organic compounds, bio-oil, derived from the extensive thermal decomposition of biomass, is recognized as a promising source for hydrogen generation. Notwithstanding its substantial organic nature and considerable molecular weight, PL has also been explored as an initial substrate for hydrogen synthesis. For instance, Zeng et al. (2015) investigated the potential conversion of the high molecular weight fraction of bio-oil, primarily composed of PL, into hydrogen gas. The pyrolysis process of cotton stalks yielded the high molecular weight fraction of bio-oil, which was subsequently separated from the low molecular weight phase after a few days of aging. The reaction occurred in a batch reactor with temperatures ranging from 850°C to 1,000°C, utilizing Fe_2O_3/Al_2O_3 as the catalyst. At 850°C, the analysis indicated that high molecular weight phenolic compounds underwent conversion into a gaseous substance containing 30.35 wt% hydrogen. Elevating the temperature to 1,000°C resulted in an increased hydrogen concentration of 43.37 wt%, attributed to the further fragmentation of C-H bonds in the hydrocarbon gases in the gaseous state at higher temperatures. The introduction of steam into the reaction environment significantly enhanced hydrogen production at a steam-to-heavy bio-oil mass ratio of 2:1. In a separate study. Xiao et al. (2014) utilized a substantial portion of bio-oil from cotton stalks for hydrogen generation using an ilmenite catalyst. The reaction occurred in a fixed bed reactor maintained at temperatures ranging from 850°C to 1,000°C, producing a gas containing 30% by weight of H_2 at 850°C under the influence of the ilmenite catalyst.

In a related study, Zeng et al. (2016) used steam injection to reduce coke generation and increase hydrogen output in the reaction environment. The findings showed that the H_2 concentration of the heavy bio-oil rose from 470 to 870 mL/mL when the steam-to-heavy bio-oil mass ratio was adjusted from 0 to 2.5:1. In addition, the gaseous product showed hydrogen yields of 13% and 84%, respectively, when comparing steam- and steam-free reaction conditions at the same reaction time and temperature. This demonstrates how vital steam is for preventing coking reactions from starting.

3.5.2 Phenolics and Aromatics

The wide range of uses for PL is shown in Figure 3.3, and its potential to be converted into alkylphenols and aromatics might significantly increase its market worth. As aromatic rings are abundant in PL structures and aromatic hydrocarbons and phenolics are expensive, many methods have been devised to synthesize molecules or chemical groups from PL selectively (De Wild et al., 2017). There has been much research on hydrotreatment as a method for transforming PL into aromatics. Compounds might also be grouped into aromatics, cyclic alkanes, and linear alkanes. The potential of hydrotreatment to convert PL into valuable chemicals is highlighted by this. The conversion of PL to hexamethylbenzene was the subject of an independent investigation by Wang et al. (2019). Under an N2 environment at 350°C, the experiment used a fixed bed reactor with γ-Al$_2$O$_3$ as the catalyst. A bio-oil made from pinewood was used to make PL. First, the cyclohexadienyl cation produced hexamethylbenzene by reacting pentamethylphenol with a hydrion. The result was cyclohexane-2,4-dienol, formed when hydride ions were transported from the cyclohexadienyl cation to methanol. Dehydration converted the molecule to pentamethylbenzene, and alkylation with methanol was a step in producing hexamethylbenzene. PL successfully produced hexamethylbenzene with a purity level of 99%.

Diverse catalysts have generated distinct types and quantities of aromatics from pyrolytic lignin PL. In a study by Zhang et al. (2018) HZSM-5, α-Al$_2$O$_3$, and MoO$_3$

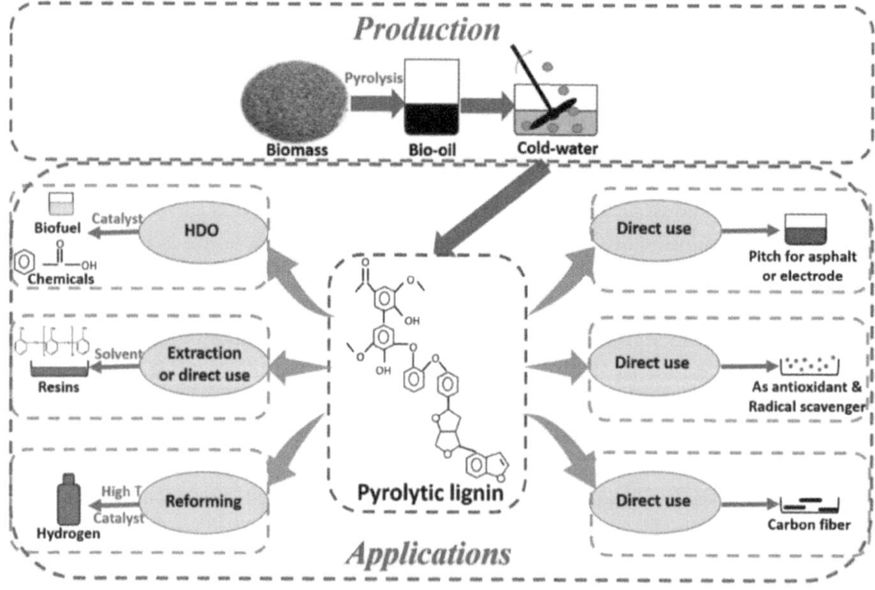

FIGURE 3.3 Various applications of PL. Adopted from Zhang et al. (2021).

were utilized to hydrotreat PL from the cold-water precipitation of bio-oil extracted from pine wood. The primary products obtained from this process were phenolics and aromatics. The presence of accessible oxygen vacancies on the surface of MoO_3 and oxygen atoms in the PL structure induced the formation of unpaired electrons and coordination bonds, thereby weakening the strength of the C-O bond in PL. The weakened C-O bonds underwent dissociation under the aggressive assault of hydrogen radicals, forming various types of aromatics. Guaiacols acted as primary intermediate compounds, undergoing demethoxylation to produce phenolics and alkylphenolics or demethylation to yield bisphenols. Aromatics were also generated through dehydration and subsequent cracking reactions. Additionally, the inclusion of $\alpha\text{-}Al_2O_3$ resulted in the formation of phenolics in lower quantities, primarily due to its limited effectiveness in catalyzing deoxygenation and cracking events.

3.5.3 Antioxidants and Radical Scavengers

A wide range of compounds with fused-ring and single-ring architectures are included in PL. These outstanding chemicals contain vanillin, 2-phenylpropionic acid, benzyl benzoate, phenyllactic acid, and fisetinidol. According to Kadarwati et al. (2016), these chemicals might be used as antioxidants or radical scavengers in the food business. Research on the antioxidant properties of phenolic compounds found in PL was carried out by Qazi et al. (2017). Phenyllactic acid and vanillin, two compounds with radical scavenging and antioxidant properties, were identified as antibacterial and antioxidant in their study. A synergistic mix of electron and H-atom transfer mechanisms gave rise to PL's antioxidant properties. In contrast, its proton or electron transfer processes were thought to be responsible for its significant scavenging activities.

Antioxidants derived from PL hold potential applications in biodiesel and bio-based lubricants. In a study by Chandrasekaran et al. (2016), bio-oil was generated through the gradual pyrolysis of birch wood and commercial KL at a temperature of 450°C. Subsequently, PL was obtained through alkali extraction of the bio-oil samples, and the phenolic-rich portion was utilized as a substitute for fossil-based antioxidants in soy biodiesel and lubricants. The antioxidant properties of the phenolic-rich phase from bio-oil exhibited comparable results to those of butylated hydroxytoluene. Notably, the antioxidant activity was attributed to phenolics with molecular weights of 302, 316, 330, and 344. This underscores the potential of utilizing PL-derived antioxidants as sustainable alternatives in biodiesel and lubricant applications.

3.5.4 Syngas

Theoretically, the gasification of 1 kg of isolated lignin with steam yields approximately 62 moles of hydrogen and 53 moles of carbon monoxide, resulting in a hydrogen-to-carbon monoxide ratio of 1.2. However, it's important to note that this gasification reaction is highly endothermic, requiring the addition of external heat to progress. In the case of autothermal reforming, the necessary heat for the process is generated through the partial oxidation of lignin. The gas mixture obtained from lignin gasification holds potential applications in producing methanol and Fischer–Tropsch fuels. Moreover, the syngas can be utilized in gas turbines to generate heat and electricity, making it a versatile and valuable resource in various energy applications (Azadi et al., 2013).

3.6   FAST PYROLYSIS OF LIGNIN FOR PRODUCING HYDROGEN

In their study, Baumline et al. (2006) examined the effects of regulated heat flux densities produced by short bursts of focused radiation on the flash pyrolysis of lignin samples. Two types of lignins, namely Kraft and Organocell lignins (OCL), were used in the investigation. Before char, vapors, and gases developed, the investigations showed that an intermediate liquid compound formed. The study measured the two lignin types' mass loss and product production rates. There was also a comparison of the results with those from earlier research using cellulose and those from the two different kinds of lignin.

Assuming that the organic component of KL undergoes complete pyrolysis, the elemental composition study suggests that 1 kg of KL organic content may theoretically yield up to 55×10^{-3} kg of gaseous H_2. Assuming all other reaction products are composed entirely of carbon and oxygen, this maximum limit fraction of 5.5 wt% is based on that assumption. This theoretical value exceeds 60×10^{-3} kg for OCL. The following is the approximate product mass distribution using a 1-second flash period with KL: 31% carbon monoxide, 28% aerosols and vapors trapped, and 41% gas. The research found that the reacting lignin had around half a mole of hydrogen, the same as the 2.8% mass fraction. According to the results, H_2 is produced from around half of the hydrogen in lignin. Light hydrocarbons, such as methane and condensed vapors, contain the remaining hydrogen. The average vapor's chemical formula will be C6H10.4O2.5 if the char is all carbon. According to elemental analysis, carbon makes up 72% of the organic component's mass in the char (Baumlin et al., 2006).

Supplementary gas treatments might achieve a much greater hydrogen %.

- The leftover vapors may crack at high temperatures using catalytic methods in the gas phase. Nevertheless, it is advisable to refrain from engaging in maturation responses.
- Steam reforming of light hydrocarbons and vapors produces CO and H_2.
- The water-gas shift process converts the high carbon monoxide percentages into hydrogen.
- Char is subjected to steam gasification.

The total process resulting in the theoretical maximum hydrogen output may be expressed as:

$$C_6H_{6.3}O_{2.2} + 9.8H_2O \rightarrow 12.95H_2 + 6CO_2 \tag{3.1}$$

3.6.1 H_2 PRODUCING MECHANISM

According to the literature, the amounts of H_2 fractions in the gases generated by cellulose, lignins, and wood pyrolysis are influenced by the operating circumstances (Broust et al., 2002). Under the specified heating conditions, the characteristics of biomass (OCL, KL, wood, or cellulose) also vary significantly. Experiments investigating ablative pyrolysis, which involves the interaction of a wood rod with a spinning hot disk, have been conducted at heat flux densities comparable to those seen in the image furnace. The molar percentages of H_2 in these circumstances do not surpass 4%, regardless of the disk temperature ranging from 873 to 1,173 K (Boutin et al., 1997). Under identical circumstances, the ablative rapid pyrolysis of biomass conducted in a cyclone reactor exhibits a range of 10%–22% depending on the wall temperature (Lédé et al., 2002).

Increasing the reactor temperature and keeping the gas-phase residence times longer might amplify secondary gas-phase thermal cracking processes, which are often thought to be responsible for a higher H_2 concentration (Baumlin et al., 2005). In studies conducted at low temperatures or when the initial products are rapidly cooled or removed from the high-temperature reaction environment, it is more probable that small volumes of hydrogen gas will be created. In ablative pyrolysis experiments, for example, this happens after rapidly removing the first liquids from the disk/biomass interface. In many cases, the gas phase is even hotter than the disk phase (Lédé, 2003).

A benefit of using focused radiation as a heat source in this study is that direct radiation absorption does not substantially heat the gasses. Furthermore, the rate at which gaseous species are generated by the sample undergoing pyrolysis at the reaction temperature is far lower than that of the gas used to transport them. Consequently, the resultant mixture of gases is almost at ambient temperature. Furthermore, the gas residence periods vary from 0.1 to 0.3 seconds and are brief. Consequently, the occurrence of further thermal reactions in the gas phase inside the reactor is improbable. Therefore, they may anticipate a high level of quenching efficiency, thereby halting any more thermal reactions inside the gas phase. Thus, these methods cannot account for the formation of large H_2 fractions with KL. In the case of lignin, prolonged exposure to high temperatures might likely lead to secondary cracking and interactions inside the thick layer of hot char. These processes may contribute to the creation of hydrogen gas, mainly due to the influence of the image furnace radiation. Nevertheless, this explanation fails to account for the observation that, in the case of KL, the H_2 percentage remains constant regardless of char formation and accumulation. Furthermore, while doing cellulose studies in the same reactor, the gases produced consist of H_2, with no char formation (Broust et al., 2002).

In the first steps of biomass pyrolysis, a layer of liquid chemicals, including cellulose and lignin, formed. According to the suggested procedure, hydrogen gas is created within the reacting sample inside this layer. Consistent with the lignin interpretation given by Ferdous et al. (2002), the direct fragmentation mechanism previously reported for cellulose is also suggested here. Hydrogen is released at high temperatures during the breakdown of lignin subunits and the rearrangement of aromatic rings. The results of the experiments performed with the electric furnace equipment corroborate this viewpoint, showing that the concentration of H_2 increases as a function of temperature. Before the incoming heat flow reaches the sample's virgin surface, it is partially blocked by the liquid layer. According to a simulation of the erosion of a cellulose cylinder subjected to intense radiation, the ILCC layer might potentially reach temperatures above 920 K. This is much higher than the temperature at which primary pyrolysis of cellulose occurs, which is around 740 K (Boutin et al., 2002). At temperatures comparable to those at which controlled trials in their electric furnace reveal substantial increases in H_2 concentration, the observed high temperature is consistent with those results.

The Broido Shafizadeh (BS)-type model remains often used in the literature to represent biomass rapid pyrolysis. This model is commonly linked to a gas-phase vapor thermal cracking process that produces CO, H_2, and hydrocarbons. The findings of this study indicate that the current paradigm is inadequate in fully explaining the gas production process. It is recommended to consider an extra reaction that takes place directly at the ILC level, as Lédé (2003) and Diebold (1994) proposed for cellulose. This reaction would compete with the evaporation of the short-lived liquid intermediate, as shown in the BS model. This conclusion is consistent with the observations and interpretations previously reported by Hopkins et al. (1984). Indeed, when the gas-phase temperatures are elevated, and the vapors remain in the system for extended periods, conventional thermal cracking may also occur, resulting in extra production of H_2.

3.7   CONCLUSION

Lignin comprises about 25% of the lignocellulosic biomass and is recognized as the optimal biomass for producing sustainable fuels and chemicals. Lignin has a greater energy density per unit mass than carbohydrates and is the primary source of renewable aromatics on Earth. Nevertheless, the intricate composition and diverse connections among the monomers transform lignin into valuable substances significantly more complicated than carbohydrates. By subjecting the syngas to a water-gas shift reaction and then separating the gases using pressure swing adsorption, it is possible to manufacture pure hydrogen. This process leads to an 80% increase in the total amount of hydrogen obtained compared to utilizing the raw syngas directly. Purified hydrogen has several applications, including its usage as a clean energy carrier, for enhancing fuel quality in petroleum and biorefineries, for synthesizing ammonia, and in fuel cells. Based on a Baumlin study (Baumlin et al., 2006), lignin has a remarkable capacity to generate gases with a very high proportion of syngas (up to 87%) and H_2 (about 50%). In the case of KL, around 40% of the hydrogen in the raw material undergoes conversion into H_2. The H_2 fractions may be enhanced by several methods, such as raising the temperature and residence duration of the gas in the reactor, implementing secondary thermal and catalytic cracking, steam reforming of vapors, water gas shift reaction of CO, and steam gasification of the charry products.

ABBREVIATION AND SYMBOLS

BTG	Biomass Technology Group
BS	Broido Shafizadeh
FTIR	Fourier transforms infrared spectroscopy
GC/MS	Gas chromatography/mass spectrometry
KL	Kraft lignin
ILCL	Ionic Liquid compound for cellulose
OCL	Organocell lignin
PL	Pyrolytic lignin
TG	Thermogravimetric analysis

REFERENCES

Amen-Chen, C., Pakdel, H. & Roy, C. 2001. Production of monomeric phenols by thermochemical conversion of biomass: A review. *Bioresource Technology*, 79, 277–299.

Ansari, K. B., Arora, J. S., Chew, J. W., Dauenhauer, P. J. & Mushrif, S. H. 2019. Fast pyrolysis of cellulose, hemicellulose, and lignin: Effect of operating temperature on bio-oil yield and composition and insights into the intrinsic pyrolysis chemistry. *Industrial & Engineering Chemistry Research*, 58, 15838–15852.

Azadi, P., Inderwildi, O. R., Farnood, R. & King, D. A. 2013. Liquid fuels, hydrogen and chemicals from lignin: A critical review. *Renewable and Sustainable Energy Reviews*, 21, 506–523.

Azuma, J.-I. & Tetsuo, K. 1988. Lignin-carbohydrate complexes from various sources. In *Methods in Enzymology*. (Vol. 161, pp. 12–18). Academic Press: Elsevier.

Baumlin, S., Broust, F., Bazer-Bachi, F., Bourdeaux, T., Herbinet, O., Toutie Ndiaye, F., Ferrer, M. & Lédé, J. 2006. Production of hydrogen by lignins fast pyrolysis. *International Journal of Hydrogen Energy*, 31, 2179–2192.

Baumlin, S., Broust, F., Ferrer, M., Meunier, N., Marty, E. & Lédé, J. 2005. The continuous self stirred tank reactor: Measurement of the cracking kinetics of biomass pyrolysis vapours. *Chemical Engineering Science*, 60, 41–55.

Binod, S., Thierry, G., Sébastien, L., Vincent, C., Frédéric, A., Sandrine, H., Philippe, M., Steve, P., Nicolas, B. & Anthony, D. 2017. *A Multitechnique Characterization of Lignin Softening and Pyrolysis.*

Boutin, O., Ferrer, M. & Lédé, J. 2002. Flash pyrolysis of cellulose pellets submitted to a concentrated radiation: Experiments and modelling. *Chemical Engineering Science*, 57, 15–25.

Boutin, O., Lede, J., Li, H. Z. & Kiener, P. 1997. Temperature of ablative pyrolysis of wood. *Comparison of Spinning Disc and Rotating Cylinder Experiments*. Biomass gasification and pyrolysis. State of the art and future prospects, pp.336–344. Pyne CPL Press, UK.

Broust, F., Ferrer, M. & Lédé, J. 2002. Fast pyrolysis of biomass in a multifunctional cyclone reactor. In *AIChE Annual Meeting*, 61–64.

Bu, Q., Lei, H., Zacher, A. H., Wang, L., Ren, S., Liang, J., Wei, Y., Liu, Y., Tang, J. & Zhang, Q. 2012. A review of catalytic hydrodeoxygenation of lignin-derived phenols from biomass pyrolysis. *Bioresource Technology*, 124, 470–477.

Chakar, F. S. & Ragauskas, A. J. 2004. Review of current and future softwood kraft lignin process chemistry. *Industrial Crops and Products*, 20, 131–141.

Chandrasekaran, S. R., Murali, D., Marley, K. A., Larson, R. A., Doll, K. M., Moser, B. R., Scott, J. & Sharma, B. K. 2016. Antioxidants from slow pyrolysis bio-oil of birch wood: Application for biodiesel and biobased lubricants. *ACS Sustainable Chemistry & Engineering*, 4, 1414–1421.

Chua, Y. W., Yu, Y. & Wu, H. 2019. Structural changes of chars produced from fast pyrolysis of lignin at 100-300 C. *Fuel*, 255, 115754.

Davda, R. R., Shabaker, J. W., Huber, G. W., Cortright, R. D. & Dumesic, J. A. 2005. A review of catalytic issues and process conditions for renewable hydrogen and alkanes by aqueous-phase reforming of oxygenated hydrocarbons over supported metal catalysts. *Applied Catalysis B: Environmental*, 56, 171–186.

De Wild, P. J., Huijgen, W. J. J., Kloekhorst, A., Chowdari, R. K. & Heeres, H. J. 2017. Biobased alkylphenols from lignins via a two-step pyrolysis-hydrodeoxygenation approach. *Bioresource Technology*, 229, 160–168.

Diebold, J. P. 1994. A unified, global model for the pyrolysis of cellulose. *Biomass and Bioenergy*, 7, 75–85.

Doherty, W. O. S., Mousavioun, P. & Fellows, C. M. 2011. Value-adding to cellulosic ethanol: Lignin polymers. *Industrial Crops and Products*, 33, 259–276.

Dutta, T., Papa, G., Wang, E., Sun, J., Isern, N. G., Cort, J. R., Simmons, B. A. & Singh, S. 2018. Characterization of lignin streams during bionic liquid-based pretreatment from grass, hardwood, and softwood. *ACS Sustainable Chemistry & Engineering*, 6, 3079–3090.

Effendi, A., Gerhauser, H. & Bridgwater, A. V. 2008. Production of renewable phenolic resins by thermochemical conversion of biomass: A review. *Renewable and Sustainable Energy Reviews*, 12, 2092–2116.

Faravelli, T., Frassoldati, A., Migliavacca, G. & Ranzi, E. 2010. Detailed kinetic modeling of the thermal degradation of lignins. *Biomass and Bioenergy*, 34, 290–301.

Ferdous, D., Dalai, A. K., Bej, S. K. & Thring, R. W. 2002. Pyrolysis of lignins: Experimental and kinetics studies. *Energy & Fuels*, 16, 1405–1412.

Gholizadeh, M., Gunawan, R., Hu, X., Kadarwati, S., Westerhof, R., Chaiwat, W., Hasan, M. M. & Li, C.-Z. 2016. Importance of hydrogen and bio-oil inlet temperature during the hydrotreatment of bio-oil. *Fuel Processing Technology*, 150, 132–140.

Hopkins, M. W., Antal Jr, M. J. & Kay, J. G. 1984. Radiant flash pyrolysis of biomass using a xenon flashtube. *Journal of Applied Polymer Science*, 29, 2163–2175.

Hu, X., Wang, Y., Mourant, D., Gunawan, R., Lievens, C., Chaiwat, W., Gholizadeh, M., Wu, L., Li, X. & Li, C. Z. 2013. Polymerization on heating up of bio-oil: A model compound study. *AIChE Journal*, 59, 888–900.

Kadarwati, S. 2016. Coke Formation during the Hydrotreatment of Bio-oil (Doctoral dissertation, Curtin University)..

Kadarwati, S., Oudenhoven, S., Schagen, M., Hu, X., Garcia-Perez, M., Kersten, S., Li, C.-Z. & Westerhof, R. 2016. Polymerization and cracking during the hydrotreatment of bio-oil and heavy fractions obtained by fractional condensation using Ru/C and NiMo/Al2O3 catalyst. *Journal of Analytical and Applied Pyrolysis*, 118, 136–143.

Kiani, P., Meshksar, M. & Rahimpour, M. R. 2023. Biogas reforming over La-promoted Ni/SBA-16 catalyst for syngas production: Catalytic structure and process activity investigation. *International Journal of Hydrogen Energy*, 48, 6262–6274.

Klemetsrud, B., Eatherton, D. & Shonnard, D. 2017. Effects of lignin content and temperature on the properties of hybrid poplar bio-oil, char, and gas obtained by fast pyrolysis. *Energy & Fuels*, 31, 2879–2886.

Lédé, J. 2003. Comparison of contact and radiant ablative pyrolysis of biomass. *Journal of Analytical and Applied Pyrolysis*, 70, 601–618.

Lédé, J., Ferrer, M. & Broust, F. 2002. Fast pyrogasification and/or pyroliquefaction of biomass in a cyclone reactor for the production of either liquids or gases. 12th European Biomass Conference, Amsterdam, NL: ETA-Florence and WIP-Munich. 191–201.

Lehto, J., Oasmaa, A., Solantausta, Y., Kytö, M. & Chiaramonti, D. 2014. Review of fuel oil quality and combustion of fast pyrolysis bio-oils from lignocellulosic biomass. *Applied Energy*, 116, 178–190.

Li, J., Bai, X., Fang, Y., Chen, Y., Wang, X., Chen, H. & Yang, H. 2020. Comprehensive mechanism of initial stage for lignin pyrolysis. *Combustion and Flame*, 215, 1–9.

Liu, C., Hu, J., Zhang, H. & Xiao, R. 2016. Thermal conversion of lignin to phenols: Relevance between chemical structure and pyrolysis behaviors. *Fuel*, 182, 864–870.

Lu, X. & Gu, X. 2022. A review on lignin pyrolysis: Pyrolytic behavior, mechanism, and relevant upgrading for improving process efficiency. *Biotechnology for Biofuels and Bioproducts*, 15, 106.

Lu, X., Zhu, X., Guo, H., Que, H., Wang, D., Liang, D., He, T., Hu, C., Xu, C. & Gu, X. 2020. Investigation on the thermal degradation behavior of enzymatic hydrolysis lignin with or without steam explosion treatment characterized by TG-FTIR and Py-GC/MS. *Biomass Conversion and Biorefinery*, Springer, 1–10.

Mosallanezhad, S., Rahmatmand, B. & Rahimpour, M. R. 2023. Photobiological process of biomass. *Reference Module in Earth Systems and Environmental Sciences*. Elsevier.

Mu, W., Ben, H., Ragauskas, A. & Deng, Y. 2013. Lignin pyrolysis components and upgrading-technology review. *Bioenergy Research*, 6, 1183–1204.

Oh, S., Lee, J. H., Choi, I.-G. & Choi, J. W. 2020. Enhancement of bio-oil hydrodeoxygenation activity over Ni-based bimetallic catalysts supported on SBA-15. *Renewable Energy*, 149, 1–10.

Qazi, S. S., Li, D., Briens, C., Berruti, F. & Abou-Zaid, M. M. 2017. Antioxidant activity of the lignins derived from fluidized-bed fast pyrolysis. *Molecules*, 22, 372.

Raveendran, K., Ganesh, A. & Khilar, K. C. 1996. Pyrolysis characteristics of biomass and biomass components. *Fuel*, 75, 987–998.

Schmitt, C. C., Raffelt, K., Zimina, A., Krause, B., Otto, T., Rapp, M., Grunwaldt, J.-D. & Dahmen, N. 2018. Hydrotreatment of fast pyrolysis bio-oil fractions over nickel-based catalyst. *Topics in Catalysis*, 61, 1769–1782.

Sukhbaatar, B., Steele, P. H. & Kim, M. G. 2009. Use of lignin separated from bio-oil in oriented strand board binder phenol-formaldehyde resins. *BioResources*, 4, 789–804.

Wang, C., Li, M. & Fang, Y. 2019. Upgrading of pyrolytic lignin into hexamethylbenzene with high purity: Demonstration of the "all-to-one" biochemical production strategy in thermo-chemical conversion. *Green Chemistry*, 21, 1000–1005.

Wang, S., Dai, G., Yang, H. & Luo, Z. 2017. Lignocellulosic biomass pyrolysis mechanism: A state-of-the-art review. Progress *in Energy and Combustion Science*, 62, 33–86.

Wang, Y., Li, X., Mourant, D., Gunawan, R., Zhang, S. & Li, C.-Z. 2012. Formation of aromatic structures during the pyrolysis of bio-oil. *Energy & Fuels*, 26, 241–247.

Wünning, P. 2001. Applications and use of lignin as raw material. In Alexander Steinbüchel and Martin Hofrichter (eds), *Biopolymers. Lignin, Humic Substances and Coal*. Wiley-VCH, Weinheim, 117–127.

Xiao, R., Zhang, S., Peng, S., Shen, D. & Liu, K. 2014. Use of heavy fraction of bio-oil as fuel for hydrogen production in iron-based chemical looping process. *International Journal of Hydrogen Energy*, 39, 19955–19969.

Yang, J., Wang, X., Shen, B., Hu, Z., Xu, L. & Yang, S. 2020. Lignin from energy plant (Arundo donax): Pyrolysis kinetics, mechanism and pathway evaluation. *Renewable Energy*, 161, 963–971.

Zakzeski, J., Bruijnincx, P. C. A., Jongerius, A. L. & Weckhuysen, B. M. 2010. The catalytic valorization of lignin for the production of renewable chemicals. *Chemical Reviews*, 110, 3552–3599.

Zeng, D.-W., Xiao, R., Huang, Z.-C., Zeng, J.-M. & Zhang, H.-Y. 2016. Continuous hydrogen production from non-aqueous phase bio-oil via chemical looping redox cycles. *International Journal of Hydrogen Energy*, 41, 6676–6684.

Zeng, D.-W., Xiao, R., Zhang, S. & Zhang, H.-Y. 2015. Bio-oil heavy fraction for hydrogen production by iron-based oxygen carrier redox cycle. *Fuel Processing Technology*, 139, 1–7.

Zhang, L., Zhang, S., Hu, X. & Gholizadeh, M. 2021. Progress in application of the pyrolytic lignin from pyrolysis of biomass. *Chemical Engineering Journal*, 419, 129560.

Zhang, X., Chen, Q., Zhang, Q., Wang, C., Ma, L. & Xu, Y. 2018. Conversion of pyrolytic lignin to aromatic hydrocarbons by hydrocracking over pristine MoO3 catalyst. *Journal of Analytical and Applied Pyrolysis*, 135, 60–66.

Zhang, Y., Monnier, J. & Ikura, M. 2020. Bio-oil upgrading using dispersed unsupported MoS2 catalyst. *Fuel Processing Technology*, 206, 106403.

Zhao, S., Liu, M., Zhao, L. & Zhu, L. 2018. Influence of interactions among three biomass components on the pyrolysis behavior. *Industrial & Engineering Chemistry Research*, 57, 5241–5249.

Zheng, Q., Zhang, D., Fu, P., Wang, A., Sun, Y., Li, Z. & Fan, Q. 2022. Insight into the fast pyrolysis of lignin: Unraveling the role of volatile evolving and char structural evolution. *Chemical Engineering Journal*, 437, 135316.

4 Pyrolysis Process for Conversion of Sewage Sludge to Hydrogen

Sergio Nogales-Delgado, Beatriz Ledesma Cano, Carmen María Álvez-Medina, and Adrián Bocho-Roas

4.1 INTRODUCTION

Due to the increasing population in cities (which comprise a considerable and growing percentage of **the** worldwide population), including urban megacities like Tokyo, Delhi, Manila, and Mexico, among others, sewage sludge is one of the most troublesome and abundant wastes, requiring the implementation of wastewater treatment plants in populated urban areas in order to limit the heavy metal concentration in untreated sewage sludge and to avoid the direct use of this waste in soils according to current international policies (Statista, 2023). Apart from the evident environmental impact of wastewater treatment, these plants require a lot of energy with the subsequent energy costs, implying up to 40% of operating costs and around 2 kW/hm^3 of treated wastewater (Hu et al., 2022). The direct use of untreated sewage sludge as landfilling or land-farming could present a negative impact on the environment, due to the presence of heavy metals or nutrient surpluses, which can affect microorganisms in agricultural soil. In that sense, biochar generation from sewage sludge could be an interesting way to enhance this waste in order to promote sustainable agriculture (Ghorbani et al., 2022).

Regarding thermochemical processes, the main objective is energy production from organic matter like sewage sludge with the lowest environmental impact possible. In that sense, pyrolysis presents some advantages like waste volume reduction (up to 50%), stabilization of organic matter (with a heavy metal concentration in the char), and generation of interesting products or fuels like biochar, liquids, or tar and synthesis gas. Pyrolysis, as a thermal decomposition (from 500°C to 1,000°C) in an inert atmosphere (usually nitrogen) of a wide range of materials, usually produces interesting compounds in the three physical states: solid (char), liquid (condensable gas), and gas (non-condensable gas). This technique is versatile and, depending on the reaction conditions (such as heating rate), the solid, liquid, or gas yield can vary. For instance, at high heating rates, which can be considered as flash pyrolysis, liquid yield is promoted. For liquid production during pyrolysis, it is an interesting alternative as the liquids obtained during this process can be easily stored or shipped, with subsequent energy use.

DOI: 10.1201/9781003382270-6

To sum up, two different and complementary trends are observed when it comes to sewage sludge treatment. On the one hand, it is necessary to manage such a pollutant waste in order to avoid a negative environmental impact, but on the other hand, it is necessary to absorb some of the energy costs related to a wastewater plant. In that sense, the energy conversion of sewage sludge, as well as its conversion to other interesting materials such as active carbons, could be an interesting way to fulfill both objectives, contributing to sustainable development, green chemistry, and a circular economy.

Based on these considerations, it is not surprising that the management of this waste has risen in the interest of **the** scientific community. This fact can be observed in Figure 4.1, where the evolution of scientific articles published in impact journals in the last two decades is included. Thus, there has been a continuous increase in this kind of publications, especially since 2010, when a considerable rise was observed (from around 50 to above 300 articles published per year), proving the scientific interest in this subject, as sewage sludge management (and especially energy management, such as pyrolysis) presents a wide range of technical and scientific opportunities.

Also, from an industrial point of view, it should be noted that there are many companies that are focused on sewage sludge pyrolysis, like Pyrocal (a company in Spain that uses pyrolysis technology to convert sewage sludge into activated carbon, biochar, and bio-oil), Biogreen (whose plant is in the UK, they use pyrolysis technology

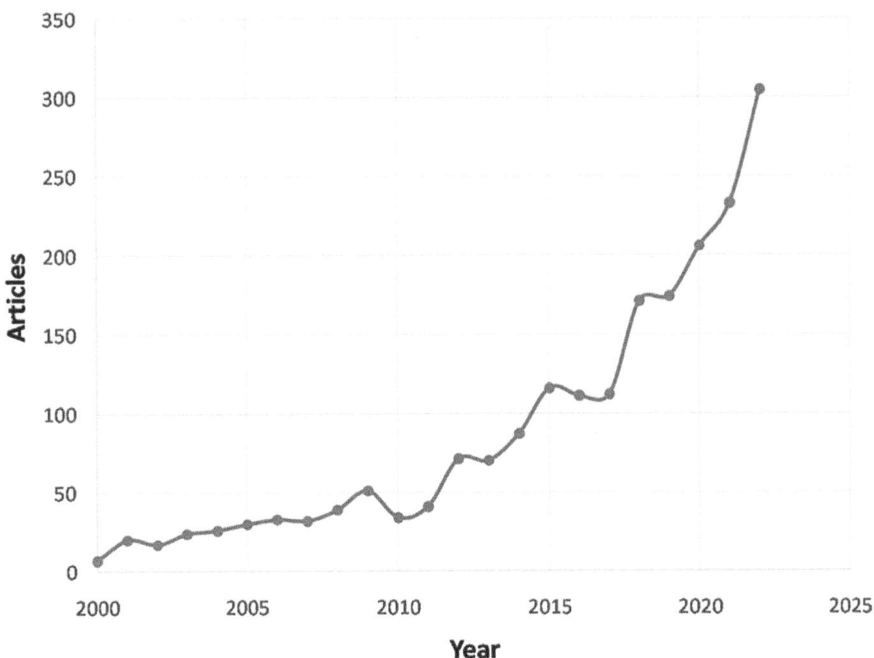

FIGURE 4.1 Articles related to sewage sludge pyrolysis (produced from Scopus).

to convert sewage sludge into biochar and bio-oil), or Infinis (located in Scotland, they use pyrolysis technology to convert sewage sludge into biochar and biogas). It is important to note that there are many other pyrolysis plants that work with sewage sludge around the world. Each plant may have different capacities, technologies, and end products. As a result, sewage sludge pyrolysis is a promising and versatile technology currently targeted at both industrial and research levels.

4.2   SEWAGE SLUDGE PRODUCTION AND CHARACTERISTICS

In this scenario, it is important to put in context sewage sludge production in a wastewater purification plant, whose stages are vital to understanding the role of pyrolysis. Considering that pyrolysis will produce different products from sewage sludge (like syngas or char), it will be interesting to consider which ones will have a direct use in specific applications (agriculture or energy production) or can be reused in the same wastewater plant at different treatment stages. Thus, the stages typically covered in a wastewater plant are primary, secondary, and tertiary treatments, which will be briefly explained as follows (see Figure 4.2).

To remove the pollutant load contained in the wastewater, it passes through a series of physical and biological processes consisting of the following stages:

4.2.1   PRETREATMENT

In this stage, suspended solids (a floating charge, sand, gravel, etc.) that can cause problems in subsequent treatments (due to their nature or size) are removed. It consists of these stages:

- Separation: Large solids and large quantities of sand are removed using a separation system (pit), located at the inlet of the sewage treatment plant collector.
- Roughing: Large objects capable of causing blockages in the different units of the installation are removed by passing the wastewater through grates with different mesh sizes.

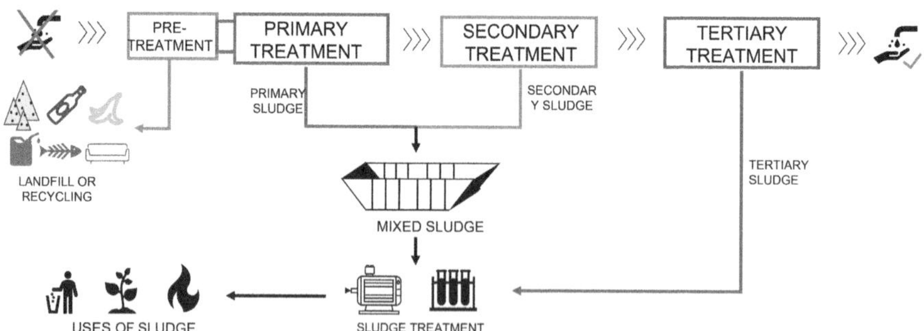

FIGURE 4.2   General overview of a wastewater plant.

- Screening: It implies a filtration on a thin support. Depending on the size of the hole, a distinction is made between macro and micro-screening.
- Dilaceration: The aim of this stage is to grind the solid matter entrained by the water.
- Dewatering: The objective is to eliminate particles with a size greater than 200 μm, to avoid sedimentation on pipes. In this way, pumps and other equipment are protected.
- De-oiling and degreasing: Fats, oils, grease, and other floating materials lighter than water, are removed. De-oiling is a liquid-liquid separation, whereas degreasing is a solid-liquid separation. In both cases, air bubbling is used to emulsify fats and improve buoyancy.
- Pre-aeration: Improving water treatability, odor control, fat separation, flocculation, etc.

4.2.2 PRIMARY TREATMENT

The objective is the separation by physical means of suspended particles that could not be retained in the pretreatment stage. During this stage, the removal of organic matter is practically negligible. The processes corresponding to this stage are (see Figure 4.3):

- Sedimentation (or primary settling): Its aim is the reduction of suspended solids in wastewater through gravity. Therefore, only settleable solids and floating matter can be removed. Sedimentation occurs for several hours in primary settling tanks (clarifiers). Three different phases take place in these tanks:

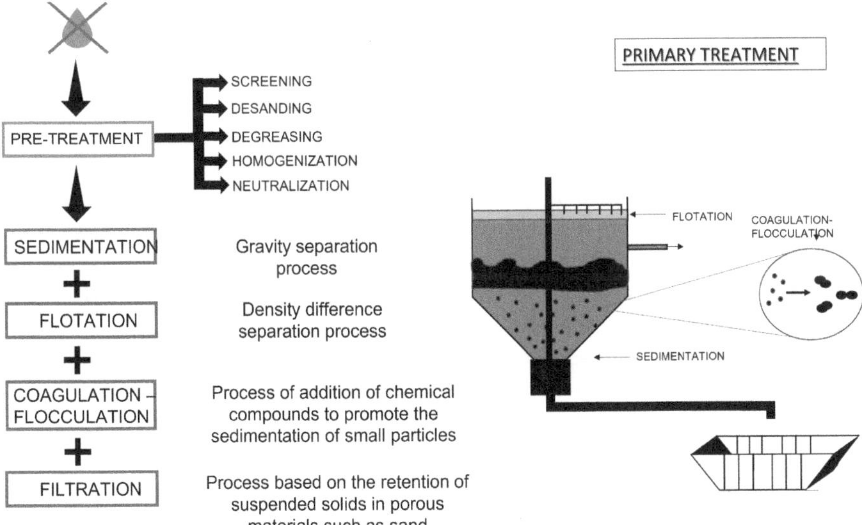

FIGURE 4.3 Primary treatment, with the subsequent primary sewage sludge generation.

- Coagulation: It aims to destabilize, by using coagulating substances and colloidal suspensions by breaking the repulsive forces due to the presence of charges of the same sign on the surface of colloids. At this point, these particles can interact with each other by agglomerating first into micro-flakes and then into flakes of increasing size.
- Flocculation: It consists of further destabilization of colloidal substances and agglomeration of destabilized particles resulting in micro-flakes and, subsequently, flakes that settle out. A combination of substances called coagulating aids or flocculants are usually added to the water.
- When the flocs reach such dimensions that they can be precipitated, the last sedimentation phase begins, in a tank where they can be removed from water.
- Flotation: Dissolved air flotation (DAF), apart from removing solid and/or liquid matter of lower density than water, allows the removal of higher-density solids. DAF consists of a system that creates air micro-bubbles within wastewater, which bind to the particles to be removed, forming aggregates capable of floating because they are less dense than water. Therefore, it can be said that the objective of this process in the primary treatment is twofold: a reduction of floating matter and a reduction of suspended solids.
- Gravity separation: This method is effective, especially for insoluble liquids such as oils.
- Sludge evacuation: A mechanical sludge collector scrubs the sludge settled at the bottom of a hopper in the lower part of the clarifier. The sludge is then pumped from this cavity to the sludge treatment system.
- Evacuation of the top layer: In some plants, the sludge scrubbers will lead the top layer (with impurities) toward their evacuation system.

4.2.3 Secondary Treatment

At this point (see Figure 4.4), the organic matter present in water is removed or reduced by the action of microorganisms, aerobic and anaerobic, which transform it into settleable solids that can be easily separated. Anaerobic digestion is a resourceful technique for waste sludge management (Pavicic et al., 2022). A secondary treatment tank (secondary clarifier) receives the wastewater from the primary aerator and clarifier after the initial removal of sludge and surface impurities takes part. At that time, 40%–60% of the solids have already been removed. After this stage, 90% of the pollutants will be removed from the water. For effective wastewater treatment, a balance must be achieved between the level of organic waste, dissolved oxygen, and bacterial levels. The processes that take place in this phase are:

- Aeration: It supplies large amounts of oxygen to wastewater for aerobic bacteria and other microorganisms that help break down harmful organic material in the sewage. The resulting clumps at the bottom of the tank called activated sludge, settle to the bottom of the wastewater. There are several types of aerators such as surface or dispersed aerators.

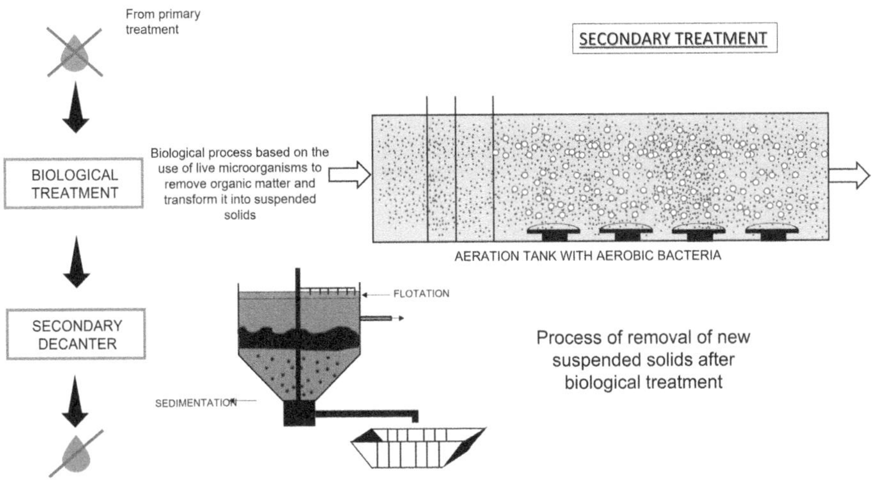

FIGURE 4.4　Secondary treatment, with secondary sewage sludge generation.

- Secondary Sedimentation or Clarification: It is often combined with aeration in a large tank. Aeration will occur at the top surface whereas sludge settling will occupy the bottom of the tank. The core of the system is the activated sludge from the aeration process. This material is rich in bacteria and other microbes responsible for the breakdown of organic materials and the formation of flocs for additional removal of solids, oil, and other waste. The secondary sedimentation stage is necessary to allow the flocs to settle and remove additional surface impurities from the surface. This final product is low in organic content.

4.2.4　TERTIARY TREATMENT

Tertiary treatment (see Figure 4.5) is the most complete procedure for treating wastewater content, but it has not been widely adopted because it is too expensive. It aims to remove the residual organic load and other pollutants not previously removed, such as P and N.

4.2.5　SEWAGE SLUDGE CHARACTERISTICS

It should be noted that, depending on the process or stage, different kinds of sewage sludge can be obtained, proving its heterogeneity. Sewage sludge obtained from wastewater treatment plants usually contains a wide range of compounds, like nontoxic compounds, toxic pollutants, compounds with N and P, pathogens and microbiological pollutants, inorganic compounds, and a considerable amount of water. Its composition and characteristics might vary depending on many factors, like the wastewater treatment plant location, the treatment processes that take place, the time of year (or even the time of day), etc. That is the reason why the components included

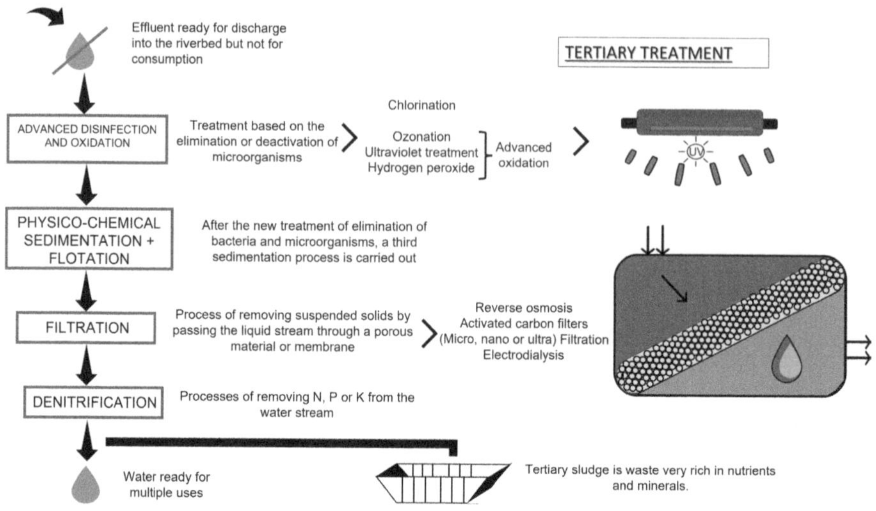

FIGURE 4.5 Tertiary treatment with tertiary sludge generation.

TABLE 4.1

Main Characteristics of Sewage Sludge (Dry Basis Except for Moisture) (Hu et al., 2022; Nunes et al., 2021)

Ultimate Analysis	%
C	23–40
H	4–6
N	2.5–8
S	0.8–2.0
O	20–24
Proximate Analysis	**%**
Moisture	70–85
Ash content	23–55
Volatile matter	35–65
Fixed carbon	1–20

in Table 4.1 present such a wide range, according to data obtained in the literature (Hu et al., 2022; Nunes et al., 2021). Notable is the high moisture content in sewage sludge, which is usually classified as free water (the majority, which can be easily removed), interstitial water (which can be removed through strong mechanical forces), vicinal water (removed by heating), and bound water (whose removal is difficult, requiring harsh pretreatments). Water content should be considered in some thermal treatments like pyrolysis, and its removal is required (which could imply additional energy costs).

As observed, considerable amounts of N, S, and O are included in sewage sludge, which can imply an increase in evolved pollutants during some treatments like combustion or pyrolysis, with NO_x or H_2S emissions, among others. These pollutants can represent a considerable threat to the environment and public health, apart from the fact that hydrogen sulfide can promote corrosion in pyrolysis facilities. Also, sewage sludge can contain some elements like heavy metals or Fe, Ca, K, and Mg, which usually act as catalysts. As a consequence, solid waste after sewage sludge thermal treatment (such as ash) might be equally valuable if they are suitably managed (Hu et al., 2022). As mentioned above, sewage sludge can imply a challenge and at the same time an opportunity, as its management is mandatory to avoid environmental problems, but also some interesting conversions to energy or other products can be achieved. Thus, the role of pyrolysis could be important, not only for syngas production but also for the improvement of hydrogen generation in other processes, apart from the evident indirect benefits that could be implied in a wastewater plant. These aspects will be covered in the following sections, as every detail counts for a more efficient pyrolysis process.

4.3 TREATMENTS TO MANAGE SEWAGE SLUDGE

There are plenty of treatments to manage sewage sludge, some traditional, some innovative, but most of them easily adapted to a wastewater plant. Most of them present advantages and disadvantages (as we will explain for pyrolysis), depending on the nature of sewage sludge or the characteristics of the product obtained. Some of the most interesting treatments are as follows:

Incineration: It has been a traditional way to manage many types of waste. It implies a combustion process where the aim is the destruction or reduction of waste. Thus, through combustion processes, the organic fraction of sewage sludge could be destroyed or decomposed in an oven at temperatures around 500°C, where its combustion takes place with excess air, converting the waste into ash, with a high inorganic content. The generated heat is used for electricity production, although this method might be expensive and imply environmental concerns, which makes it not suitable for current sustainable practices and policies (Di Giacomo and Romano, 2022; Siddiqui et al., 2023).

Gasification: It is the process that involves the thermochemical transformation of biomass, at temperatures above 700°C and with a controlled process involving heat, steam, and oxygen to convert biomass to hydrogen and other products (carbon monoxide, hydrogen, and carbon dioxide), without combustion. Then, carbon monoxide reacts with water to form carbon dioxide and more hydrogen via a water-gas shift reaction. Adsorbers or special membranes can separate hydrogen from this gas stream. The proportion of hydrogen that can be obtained from the synthesis gas generated is approximately 40%, in a process that is less carbon intensive than the production of gray or blue hydrogen (from natural gas).

In addition, this process is less expensive than the electrolysis used for green hydrogen, for the moment, and allows the continuous production of energy thanks to the availability of biomass. However, it is still a complex process in which many reactions are involved and unwanted products are generated, such as carbon monoxide (CO) and coke. Also, the process requires severe reaction conditions, followed by additional steps of separation and purification of the hydrogen produced to ensure it is of high quality, which complicates its implementation in small-sized industries.

Hydrothermal carbonization (HTC): It can be defined as combined hydrolysis, condensation, aromatization, dehydration, and decarboxylation reactions of waste in suspension with water applying mild temperatures (170°C–250°C) under saturated pressure for short residence times (between 5 and 240 minutes). It is presented as an alternative for pyrolysis because, unlike the latter, HTC does not require prior sludge drying, which involves costly pretreatments. Precisely, sewage sludge, as already mentioned, has a high moisture content. Therefore, the HTC process can improve the properties of sewage sludge into a carbon-rich solid product (hydrochar, HC) comparable to coal, thereby increasing its calorific value. Then, HTC's main products are HC and a liquid phase with a high content of organic matter, as well as a little gaseous part (5%) that is usually made up of CO_2. Therefore, it is an efficient method for sewage sludge with energy and chemical recovery. To sum up, HTC technology represents an economical, practical, and rapid treatment method for efficient sewage sludge treatment and recovering energy and chemicals.

Pyrolysis: It is a mature technology of interest to researchers for the treatment of sewage sludge since it is considered environment-friendly and generates fewer emissions than incineration (Manara and Zabaniotou, 2012). Pyrolysis consists of thermal degradation of a waste in the absence of oxygen, without combustion. The only oxygen present is the content in the waste to be treated, so these substances are decomposed by heat (300°C–800°C), without combustion reactions taking place. The resulting products are three phases which were briefly described in Figure 4.6:

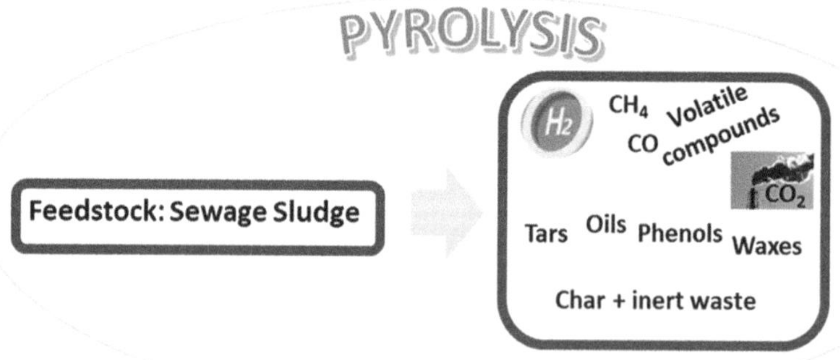

FIGURE 4.6   Products from the pyrolysis process for sewage sludge.

- **Gas:** CO, CO_2, H_2, CH_4, and more volatile compounds from the cracking of organic molecules, together with those already existing in the waste. This gas is very similar to the synthesis gas obtained in gasification, but there is a greater presence of tars, waxes, etc. to the detriment of gases because pyrolysis works at temperatures below gasification. In order to obtain more hydrogen, CO and CH_4 can be further processed by the steam reforming method and water-gas shift reaction (WGS, equation (4.1)), through whose process it is possible to eliminate greenhouse gases. As explained later on, pyrolysis could play an indirect role in the improvement of the performance of biogas or methane steam reforming (Fremaux et al., 2015; Nipattummakul et al., 2010).

$$CO + H_2O \leftrightarrow H_2 + CO_2 \tag{4.1}$$

- **Liquid:** Long-chain hydrocarbons such as tars, oils, phenols, and waxes formed during condensation at room temperature, which can be used for different purposes (Manara and Zabaniotou, 2012).
- **Solid:** High carbon material and other inert waste components like heavy metals which can be used in a wide variety of industrial applications, especially as adsorbent or catalyst support, as explained in the following sections (Nipattummakul et al., 2010).

According to TG (thermogravimetric) and DTG (derivative thermogravimetric) curves obtained from thermogravimetric studies in an inert atmosphere (see Figure 4.7), the thermal decomposition behavior during pyrolysis suggests that there are three stages

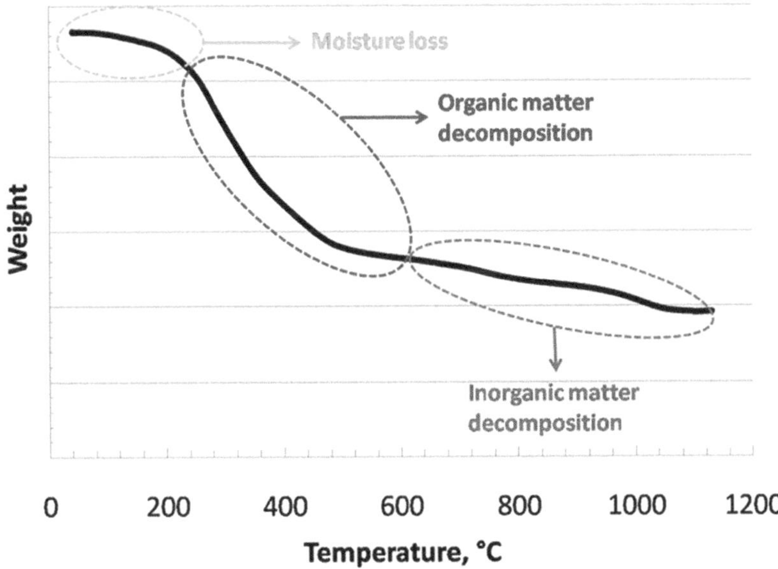

FIGURE 4.7 The typical thermogravimetric curve for sewage sludge during pyrolysis.

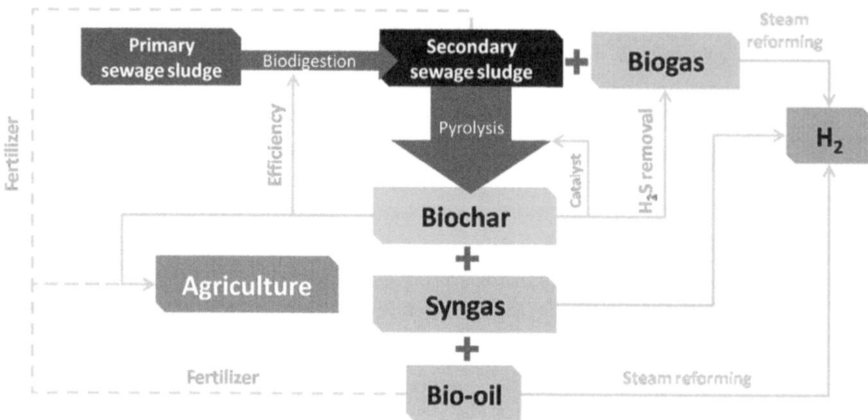

FIGURE 4.8 Pyrolysis and its relationship with wastewater process.

during pyrolysis: from 25°C to 200°C, where moisture loss during drying takes place; from 200°C to 600°C where the decomposition of biodegradable organic matter (proteins, carboxylic acids or cellulose) and biodegradable organic matter (aromatics, saturated aliphatic and long-chain aliphatic amides) takes place; and above 600°C, where decomposition of inorganic matter such as calcium carbonate occurs (Naqvi et al., 2021).

This way, pyrolysis could be an interesting starting point for sewage sludge treatment, offering interesting products explained in this chapter. Nevertheless, the most interesting point about pyrolysis applied to sewage sludge is its possible implementation in a wastewater plant, offering many opportunities to improve the efficiency of both pyrolysis and wastewater treatment, as observed in Figure 4.8.

As observed in this figure, depending on the products obtained through sewage sludge pyrolysis, different applications can be obtained, which are directly related to wastewater plants and, therefore, make pyrolysis more efficient in this context.

4.3.1 Sewage Sludge Pyrolysis for Hydrogen Production

Possibly due to the fact that sewage sludge presents high moisture levels (among other factors that are taken into account later on), the use of pyrolysis for hydrogen production from sewage sludge has not been widely studied in the scientific literature. Nevertheless, due to the increasing interest in converting wastes with difficult management to interesting energy sources, attention to pyrolysis could be increased in the future, especially when efficiency problems related to sewage sludge pretreatments are solved in the future.

In that sense, the most interesting advantages of treating sewage sludge using pyrolysis (mainly in the temperature range of 300°C–900°C) lie in the reduction of its volume, heavy metals removal, the destruction of pathogens and microorganisms (as explained in previous sections), and production of biogas, tar, and char with a certain utilization value. Hydrogen is considered the fuel of the future as an alternative to fossil fuels and greenhouse gas emitters (Nikolaidis and Poullikkas, 2017).

Despite its enormous potential, it is a difficult fuel to obtain, since it is not found isolated in nature, and precisely, the method by which this element is obtained determines whether it is a clean and sustainable fuel. However, attention should be focused on how all these technologies can most quickly and effectively reduce greenhouse gases. In the search for methods to produce clean energy, and thereby add value to sewage sludge, the production of hydrogen from its pyrolysis has been investigated. According to several scientific studies, sewage sludge is considered a promising material to produce hydrogen through pyrolysis, where high concentrations of hydrogen in pyrolysis gas (up to 30%–36% v/v) can be obtained. On the contrary, lower hydrogen levels (around 20%) were obtained in other studies, with high reaction temperatures (900°C) and low heating rates (10°C/min) (Ghodke et al., 2021). In that sense, these differences point out that many factors can be vital to obtaining high hydrogen yields in sewage sludge pyrolysis. For instance, sewage sludge composition, its moisture level, or the operating conditions play an important role in hydrogen generation during pyrolysis. The operating conditions or variables that have the most influence on hydrogen production are temperature, heating rate, residence time, the nature of raw material, the moisture content in the sample, and the catalyst addition. Some authors observed that increasing the moisture content in the sample, as well as the reaction temperature at 50°C/min plays a fundamental role in hydrogen production (Xu et al., 2014). Other authors (Domínguez et al., 2006) experimented with hydrogen-rich fuel gas production from the pyrolysis of wet sewage sludge with moisture content in the range of 70–80 wt. %. On the other hand, some studies obtained similar hydrogen productions using a bench-scale fixed bed reactor with less sewage sludge moisture and found that increasing the moisture above 43% lowers hydrogen production. They observed that H_2 content reaches the highest value of 7.76 mol/kg dry basis wet sludge and 42.13 vol%, at a moisture of 43.38%. In the case of temperature, they found a greater production of hydrogen when going from 600°C to 850°C (Luo et al., 2017). Similarly, other research work found the highest values of 51.48% in hydrogen production at 800°C (Wang et al., 2022), whereas the highest yield for hydrogen production found that the yield of H_2 increased from 1.26 mmol/g at 600°C to 9.07 mmol/g at 900°C (Liu et al., 2021). These authors not only concluded that at higher temperatures (800°C and 900°C) hydrogen production increases, but that it is also the dominant gaseous product. In another study (Beik et al., 2023), higher H_2 concentrations were found when this variable was increased. This fact can be explained by taking into account that an increase in temperature favors the three types of reactions that benefit the production of H_2, as can be seen, reflected in Figure 4.9, which is adapted from the findings obtained by other authors (Karaca et al., 2018; Xin et al., 2019).

As observed in this figure, the first decomposition stage takes place at low temperatures, with the subsequent generation of tar and char. At higher temperatures, apart from the abovementioned products, gas compounds are obtained: mainly hydrogen, carbon monoxide, carbon dioxide, methane, and other organic fragments. All these gaseous compounds have to a certain extent energy application, especially when they are converted to hydrogen through different methods. Again, as explained in later sections (where solid and liquid products obtained through pyrolysis were considered as hydrogen enhancers in indirect ways), sewage sludge pyrolysis can directly

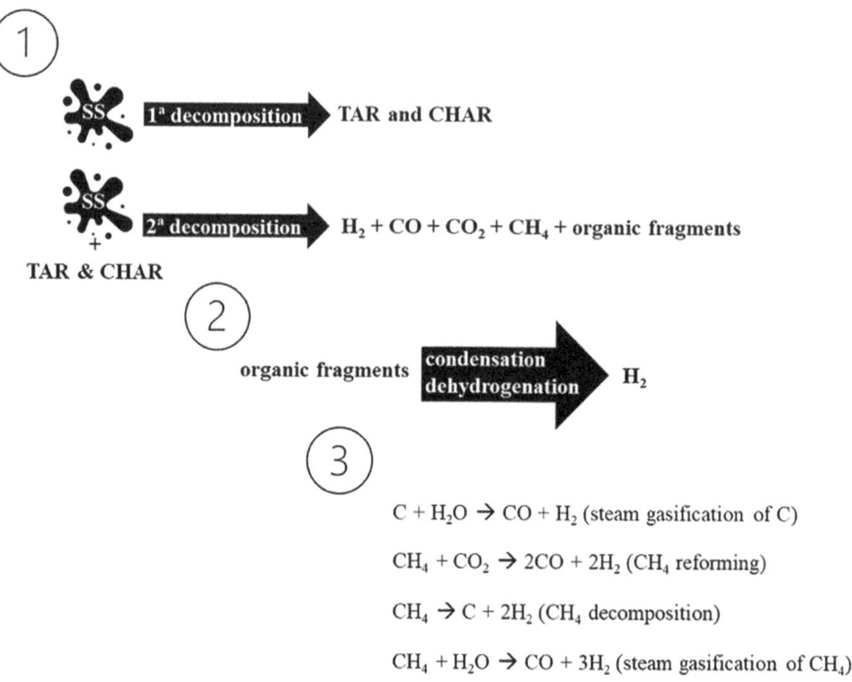

FIGURE 4.9 Main stages depend on the temperature concerning sewage sludge (SS) pyrolysis and possibilities for hydrogen production.

produce hydrogen, but also the other by-products obtained during this process can contribute to additional hydrogen production through different chemical routes. This way, many compounds obtained during pyrolysis could react in different ways to produce hydrogen, such as methane steam reforming or decomposition.

In any case, regarding the effect of temperature on hydrogen production through pyrolysis, some studies found a greater production of hydrogen when going from 600°C to 850°C at a specific humidity (Luo et al., 2017). In another research work, a low flow of carrier gas was used, so that the steam, organic vapors, and gases released during the process reacted at elevated temperatures before leaving the reactor and the authors found that the presence of water increased the production of gases at elevated temperatures and high heating rates (Domínguez et al., 2006). The steam generated during the treatment reacted with both the vapors (steam reforming) and the solid residue produced (steam gasification), resulting in increased hydrogen production. Another influential factor could be the flocculant that is added to the sewage sludge and could accelerate pyrolysis when working at high heating rates as reported elsewhere (Chu et al., 2001). Regarding catalysis addition, some authors (Yang et al., 2018), studied the catalytic effects of three main components of SiO_2, Al_2O_3, and Fe_2O_3 on the pyrolysis of sewage sludge and improvement in hydrogen production was attributed to the catalytic effects of Fe_2O_3 and Al_2O_3.

Syngas is a high-quality product, and its production requires long reaction times, high reaction temperatures, and rates, requiring effective pyrolysis technology for a quick thermal decomposition. That is the reason why syngas obtained through traditional pyrolysis present low-quality syngas generation. As an alternative, microwave pyrolysis could contribute to a faster heating rate compared to conventional pyrolysis, with better heat transfer (temperature increases from the inside of sewage sludge), being suitable for fast and flash pyrolysis. This way, different studies have covered the possible use of microwave pyrolysis of sewage sludge, optimizing the conditions to increase H_2 and CO fraction and proving this technique as an alternative to increasing syngas yield in sewage sludge pyrolysis (Oh et al., 2023).

4.3.2 SEWAGE SLUDGE PYROLYSIS FOR BIO-OIL PRODUCTION

Sewage sludge pyrolysis for liquid production is equally an interesting way to manage this waste (Fonts et al., 2012). Apart from syngas (including hydrogen), other products can be obtained during sewage sludge pyrolysis, which can make this process more efficient. Some researchers have studied bio-oil production, although these processes present high costs and obtain low product quality (Oh et al., 2023). Alternatively, the use of bio-oils or other liquid fractions could be suitable as fertilizers due to their nutrient content.

4.3.3 SEWAGE SLUDGE PYROLYSIS FOR BIOCHAR PRODUCTION: CONTRIBUTION FOR HYDROGEN PRODUCTION

As previously explained, one of the most interesting phases obtained during sewage sludge pyrolysis is the solid one, that is, biochar. In that sense and depending on many factors like heating rate or the use of further heating treatments like activation through carbon dioxide, steam or, oxygen, activated carbons with interesting characteristics (such as high porosity and surface area) can be obtained for different purposes related to sewage sludge management. Mainly, there are three ways where biochar can be directly or indirectly used as a value-added product, enhancing the performance of sewage sludge plants acting as an active carbon for different purposes: improvement of soils in agriculture, microorganism hosting for a better performance of biogas production, H_2S removal from fuel gas mixtures like biogas or syngas, and its use as catalyst support.

- **Improvement of Soils:** Depending on pyrolysis production (slow and fast pyrolysis, among others), the characteristics of biochar can be suitable for sustainable purposes related to soil and agricultural practices. Thus, with slow pyrolysis (with heating rates of 5–10°C/min), products with high surface area can be used to adsorb pollutants in air, water, and soil. In this case, some functional groups can interact with heavy metals through H-bonding and complexation, and due to their microporosity, biochar from sewage sludge can adsorb organic pollutants in soil water. Also, biochar could take part in soil remediation, reducing greenhouse gas emissions through solid carbon sequestration in soils, improving agricultural characteristics like fertility, pH and soil microbial activity, and retention of nutrients, with the subsequent improvement in agricultural productivity (Hu et al., 2022).

- **Microorganism Hosting:** The conversion of sludge (and its possible combination with food waste) into biogas through anaerobic digestion has gained attention in recent years, being a resourceful stage during wastewater treatment. As a consequence, interesting fuel gas mixtures (like biogas) are obtained, which will imply a better energy yield if the amount of gas and its quality (with higher CH_4 ratios) are good enough. Anaerobic digestion implies stages like hydrolysis, acidogenesis. or methanogenesis, obtaining methane and other compounds like H_2S (which will be covered in the following section). The possible use of biochar obtained from sewage sludge pyrolysis to improve anaerobic digestion might be interesting, as this porous material could be used as a host for microorganisms. Subsequently, sewage sludge could be doubly important in this process, acting as raw material and an additive to improve anaerobic digestion performance. Concerning biochar obtained through pyrolysis, its addition during anaerobic digestion could enhance the removal of chemical oxygen demand, reducing the lag phase of methanogenesis and therefore increasing CH_4 production, improving the elemental composition of solid digestate, which could be used as fertilizer. Moreover, some contaminants or by-products that could inhibit biodigestion can be adsorbed by surface functional groups in biochar. On the other hand, biochar addition can increase alkalinity in an anaerobic digestion medium, reducing ammonia inhibition and acid stress to microorganisms implied in the process, and the porous nature of biochar promotes microorganism colonization (such as bacteria and archaea). Finally, functional groups in biochar can enhance methane yield through direct or indirect electron transfer among anaerobic microbes (Kumar et al., 2021).
- **H_2S Removal:** It is well known that H_2S generation during industrial processes such as biogas production can be a matter of concern, as it is a poisonous gas that can be deadly at around 1,000 ppm. Additionally, in the presence of water, it generates acidic solutions which can provoke corrosion in industrial facilities. Therefore, its removal once generated is an important task, with different techniques like the use of biofilters, absorption through alkanolamines, ionic liquids, or deep eutectic solvents (among others), and adsorption being interesting (Pudi et al, 2022). Regarding the latter, adsorption is a surface-based process where a molecule from a liquid or gas is transferred to a solid surface (such as activated carbon). There are different interactions that can contribute to the adsorption of molecules, mainly weak interactions like Van der Waals forces or strong interactions, implying chemisorptions. Thus, and apart from the kind of interaction, other factors can be vital to consider a specific porous material for this purpose, such as porosity (pore size and volume, pore size distribution, etc.), capacity and selectivity toward target molecules, chemical and thermal stability and reusability or service life, among others. Specifically, regarding the subject of this book chapter, the possible use of activated carbons for H_2S removal could be an interesting point, especially considering the possibility of sewage sludge reintegration in water treatment plants. This way, the use of biochar from sewage sludge pyrolysis could be an interesting starting point for

activated carbon generation through further thermal treatment with carbon dioxide, steam, or oxygen in order to increase its specific surface area or pore volume. Also, and taking into account the combined use of activated carbons obtained from this waste with other agricultural or food wastes like barks, husk, or spent coffee grains, this kind of adsorbent is quite cheap as the raw material from which it is obtained is easily available carbon sources. Additionally, the combined use of activated carbons with other materials like zeolites, metal oxides (which can be obtained from alternative treatment of sewage sludge like calcinations), or porous organic polymers could be an interesting way to obtain synergistic effects on pollutant adsorption, as they can provide additional active sites for H_2S capture. On the other hand, the use of active carbons as hosts of some microorganisms could be suitable for hydrogen sulfide conversion to S (Pavicic et al., 2022).

- **Use as Catalyst Support:** Due to some of its main properties, like high surface area and adjustable pore size and volume (through pyrolysis or additional activation), biochar can be useful as catalyst support in heterogeneous catalysis of a wide range of chemical reactions (from biodiesel production through transesterification of vegetable oils to tar steam reforming), due to its low-cost production and environmental friendliness. Consequently, catalyst design from sewage sludge biochar could be a subject with endless possibilities, and future research could be focused on its re-use in sewage sludge energy conversion (especially during pyrolysis or biogas steam reforming).

Having said that, it is clear that biogas purification and upgrading are important in steam reforming (see Figure 4.10), which can be carried out through different means (Pavicic et al., 2022):

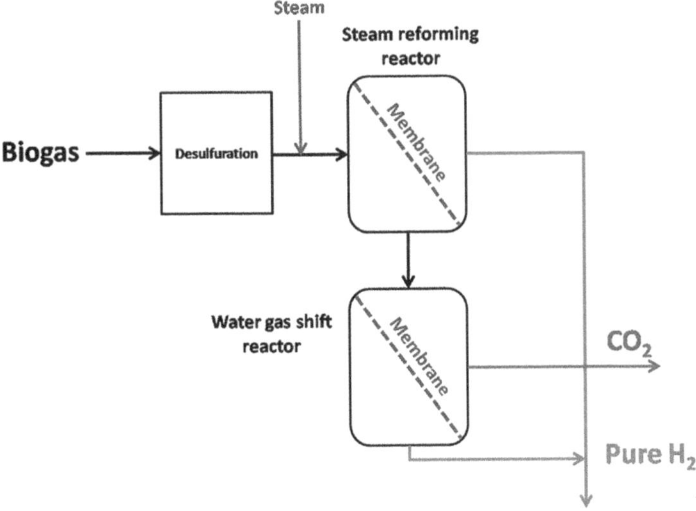

FIGURE 4.10 Hydrogen production from biogas through steam reforming.

- H_2S removal, where the use of adsorbents can be useful, like char produced during pyrolysis, to avoid corrosion in facilities.
- CH_4 yield improvement, which is essential to increase the efficiency of steam reforming reaction, reducing other inconvenient compounds like H_2S that usually promote coke deposition on catalysts or Pd/Au membranes, worsening the reusability or self-life of these components.

4.4 CHALLENGES RELATED TO SEWAGE SLUDGE PYROLYSIS

Although sewage sludge pyrolysis presents many advantages and seems to be a solid alternative to traditional sewage sludge management, there are some disadvantages that should be addressed in order to make this process more efficient and competitive, like the correct implementation of these techniques (including upgrading steps, for instance) in industry to obtain commercial products.

In the case of pyrolysis, drying processes in sewage sludge is an important requirement to improve the performance of this waste during this thermal treatment, which implies a lower efficiency of the whole process. This is especially worrying when secondary sewage sludge after centrifugation usually presents moisture values around 70%–80%, which can be reduced to approximately 30% by additional sustainable drying processes such as solar drying systems. An interesting alternative for wet and sludgy biogenic residues is hydrothermal carbonization, which presents a high energy efficiency (Reißmann et al., 2021).

Thus, one of the main challenges related to the energy use of sewage sludge through pyrolysis is the requirement of dry products to carry out this thermal treatment. Due to the high moisture content in sewage sludge (around 80% in sewage sludge cakes after mechanical dewatering), drying methods are needed to make this waste suitable for this process, which usually implies a lot of energy and subsequent high economic cost (Hu et al., 2022). In order to solve this problem, some alternatives have been considered in the literature, like the use of solar drying systems to cheapen the drying process, co-pyrolysis with other interesting products, or the use of hydrothermal carbonization, where the use of wet wastes is suitable, as it is a chemical reaction where water at high pressure and low temperature (around 180°C–260°C) is used.

Apart from that, there are environmental concerns, such as how to manage evolved pollutants like NH_3, SO_2, or HCl, among others, to obtain cleaner syngas. Also, the catalysts used in sewage sludge pyrolysis (or other abovementioned processes such as steam reforming) usually undergo deactivation due to several factors (poisoning, coke deposition, sintering, etc.). Therefore, lengthening the activity of these catalysts, along with catalyst regeneration, is another interesting research line.

4.5 CONCLUSION

In this chapter, a thorough review of pyrolysis applied to sewage hydrogen production has been carried out. Several conclusions can be reached, like the following: Sewage sludge production is a challenging and increasing waste, which is a continuous concern for urban areas that should be correctly managed from efficiency and environmental points of view. According to sewage sludge production and management, pyrolysis can play an important role as a direct and efficient way to manage this waste to obtain syngas. In that sense, the use of microwave pyrolysis seems to be a promising way to improve syngas yield, as the heating rate is considerably improved. Additionally, pyrolysis can present indirect advantages that can improve the efficiency of many aspects covered in a sewage sludge plant, focused on indirect hydrogen production. Thus, the different phases obtained during sewage sludge pyrolysis (especially biochar) can contribute to a better performance of water treatment, both in biogas production (improving the performance of microorganisms) or purification (acting as active carbon to remove H_2S, avoiding corrosion in facilities). However, there are some challenges like the high moisture in sewage sludge, which is a disadvantage for pyrolysis efficiency. In that sense, there are alternatives like the use of solar dryers or the use of hydrothermal carbonization to avoid high energy costs. Additionally, equivalent processes like hydrothermal carbonization could complement traditional pyrolysis, depending on the kind of product to be produced. In general, the techniques explained in this book chapter can be complementary, easily adapted to the specific circumstances of sewage treatment plants, and contribute to a circular economy, as most by-products from wastewater can be reused during its treatment in wastewater plants in order to improve efficiency and sustainability in such a necessary facility in urban areas.

REFERENCES

Beik, F, Williams, L, Brown, T, and Wagland, ST. 2023. "Development and prototype testing of a novel small-scale pyrolysis system for the treatment of sanitary sludge." *Energy Conversion and Management*, 277, 116627. https://doi.org/10.1016/j.enconman.2022.116627

Chu, CP, Lee, DJ, and Chang, CY. 2001. "Thermal pyrolysis characteristics of polymer flocculated waste activated sludge." *Water Research*, 35, 49e56. https://doi.org/10.1016/S0043-1354(00)00235-9

Di Giacomo, G., and Romano, P. 2022. Evolution and Prospects in Managing Sewage Sludge Resulting from Municipal Wastewater Purification. Energies, 15(15). https://doi.org/10.3390/en15155633

Domínguez, A, Menéndez, JA, Inguanzo, M, and Pís, JJ. 2006. "Production of bio-fuels by high temperature pyrolysis of sewage sludge using conventional and microwave heating." *Bioresource Technology*, 97, 1185–1193.

Fonts, I, Gea, G, Azuara, M, Ábrego, J, and Arauzo, J. 2012. "Sewage sludge pyrolysis for liquid production: A review." *Renewable and Sustainable Energy Reviews*, 16(5), 2781–2805. https://doi.org/10.1016/j.rser.2012.02.070

Fremaux, S, Beheshti, SM, Ghassemi, H, and Shahsavan-Markadeh, R. 2015. "An experimental study on hydrogen-rich gas production via steam gasification of biomass in a research-scale fluidized bed." *Energy Conversion Management*, 91, 427–432.

Ghodke, PK, Sharma, AK, Pandey, JK, Chen, W-H, Patel, A, and Ashokkumar, V. 2021. "Pyrolysis of sewage sludge for sustainable biofuels and value-added biochar production." *Journal of Environmental Management*, 298, 113450. https://doi.org/10.1016/j.jenvman.2021.113450

Ghorbani, M, Konvalina, P, Walkiewicz, A, Neugschwandtner, RW, Kopecký, M, Zamanian, K, Chen, WH, and Bucur, D. 2022. "Feasibility of biochar derived from sewage sludge to promote sustainable agriculture and mitigate GHG Emissions—A review." *International Journal of Environmental Research and Public Health*, 19(19). https://doi.org/10.3390/ijerph191912983

Hu, M, Hu, H, Ye, Z, Tan, S, Yin, K, Chen, Z, Guo, D, et al. 2022. "A review on turning sewage sludge to value-added energy and materials via thermochemical conversion towards carbon neutrality." *Journal of Cleaner Production*, 379(P1), 134657. https://doi.org/10.1016/j.jclepro.2022.134657

Karaca, C, Sözen, S, Orhon, D, and Okutan, H. 2018. "High temperature pyrolysis of sewage sludge as a sustainable process for energy recovery." *Waste Management*, 78, 217–226. https://doi.org/10.1016/j.wasman.2018.05.034

Kumar, M., Dutta, S., You, S., Luo, G., Zhang, S., Show, P. L., Sawarkar, A. D., Singh, L., and Tsang, D. C. W. 2021. A critical review on biochar for enhancing biogas production from anaerobic digestion of food waste and sludge. Journal of Cleaner Production, 305. https://doi.org/10.1016/j.jclepro.2021.127143

Liu, Y, Chen, T, Gao, B, Meng, R, Zhou, P, Chen, G, Zhan, Y, Lu, W, and Wang, H. 2021. "Comparison between hydrogen-rich biogas production from conventional pyrolysis and microwave pyrolysis of sewage sludge: Is microwave pyrolysis always better in the whole temperature range?" *International Journal of Hydrogen Energy*, 46(45), 23322–23333. https://doi.org/10.1016/j.ijhydene.2020.05.165

Luo, S, Guo, J, and Feng, Y. 2017. "Hydrogen-rich gas production from pyrolysis of wet sludge in situ steam agent." *International Journal of Hydrogen Energy*, 42, 18309–18314.

Manara, P, and Zabaniotou, A. 2012. "Towards sewage sludge based biofuels via thermochemical conversion—A review." *Renewable Sustainable Energy Review*, 16, 2566e82. https://doi.org/10.1016/j.rser.2012.01.074

Naqvi, SR, Tariq, R, Shahbaz, M, Naqvi, M, Aslam, M, Khan, A, Mackey, H, Mckay, G, and Al-Ansari, T. 2021. "Recent developments on sewage sludge pyrolysis and its kinetics: Resources recovery, thermogravimetric platforms, and innovative prospects." *Computers and Chemical Engineering*, 150, 107325. https://doi.org/10.1016/j.compchemeng.2021.107325

Nikolaidis, P, and Poullikkas, A. 2017. "A comparative overview of hydrogen production processes." *Renewewable Sustainable Energy Review*, 67, 597e611. https://doi.org/10.1016/j.rser.2016.09.044

Nipattummakul, N, Ahmed II, Kerdsuwan, S, and Gupta, AK. 2010. "Hydrogen and syngas production from sewage sludge via steam gasification." *International Journal of Hydrogen Energy*, 35, 11738e45. https://doi.org/10.1016/j.ijhydene.2010.08.032

Nunes, N, Ragonezi, C, Gouveia, C, and Pinheiro de Carvalho, M. 2021. "Review of sewage sludge as a soil amendment in relation to current international guidelines: A heavy metal perspective." *Sustainability*, 13, 2317.

Oh, DY, Kim, D, Choi, H, and Park, KY. 2023. "Syngas generation from different types of sewage sludge using microwave-assisted pyrolysis with silicon carbide as the absorbent." *Heliyon*, 9, e14165.

Pavicic, J, Mavar, KN, Brkic, V, and Simon, K. 2022. "Biogas and biomethane production and usage: Technology development, advantages and challenges in Europe." *Energies,* 15, 2940.

Pudi, A., Rezaei, M., Signorini, V., Andersson, M. P., Baschetti, M. G., and Mansouri, S. S. 2022. Hydrogen sulfide capture and removal technologies: A comprehensive review of recent developments and emerging trends. Separation and Purification Technology, 298, 121448. https://doi.org/10.1016/j.seppur.2022.121448

Reißmann, D, Thrän, D, Blöhse, D, and Bezama, A. 2021. "Hydrothermal carbonization for sludge disposal in Germany: A comparative assessment for industrial-scale scenarios in 2030." *Journal of Industrial Ecology*, 25(3), 720–734. https://doi.org/10.1111/jiec.13073.

Scopus. "Scopus." https://www.scopus.com/home.uri

Siddiqui, M. I., Rameez, H., Farooqi, I. H., and Basheer, F. 2023. Recent Advancement in Commercial and Other Sustainable Techniques for Energy and Material Recovery from Sewage Sludge. Water (Switzerland), 15, (5). https://doi.org/10.3390/w15050948

Statista. 2023. https://es.statista.com/

Wang, B, Yinhe, L, Guan, Y, and Feng, Y. 2022. "Characteristic of the production of hydrogen-rich combustible gas by pyrolysis of high-ash sewage sludge." *Journal of Cleaner Production*, 334, 130224. https://doi.org/10.1016/j.jclepro.2021.130224

Xin, S, Guo, L, Lifang, L, Wei, T, and Xu, Q. 2019. "Hydrogen-rich gas production from soybean straw via microwave pyrolysis under CO2 atmosphere." *Energy Sources, Part A: Recovery, Utilization and Environmental Effects.* https://doi.org/10.1080/15567036.2019.1676330

Xu, C, Chen, W, and Hong, J. 2014. "Life-cycle environmental and economic assessment of sewage sludge reatment in China." *Journal of Clean Production*, 67, 79–87.

Yang, J, Xu, X, Liang, S, Guan, R, Li, H, Chen, Y, Liu, B, Song, J, Yu, W, Xiao, K, Hou, H, Hu, J, Yao, H, and Xiao, B. 2018. "Enhanced hydrogen production in catalytic pyrolysis of sewage sludge by red mud: Thermogravimetric kinetic analysis and pyrolysis characteristics." *International Journal of Hydrogen Energy*, 43(16), 7795–7807. https://doi.org/10.1016/j.ijhydene.2018.03.018

5 Pyrolysis of Municipal Biomass Wastes for Hydrogen Production

Arash Sadeghi and Mohammad Reza Rahimpour

5.1 INTRODUCTION

The increasing global demand for energy and the need to reduce greenhouse gas emissions have increased interest in renewable and sustainable energy sources (Sadeghi et al., 2023). Hydrogen, a clean and versatile energy carrier, has gained significant attention recently (Zainal et al., 2024). It could replace fossil fuels in various sectors, including transportation, power generation, and industrial processes (Hassan et al., 2023c). However, hydrogen production from conventional sources such as natural gas and coal is associated with significant carbon emissions (Amin et al., 2022, Ishaq et al., 2022). To address this challenge, researchers and scientists have been exploring alternative methods for hydrogen production that are environmentally friendly and sustainable (Hassan et al., 2023a). One such method is the pyrolysis of municipal biomass wastes. Municipal biomass wastes, including agricultural residues, food waste, and sewage sludge, are abundant and readily available biomass sources. By converting these wastes into hydrogen through pyrolysis, we could simultaneously address the issues of waste management and hydrogen production (Matamba et al., 2023, Wang and Tester, 2023). According to the Organization for Economic Co-operation and Development (OECD) report, the world's top 10 largest municipal waste generation from 2018 to 2021 is shown in Figure 5.1.

Pyrolysis is a thermochemical process involving decomposing organic materials at high temperatures without oxygen (Rogdakis and Bitsikas, 2018). During pyrolysis, the biomass wastes are heated to temperatures ranging from 300°C to 800°C, causing them to undergo thermal decomposition. This process produces a mixture of gases, liquids, and solids, with hydrogen being one of the leading gases (Czajczyńska et al., 2017, Xu et al., 2021). Using municipal biomass wastes for hydrogen production through pyrolysis offers several advantages. Firstly, it provides a sustainable solution for waste management, as it allows for the utilization of biomass wastes that would otherwise be disposed of in landfills or incinerated (Wang and Tester, 2023, Nguyen-Thi et al., 2023). By converting these wastes into hydrogen, we could reduce the environmental impact associated with their disposal (Nandhini et al., 2022). Secondly, pyrolysis offers a carbon-neutral or even carbon-negative approach to hydrogen production. Unlike conventional methods that rely on fossil fuels, pyrolysis of biomass wastes does not release additional carbon dioxide into the atmosphere.

DOI: 10.1201/9781003382270-7

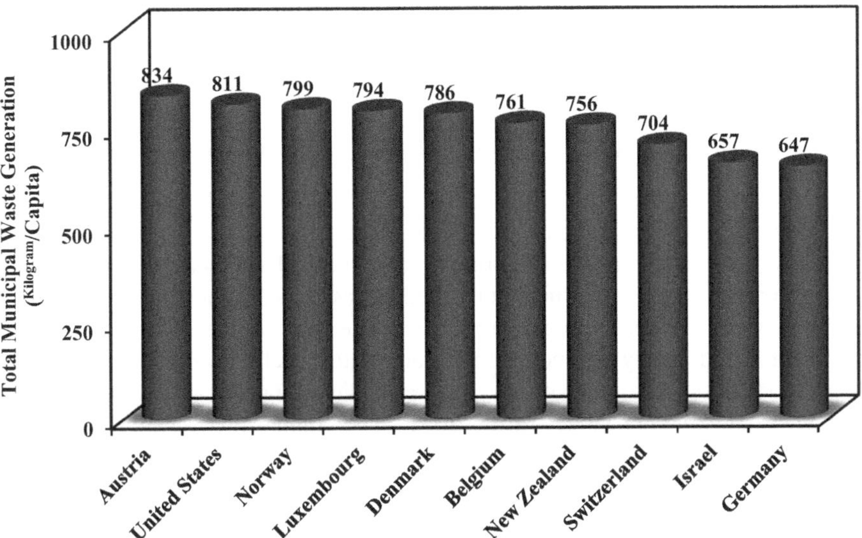

FIGURE 5.1 World's Top 10 largest municipal waste generation from 2018 to 2021 (OECD, 2014).

The carbon content in the biomass is converted into biochar, a solid residue that could be used as a soil amendment to improve soil fertility and carbon sequestration (Al-Rumaihi et al., 2022).

Furthermore, hydrogen production through pyrolysis could be integrated with existing energy systems, such as combined heat and power (CHP) plants (Yu et al., 2023, Pham, 2005). The heat generated during the pyrolysis process could be used for electricity generation or heating, increasing the system's overall energy efficiency (Li et al., 2023b). Despite its potential, there are challenges associated with utilizing municipal biomass wastes for hydrogen production through pyrolysis. The composition of biomass wastes could vary significantly, affecting the yield and quality of the hydrogen produced (Aziz et al., 2021). The presence of impurities, such as sulfur and nitrogen compounds, could also impact the pyrolysis process's efficiency and the hydrogen gas quality (Bakhtyari et al., 2018). In addition, the design and operation of pyrolysis reactors play a crucial role in determining the performance and scalability of the process. The selection of an appropriate reactor type, such as a fixed bed, fluidized bed, rotary kiln, or microwave pyrolysis system, depends on feedstock characteristics, desired product yields, and process conditions (Al-Rumaihi et al., 2022, Potnuri et al., 2022, Jouhara et al., 2018). Research efforts have focused on catalyst development, process optimization, and control strategies to overcome these challenges and optimize the pyrolysis process for hydrogen production. Catalysts could enhance the pyrolysis reaction kinetics and improve the selectivity toward hydrogen gas (Setiabudi et al., 2020, Zamri et al., 2023, Osman et al., 2023). Process optimization techniques, such as experimental design and statistical analysis, could help identify the optimal operating conditions for maximizing hydrogen yield and quality (Taylor et al.,

2023). Process control strategies, including temperature and gas flow regulation, are essential for maintaining stable and efficient pyrolysis operations (Hussain et al., 2023).

5.2 MUNICIPAL WASTES AND THEIR COMPOSITION

Municipal wastes, solid waste, or garbage, are generated from households, commercial establishments, institutions, and public places. These wastes comprise various materials, including organic and inorganic components. Understanding the composition of municipal wastes is crucial for effective waste management and utilization in pyrolysis processes for hydrogen production. The composition of municipal waste could vary significantly depending on various factors such as geographical location, population density, lifestyle patterns, and waste management practices (Abdel-Shafy and Mansour, 2018, Pheakdey et al., 2022). The composition of municipal waste generated in the United States in 2013 is schemed in Figure 5.2.

However, specific standard components are typically found in municipal wastes. Organic waste is a significant component of municipal waste and consists of biodegradable materials derived from plants and animals. This includes food waste, yard waste, paper products, and other organic materials. Food waste, such as leftover food, fruit and vegetable peels, and spoiled food, is a significant contributor to organic waste in municipal solid waste. Yard waste includes grass clippings, leaves, branches, and other garden waste. Paper products, such as newspapers, cardboard, and packaging materials, contribute to the organic fraction of municipal waste (Ayilara et al., 2020, Bandini et al., 2022). Also, organic waste is an excellent feedstock for pyrolysis processes as it contains high levels of carbon and hydrogen, which could be

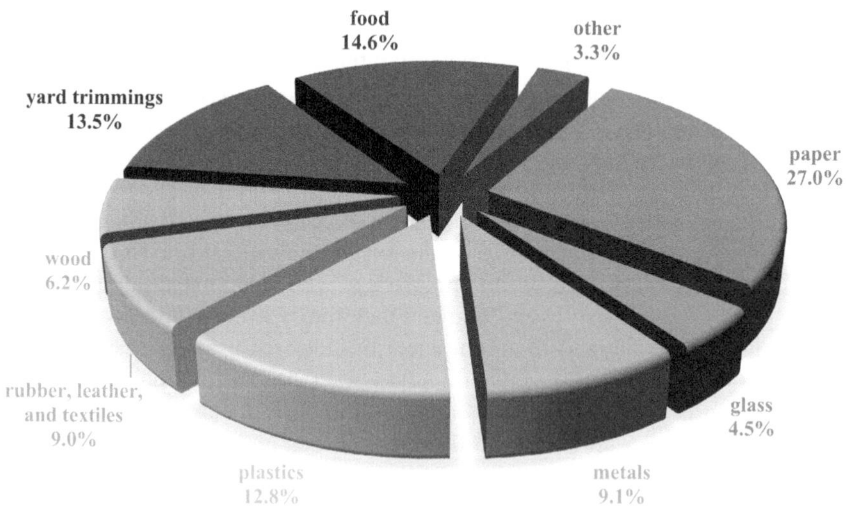

FIGURE 5.2 Municipal waste composition generated in the United States in 2013 (Environmental Protection Agency U.S., 2011).

converted into hydrogen-rich gases through thermal decomposition. The organic fraction of municipal wastes could potentially be a valuable source of hydrogen production through pyrolysis (Al-Rumaihi et al., 2022). Besides, inorganic waste refers to nonbiodegradable materials derived from minerals and synthetic sources. This includes plastics, glass, metals, ceramics, and construction debris. Plastics, including packaging materials, bottles, and bags, are a significant component of inorganic waste in municipal solid waste (Kibria et al., 2023). Glass waste consists of bottles, jars, and other glass products (Silva et al., 2017). Metals, such as aluminum and steel cans, contribute to the inorganic fraction of municipal waste. Ceramics, including broken dishes and tiles, are also part of the inorganic waste stream (Hossain and Roy, 2020). Construction debris, such as concrete, bricks, and wood waste, is another component of inorganic waste. Inorganic waste poses challenges in pyrolysis processes as it does not undergo thermal decomposition like organic waste (Aprilia et al., 2013). However, pyrolysis could convert certain plastics into valuable products like syngas or hydrogen (Shah et al., 2023). The presence of inorganic waste in municipal solid waste requires proper sorting and separation to ensure the quality and efficiency of the pyrolysis process (Zhang et al., 2022). In addition, hazardous waste refers to materials that threaten human health and the environment due to their toxic, flammable, corrosive, or reactive nature (Inglezakis and Moustakas, 2015). This includes substances such as batteries, electronic waste, chemicals, pharmaceuticals, and certain types of household cleaners. Hazardous waste requires special handling and disposal methods to prevent contamination and environmental adverse effects. Hazardous waste should be strictly avoided in pyrolysis for hydrogen production as it could release harmful emissions and by-products (Bhatt et al., 2022). Proper waste management practices should be implemented to separate and treat hazardous waste separately from municipal solid waste (Pheakdey et al., 2022). Apart from the principal components mentioned above, municipal wastes may contain other miscellaneous materials. This includes textiles, rubber, leather, ashes, and household items. Textiles, such as clothing and fabrics, contribute to the waste stream. Rubber waste includes tires and other rubber products. The leather waste consists of discarded leather products (Czajczyńska et al., 2017). Ashes from incineration processes and miscellaneous household items, such as toys and small appliances, also contribute to the composition of municipal wastes. These miscellaneous components may have varying levels of recyclability and suitability for pyrolysis processes (Lam et al., 2010). Proper waste management practices, including sorting and recycling, could help maximize the utilization of these materials and minimize their impact on the environment (Abdel-Shafy and Mansour, 2018).

5.3 SUSTAINABLE MANAGEMENT OF MUNICIPAL BIOMASS WASTES

Sustainable management of municipal biomass wastes is crucial for successfully implementing pyrolysis technology for hydrogen production. In this regard, efficient waste collection and segregation systems are essential for the sustainable management of municipal biomass wastes. Proper waste collection ensures that biomass wastes are collected from households, commercial establishments, and other sources in a timely and organized manner. This helps prevent the accumulation of waste in public spaces and reduces the risk of environmental pollution. Segregation of municipal biomass wastes is another crucial aspect of sustainable waste management. It involves separating different types of biomass wastes, such as food waste, yard waste, and sewage sludge, at the source (Wang and Tester, 2023). This allows for effectively utilizing specific biomass waste streams for different purposes, including pyrolysis for hydrogen production. Segregation also helps properly dispose of non-biomass waste components, such as plastics and metals, which could be recycled or treated separately. Various waste treatment technologies could be employed to ensure the sustainable management of municipal biomass wastes. These technologies aim to convert biomass wastes into valuable products, such as biofuels, biogas, and biochar, while minimizing their environmental impact. Common waste treatment technologies include composting, anaerobic digestion, and pyrolysis (Abdel-Shafy and Mansour, 2018). Composting is a natural process involving decomposing organic materials, such as food and yard waste, under controlled conditions. It produces compost, a nutrient-rich soil amendment that could be used in agriculture and landscaping. Composting not only reduces the volume of biomass wastes but also helps recycle nutrients and improve soil fertility (Singh et al., 2022b). Anaerobic digestion is another waste treatment technology that involves the decomposition of organic materials in the absence of oxygen. It produces biogas, a mixture of methane and carbon dioxide, which could be used as a renewable energy source. Anaerobic digestion reduces the environmental impact of biomass wastes and provides a sustainable energy source (Nhubu et al., 2022, Sarker et al., 2019). Pyrolysis is a thermochemical process that could convert biomass wastes into bio-oil, syngas, and biochar. It involves heating the biomass wastes without oxygen, which leads to the decomposition of organic compounds and the production of valuable products. Pyrolysis offers several advantages, including high energy efficiency, versatility in feedstock utilization, and the potential for hydrogen production. It could be integrated with other waste treatment technologies to maximize the utilization of biomass wastes and minimize their environmental impact. Proper waste disposal is an integral part of sustainable waste management. After the treatment process, the residual waste or by-products that cannot be further utilized or recycled need to be disposed of in an environmentally responsible manner (Durak, 2023, Ethaib et al., 2020). This may involve landfilling, incineration, or other waste disposal methods that comply with local regulations and environmental standards. It is crucial to consider the environmental impact of waste disposal methods to ensure the overall sustainability of the waste management process. Landfilling, for example, could lead to the generation of greenhouse gases, leachate contamination, and the depletion of valuable land resources (Abubakar et al., 2022, Siddiqua et al., 2022). Incineration, on the other hand, could result in air pollution and the release of harmful emissions (Domingo et al., 2020). To mitigate the environmental impact of waste disposal, it is essential to implement proper waste treatment technologies, such as pyrolysis, to minimize waste generation and maximize resource recovery. Pyrolysis offers the potential to convert biomass wastes into valuable products, such as hydrogen while reducing the environmental footprint associated with waste disposal. The sustainable management of municipal biomass wastes requires a supportive policy and regulatory framework (Kumar et al., 2023). Governments and regulatory bodies are crucial in promoting waste management practices that prioritize sustainability and resource recovery. This includes the development of waste management policies, regulations, and incentives that encourage the adoption of technologies like pyrolysis for hydrogen production. Policy measures could include the establishment of waste management targets, the implementation of waste segregation and recycling programs, and the promotion of renewable energy sources. Regulatory frameworks could ensure compliance with environmental standards, waste disposal regulations, and emissions control measures. Incentives, such as tax credits or subsidies, could encourage adopting sustainable waste management practices and investing in pyrolysis technology. A comprehensive policy and regulatory framework could provide the necessary support and guidance for the sustainable management of municipal biomass wastes. It could create a favorable environment for implementing pyrolysis technology for hydrogen production, enabling the transition toward a more sustainable and resource-efficient waste management system (Möslinger et al., 2023, Purnell et al., 2019, Masud et al., 2023).

5.4 PYROLYSIS AS A HYDROGEN PRODUCTION TECHNOLOGY

Pyrolysis is a thermochemical process that could convert municipal biomass wastes into hydrogen gas. It involves the decomposition of organic materials at high temperatures without oxygen. This process produces a mixture of gases, liquids, and solids, with hydrogen gas being one of the leading products (Sharma et al., 2015, Czajczyńska et al., 2017). The principle behind pyrolysis is the thermal decomposition of biomass wastes, such as agricultural residues, wood chips, and municipal solid waste, into smaller molecules without oxygen. The process occurs in a reactor, where the biomass is heated to temperatures typically 400°C–800°C. At these high temperatures, the complex organic compounds present in the biomass break down into simpler molecules, including hydrogen gas (Tomczyk et al., 2020, Amalina et al., 2022). The pyrolysis process can be divided into three main stages: drying, pyrolysis, and char formation. During the drying stage, the moisture content of the biomass is evaporated, preparing it for the subsequent pyrolysis reactions. In the pyrolysis stage, the biomass undergoes thermal decomposition, releasing volatile gases, such as hydrogen, carbon monoxide, methane, and other hydrocarbons (Tihay and Gillard, 2010). The remaining solid residue, char, is rich in carbon and could be further processed for various applications, including energy generation. Several pyrolysis processes could be employed for hydrogen production from municipal biomass wastes. These include fast pyrolysis, slow pyrolysis, and intermediate pyrolysis. Fast pyrolysis is a rapid heating process that operates at high temperatures and short residence times. It produces a high yield of liquid bio-oil and a smaller amount of gases and char (Hornung, 2012).

The liquid bio-oil could be further processed to obtain hydrogen gas through various upgrading techniques. Slow pyrolysis, on the other hand, involves a slower heating rate and longer residence times. This process produces a higher yield of solid char and less liquid bio-oil and gases (Lachos-Perez et al., 2023). The solid char could be used as an energy source or precursor for producing activated carbon. Intermediate pyrolysis combines fast and slow pyrolysis, operating at moderate temperatures and residence times. It produces a balanced yield of liquid bio-oil, gases, and char. This process offers flexibility in terms of product distribution and could be optimized for hydrogen production (Al-Rumaihi et al., 2022).

5.5 ADVANTAGES AND LIMITATIONS OF PYROLYSIS FOR HYDROGEN PRODUCTION

Pyrolysis is a promising technology for hydrogen production from municipal biomass wastes. It offers several advantages over other hydrogen production methods but has some limitations that must be considered (Bakhtyari et al., 2018).

5.5.1 ADVANTAGES OF PYROLYSIS FOR HYDROGEN PRODUCTION

One of the significant advantages of pyrolysis for hydrogen production is using municipal biomass wastes as feedstock. Municipal biomass wastes, such as agricultural residues, food waste, and sewage sludge, are abundant and readily available. By utilizing these wastes, pyrolysis offers a sustainable and renewable source of feedstock for hydrogen production. Also, pyrolysis has the potential to achieve a high hydrogen yield from municipal biomass wastes. During pyrolysis, the biomass is heated without oxygen, producing a mixture of gases, including hydrogen. The hydrogen yield could be further enhanced by optimizing the process parameters, such as temperature, heating rate, and residence time. In addition, pyrolysis is an energy-efficient process allowing energy recovery from municipal biomass wastes. The heat generated during the pyrolysis process could be utilized for various purposes, such as steam generation, electricity production, or heating. This energy recovery reduces the overall energy consumption and increases the economic viability of the hydrogen production process. Another advantage of pyrolysis for hydrogen production is its potential to be carbon-neutral or even carbon-negative. Municipal biomass wastes are considered carbon-neutral because the carbon dioxide released during biomass combustion is offset by the carbon dioxide absorbed during biomass growth. By converting biomass into hydrogen through pyrolysis, the carbon emissions could be significantly reduced, contributing to mitigating greenhouse gas emissions. Besides, pyrolysis provides a sustainable solution for the management of municipal biomass wastes. Instead of disposing of these wastes in landfills or incinerating them, pyrolysis offers a way to convert them into valuable products, such as hydrogen. This reduces the environmental impact of waste disposal and creates a potential revenue stream from the sale of hydrogen (Aziz et al., 2021, Igliński et al., 2023, Williams, 2021, Hosseini et al., 2015).

5.5.2 LIMITATIONS OF PYROLYSIS FOR HYDROGEN PRODUCTION

The composition of municipal biomass wastes could vary significantly, depending on the source and type of waste. This variability could affect the pyrolysis process and the hydrogen quality produced. Different feedstock compositions may require adjustments in process parameters and catalysts, which could increase the complexity and cost of the hydrogen production system. In addition, during the pyrolysis process, tar production is a common challenge. Tar is a complex mixture of organic compounds that could condense and solidify, leading to fouling and clogging of the reactor and downstream equipment. Tar formation reduces the hydrogen yield and increases the pyrolysis system's maintenance and operational costs. Therefore, effective tar removal techniques or catalysts are required to minimize tar formation and improve the overall efficiency of the process.

Furthermore, pyrolysis is an energy-intensive process that requires high temperatures to decompose biomass. The energy input for heating the reactor and maintaining the desired temperature could be significant, especially for large-scale hydrogen production. This energy requirement could affect the pyrolysis system's efficiency and economic viability. Also, using catalysts in pyrolysis could enhance the hydrogen yield and improve the selectivity of the process. However, catalyst deactivation is a common issue that could occur due to fouling, poisoning, or sintering of the catalyst. Catalyst deactivation reduces the effectiveness of the catalyst and requires frequent replacement or regeneration, which adds to the operational costs of the pyrolysis system. Besides, scaling up the pyrolysis process from laboratory-scale to commercial-scale could be challenging. The design and operation of large-scale pyrolysis reactors require careful consideration of heat transfer, mass transfer, and reaction kinetics. Integrating the pyrolysis system with other downstream processes, such as hydrogen recovery and purification, adds complexity to the scale-up process (Lopez et al., 2022, Alshareef et al., 2023, Kumar et al., 2022, Traven, 2023, Haque and Azad, 2023).

5.6 FACTORS AFFECTING THE PYROLYSIS PROCESS

The pyrolysis process is influenced by various factors that could significantly impact the efficiency and product yield. Understanding these factors is crucial for optimizing pyrolysis and maximizing hydrogen production from municipal biomass wastes (Aboelela et al., 2023).

5.6.1 FEEDSTOCK CHARACTERISTICS

The characteristics of the feedstock play a vital role in determining the pyrolysis process's outcome. Municipal biomass wastes could vary in composition, moisture content, particle size, and density, affecting the pyrolysis reaction. The moisture content of the feedstock is critical as it affects the heating value and the reaction kinetics. High moisture content could lead to incomplete pyrolysis and lower hydrogen production. Therefore, it is essential to ensure proper drying or pretreatment of the feedstock to achieve optimal results. The particle size and density of the feedstock also influence the pyrolysis process. Smaller particle sizes provide a larger surface area for heat transfer and reaction, resulting in faster pyrolysis rates. Additionally, the density of the feedstock affects the bed porosity and gas flow within the reactor, influencing the heat transfer and product distribution. Therefore, optimizing the feedstock characteristics is crucial for efficient hydrogen production through pyrolysis (Al-Rumaihi et al., 2022, Durak, 2023, Saha et al., 2023).

5.6.2 PROCESS PARAMETERS IN PYROLYSIS

Temperature is one of the most critical process parameters in pyrolysis. It directly affects the reaction kinetics and the composition of the pyrolysis products (Xie et al., 2021). The temperature range for biomass pyrolysis typically falls between 300°C and 800°C. The pyrolysis reaction is slow at lower temperatures, limiting hydrogen production (Li et al., 2023a). On the other hand, higher temperatures could lead to excessive biomass cracking, resulting in unwanted by-products forming (Aboelela et al., 2023). Therefore, finding the optimal temperature range is crucial to maximizing hydrogen production while minimizing the formation of undesirable compounds (Márquez Negro et al., 2023). The heating rate is another critical parameter that influences the pyrolysis process. It refers to the rate at which the biomass temperature increases during the heating phase (Ni et al., 2022). Higher heating rates could promote rapid pyrolysis, increasing hydrogen production. However, excessively high heating rates could also cause thermal degradation, forming char and tar. Therefore, the right balance between heating rate and hydrogen yield is essential for efficient pyrolysis (Safdari et al., 2019). Also, Residence time refers to the duration for which the biomass remains in the pyrolysis reactor. It is crucial in determining the extent of biomass decomposition and hydrogen yield. Longer residence times allow for more complete pyrolysis of the biomass, leading to higher hydrogen production (Al-Rumaihi et al., 2022).

However, excessively long residence times could also result in the formation of unwanted by-products. Therefore, optimizing the residence time is necessary to achieve the desired hydrogen yield while minimizing the formation of undesirable compounds (Solar et al., 2016). Besides, the composition of the biomass feedstock significantly influences the pyrolysis process and the hydrogen yield. Different types of biomasses have varying chemical compositions, which could affect the reaction kinetics and the formation of pyrolysis products. Biomass with high cellulose and hemicellulose content tends to produce more hydrogen, while biomass with high lignin content may form more char and tar. Therefore, selecting the appropriate feedstock composition is crucial for maximizing hydrogen production (Sridevi et al., 2023). In addition, using catalysts in pyrolysis could significantly enhance the hydrogen production efficiency. Catalysts could lower the activation energy required for the pyrolysis reaction, increase the selectivity toward hydrogen, and reduce the formation of unwanted by-products. Various catalysts, such as transition metals, zeolites, and activated carbon, have been explored for biomass pyrolysis (Hubble et al., 2022, Grams et al., 2020, Zamri et al., 2023).

The choice of catalyst and its concentration could significantly influence the pyrolysis process and hydrogen yield (Alagumalai et al., 2023, Rangel et al., 2023). Although pyrolysis is typically conducted under atmospheric pressure, the application of elevated pressures could have a significant impact on the pyrolysis process. Higher pressures could increase the yield of hydrogen and reduce the formation of by-products. The increased pressure could also enhance the heat transfer and improve the overall efficiency of the pyrolysis system. However, operating at high pressures may require additional safety measures and equipment, which could increase the cost of the process (Greco et al., 2021, Zhu et al., 2022, Zhao et al., 2023). The moisture content of the biomass feedstock could affect the pyrolysis process and the hydrogen yield. Higher moisture content could produce more water vapor during pyrolysis, diluting the hydrogen gas and reducing its purity. Therefore, it is essential to control the moisture content of the biomass to optimize hydrogen production (Eke et al., 2020). The design of the pyrolysis reactor also plays a crucial role in determining the process parameters and the efficiency of hydrogen production. Different reactor configurations, such as fixed beds, fluidized beds, rotary kilns, and microwave systems, have advantages and limitations. The choice of reactor design depends on factors such as feedstock characteristics, desired hydrogen yield, and process scalability (Guedes et al., 2018, Kulkarni et al., 2023, Attanayake et al., 2023, Raza et al., 2021).

Pretreatment of the feedstock before pyrolysis could significantly impact the process efficiency and product yields. Various pretreatment methods, such as drying, size reduction, and torrefaction, could modify feedstock properties and enhance pyrolysis (Griffin et al., 2022). Drying the feedstock reduces the moisture content, improving the heating value and reaction kinetics (Fonseca et al., 2019). Size reduction increases the surface area, promoting faster pyrolysis rates (Hong et al., 2020). Torrefaction could enhance the energy density and stability of the feedstock, resulting in improved pyrolysis performance. Proper feedstock pretreatment is essential to optimize the pyrolysis process and maximize hydrogen production. The selection of the pretreatment method depends on the feedstock's specific characteristics and the desired product distribution (Louwes et al., 2017, Chen et al., 2021).

5.7 MARKET POTENTIAL AND BUSINESS OPPORTUNITIES

The market potential for the pyrolysis of municipal biomass wastes for hydrogen production is significant and offers numerous business opportunities (Czajczyńska et al., 2017, Ni et al., 2006). The demand for hydrogen is expected to proliferate in the coming years. The transportation sector, in particular, is increasingly looking for clean and efficient energy sources to reduce greenhouse gas emissions (Singh et al., 2015). Hydrogen fuel cells are seen as a viable solution for zero-emission vehicles, creating a significant market opportunity for hydrogen production technologies like pyrolysis (Hassan et al., 2023b, Malik et al., 2023). Besides, governments worldwide are implementing renewable energy policies and providing incentives to promote the adoption of clean energy technologies. These policies often include targets for renewable hydrogen production and financial support for projects that contribute to the decarbonization of the energy sector. Pyrolysis-based hydrogen production systems could benefit from these policies and incentives, making it an attractive business opportunity (Nnabuife et al., 2023). The management of municipal biomass wastes is a pressing issue in many countries. Landfills are filling up, and the environmental impact of waste disposal is becoming increasingly evident. Pyrolysis offers a sustainable solution by converting these wastes into valuable hydrogen gas. Companies that could effectively manage and utilize municipal biomass wastes through pyrolysis could tap into the growing market for waste management and contribute to the circular economy (Li and Skelly, 2023). In addition, many industries and businesses want to diversify their energy portfolios to reduce dependence on fossil fuels and mitigate climate change risks. Pyrolysis-based hydrogen production allows companies to incorporate a clean and renewable energy source into their operations. Transportation, manufacturing, and power generation industries could benefit from integrating hydrogen produced through pyrolysis into their energy mix. Developing and deploying pyrolysis-based hydrogen production systems requires collaboration and partnerships between various stakeholders. Companies specializing in waste management, renewable energy, and hydrogen technologies could join forces to create integrated solutions. Collaboration with research institutions and universities could also help advance technology and explore new business opportunities (Marouani et al., 2023, Hassan et al., 2023c, Osman et al., 2022).

Furthermore, the market potential for pyrolysis-based hydrogen production is not limited to domestic demand. Many countries want to import hydrogen to meet their energy needs and reduce dependence on fossil fuels. Pyrolysis-based hydrogen production systems could position themselves to tap into the global hydrogen market by exporting hydrogen produced from municipal biomass wastes (Marouani et al., 2023, Squadrito et al., 2023). Pyrolysis produces hydrogen and generates valuable by-products such as biochar and bio-oil. These by-products have various applications in the agriculture, construction, and energy sectors. Companies could explore opportunities to commercialize these value-added products, further enhancing the economic viability of pyrolysis-based hydrogen production (Danesh et al., 2023). Also, pyrolysis systems could be designed to be scalable and modular, allowing for easy replication and expansion. This flexibility enables businesses to start small and gradually increase their production capacity based on market demand. The modular design also allows for deploying pyrolysis systems in various locations, including urban areas, where municipal biomass wastes are abundant. Establishing pyrolysis-based hydrogen production facilities could create employment opportunities and contribute to local economic development. The construction and operation of these facilities require skilled labor, creating jobs in engineering, manufacturing, and operations (Al-Rumaihi et al., 2022).

Additionally, developing a local supply chain for biomass feedstock could support the growth of the agricultural sector and create additional employment opportunities. Despite all this, the field of pyrolysis for hydrogen production is continuously evolving, with ongoing research and development efforts focused on improving efficiency, reducing costs, and enhancing the overall performance of the technology. Businesses that invest in research and development could stay at the forefront of technological advancements and gain a competitive edge in the market (Al-Rumaihi et al., 2022).

5.8 INTEGRATION OF PYROLYSIS WITH OTHER ENERGY SYSTEMS

Integrating pyrolysis with other energy systems is a promising approach to enhance the efficiency and sustainability of hydrogen production from municipal biomass wastes.

5.8.1 Gasification-Pyrolysis Integration

Gasification is a thermochemical process that converts carbonaceous materials into a synthesis gas (syngas) consisting mainly of hydrogen, carbon monoxide, and methane. Integrating gasification with pyrolysis offers several advantages for hydrogen production from municipal biomass wastes. Firstly, the syngas produced from gasification could be used as a heat source for the pyrolysis process, reducing the external energy requirements. This co-feeding approach enhances the system's energy efficiency and reduces operational costs. Furthermore, the syngas generated from gasification could be directly utilized for hydrogen production through a water-gas shift reaction, where carbon monoxide reacts with steam to produce additional hydrogen and carbon dioxide. This integration produces hydrogen-rich gas streams, which could be further purified and utilized for various applications. The gasification-pyrolysis integration also enables the utilization of different feedstocks, including municipal solid waste, agricultural residues, and forestry wastes. By combining these feedstocks, the system could achieve a more stable and reliable hydrogen production, as the variations in feedstock composition could be compensated by adjusting the gasification and pyrolysis parameters (Im-orb et al., 2017, Durak, 2023).

5.8.2 Combustion-Pyrolysis Integration

Combustion is a widely used energy conversion process involving the oxidation of carbonaceous materials to produce heat. The integration of combustion with pyrolysis could be beneficial for hydrogen production from municipal biomass wastes. In this integration, the char residue obtained from the pyrolysis process could be used as a combustion fuel, providing the necessary heat for the pyrolysis reaction. Combining combustion with pyrolysis could improve the system's overall energy efficiency, as the heat generated from the combustion process could be utilized for the endothermic pyrolysis reaction. This integration reduces the external energy requirements and enhances the sustainability of hydrogen production. Additionally, the combustion-pyrolysis integration allows for utilizing the by-products generated from the pyrolysis process, such as biochar. Biochar is a carbon-rich material that could be used as a soil amendment, contributing to carbon sequestration and improving soil fertility. By integrating combustion with pyrolysis, biochar could be directly utilized as a fuel, reducing the need for additional biomass resources and enhancing overall resource utilization (Hamel et al., 2001, Liu and Liu, 2005, Shen et al., 2015).

5.8.3 Anaerobic Digestion-Pyrolysis Integration

Anaerobic digestion is a biological process that converts organic materials into biogas, primarily methane and carbon dioxide. Integrating anaerobic digestion with pyrolysis offers a synergistic approach to hydrogen production from municipal biomass wastes. In this integration, the digestate obtained from the anaerobic digestion process could be utilized as a feedstock for pyrolysis, enhancing the overall energy conversion efficiency. The anaerobic digestion-pyrolysis integration allows the utilization of the organic fraction of municipal solid waste, sewage sludge, and other organic residues. By combining these feedstocks, the system could achieve a more sustainable and efficient hydrogen production, as the organic materials could be sequentially converted into biogas and then further processed through pyrolysis to produce hydrogen-rich gas streams (Ipiales et al., 2021, Pecchi and Baratieri, 2019).

Furthermore, the integration of anaerobic digestion with pyrolysis provides additional benefits, such as the reduction of organic waste volumes, the production of valuable by-products, and the mitigation of greenhouse gas emissions. The digestate obtained from the anaerobic digestion process could be used as a soil amendment, contributing to nutrient recycling and improving soil health. In addition to integrating pyrolysis with specific energy systems, developing hybrid systems that combine multiple energy conversion technologies could further enhance the efficiency and sustainability of hydrogen production from municipal biomass wastes. These hybrid systems could be designed to optimize the utilization of different feedstocks, maximize energy recovery, and minimize environmental impacts. For example, a hybrid system could integrate pyrolysis, gasification, and anaerobic digestion sequentially, where the feedstock is first subjected to pyrolysis to produce bio-oil and char, followed by char gasification to produce syngas. Finally, the remaining organic fraction is processed through anaerobic digestion to produce biogas. This sequential integration allows for efficient feedstock utilization and maximizes the production of hydrogen-rich gas streams. Integrating pyrolysis with other energy systems requires careful design and optimization to ensure the compatibility of the different processes and maximize the overall system performance. Selecting appropriate technologies, process parameters, and feedstock characteristics is crucial in achieving the desired outcomes (Singh et al., 2022a, Ghysels et al., 2020, González-Arias et al., 2020, Giwa et al., 2019, Opatokun et al., 2017).

5.9 CONCLUSION AND FUTURE PERSPECTIVES

The pyrolysis of municipal biomass wastes for hydrogen production is a promising technology that offers numerous benefits. Pyrolysis of municipal biomass wastes has shown significant potential for hydrogen production. The process could convert biomass into hydrogen-rich gases, which could be further purified and utilized for various applications. Utilizing municipal biomass wastes for hydrogen production offers a sustainable solution for waste management. Converting these wastes into valuable hydrogen gas helps reduce the environmental impact associated with waste disposal. In this regard, optimizing pyrolysis parameters, such as temperature, residence time, and feedstock composition, is crucial in maximizing hydrogen production. Also, using catalysts in pyrolysis could significantly enhance hydrogen production by promoting the decomposition of biomass and improving the selectivity toward hydrogen gas. Catalyst selection, preparation, and regeneration are essential for efficient hydrogen production. Though pyrolysis offers a sustainable solution for waste management and hydrogen production, assessing and mitigating the environmental impact associated with the process is essential. Proper emissions control and monitoring techniques should be implemented to minimize the release of harmful by-products. Besides, the economic analysis of pyrolysis systems is crucial to determine the feasibility and commercial viability of hydrogen production. Factors such as capital investment, operational costs, and market potential should be considered to ensure the economic sustainability of the technology. As discussed above, further research should optimize the pyrolysis process parameters to enhance hydrogen production efficiency. This includes investigating the effect of different feedstock compositions, temperature profiles, and residence times on hydrogen gas yield and quality. Also, developing efficient catalysts for pyrolysis is crucial to improve hydrogen production. Future research should explore novel catalyst materials and synthesis methods to enhance hydrogen gas's catalytic activity, stability, and selectivity. In addition, comprehensive environmental impact assessments should be conducted to evaluate the emissions and by-products associated with pyrolysis. This will help develop effective mitigation strategies and ensure the technology's sustainability. Besides, further economic analysis and feasibility studies should be conducted to assess the cost-effectiveness of pyrolysis systems for hydrogen production. This includes evaluating the capital and operational costs, potential revenue streams, and market opportunities.

Furthermore, research should explore the integration of pyrolysis with other renewable energy systems, such as solar and wind, to enhance the overall energy efficiency and sustainability of hydrogen production. Governments and regulatory bodies should also provide policy support and incentives to promote adopting pyrolysis technology for hydrogen production. This includes the development of favorable regulatory frameworks, financial incentives, and research funding. Finally, collaboration between researchers, industry stakeholders, and policymakers is essential to accelerate the development and deployment of pyrolysis technology. Knowledge-sharing platforms, conferences, and workshops should be organized to facilitate information exchange and collaboration.

REFERENCES

Abdel-Shafy, H. I. & Mansour, M. S. 2018. Solid waste issue: Sources, composition, disposal, recycling, and valorization. *Egyptian Journal of Petroleum*, 27, 1275–1290.

Aboelela, D., Saleh, H., Attia, A. M., Elhenawy, Y., Majozi, T. & Bassyouni, M. 2023. Recent advances in biomass pyrolysis processes for bioenergy production: Optimization of operating conditions. *Sustainability*, 15, 11238.

Abubakar, I. R., Maniruzzaman, K. M., Dano, U. L., Alshihri, F. S., Alshammari, M. S., Ahmed, S. M. S., AL- Gehlani, W. A. G. & Alrawaf, T. I. 2022. Environmental sustainability impacts of solid waste management practices in the global South. *International Journal of Environmental Research and Public Health*, 19, 12717.

Alagumalai, A., Devarajan, B. & Song, H. 2023. Unlocking the potential of catalysts in thermochemical energy conversion processes. *Catalysis Science & Technology*. DOI:10.1039/D3CY00848G

Al-Rumaihi, A., Shahbaz, M., Mckay, G., Mackey, H. & Al-Ansari, T. 2022. A review of pyrolysis technologies and feedstock: A blending approach for plastic and biomass towards optimum biochar yield. *Renewable and Sustainable Energy Reviews*, 167, 112715.

Alshareef, R., Nahil, M. A. & Williams, P. T. 2023. Hydrogen production by three-stage (i) pyrolysis,(ii) catalytic steam reforming, and (iii) water gas shift processing of waste plastic. *Energy & Fuels*, 37, 3894–3907.

Amalina, F., Abd Razak, A. S., Krishnan, S., Sulaiman, H., Zularisam, A. & Nasrullah, M. 2022. Biochar production techniques utilizing biomass waste-derived materials and environmental applications-A review. *Journal of Hazardous Materials Advances*, 7, 100134.

Amin, M., Shah, H. H., Fareed, A. G., Khan, W. U., Chung, E., Zia, A., Farooqi, Z. U. R. & Lee, C. 2022. Hydrogen production through renewable and non-renewable energy processes and their impact on climate change. *International Journal of Hydrogen Energy*, 47, 33112–33134.

Aprilia, A., Tezuka, T. & Spaargaren, G. 2013. Inorganic and hazardous solid waste management: Current status and challenges for Indonesia. *Procedia Environmental Sciences*, 17, 640–647.

Attanayake, D. D., Sewerin, F., Kulkarni, S., Dernbecher, A., Dieguez-Alonso, A. & Van Wachem, B. 2023. Review of modelling of pyrolysis processes with CFD-DEM. *Flow, Turbulence and Combustion*, 111, 355–408.

Ayilara, M. S., Olanrewaju, O. S., Babalola, O. O. & Odeyemi, O. 2020. Waste management through composting: Challenges and potentials. *Sustainability*, 12, 4456.

Aziz, M., Darmawan, A. & Juangsa, F. B. 2021. Hydrogen production from biomasses and wastes: A technological review. *International Journal of Hydrogen Energy*, 46, 33756–33781.

Bakhtyari, A., Makarem, M. & Rahimpour, M. 2018. Hydrogen production through pyrolysis. *Encyclopedia of Sustainability Science and Technology*, 1–28. DOI:10.1007/978-1-493 9-2493-6_956-1

Bandini, F., Taskin, E., Bellotti, G., Vaccari, F., Misci, C., Guerrieri, M. C., Cocconcelli, P. S. & Puglisi, E. 2022. The treatment of the organic fraction of municipal solid waste (OFMSW) as a possible source of micro-and nano-plastics and bioplastics in agroecosystems: A review. *Chemical and Biological Technologies in Agriculture*, 9, 1–17.

Bhatt, K. P., Patel, S., Upadhyay, D. S. & Patel, R. N. 2022. A critical review on solid waste treatment using plasma pyrolysis technology. *Chemical Engineering and Processing-Process Intensification*, 177, 108989.

Chen, W.-H., Lin, B.-J., Lin, Y.-Y., Chu, Y.-S., Ubando, A. T., Show, P. L., Ong, H. C., Chang, J.-S., Ho, S.-H. & Culaba, A. B. 2021. Progress in biomass torrefaction: Principles, applications and challenges. Progress *in Energy and Combustion Science*, 82, 100887.

Czajczyńska, D., Anguilano, L., Ghazal, H., Krzyżyńska, R., Reynolds, A., Spencer, N. & Jouhara, H. 2017. Potential of pyrolysis processes in the waste management sector. *Thermal Science and Engineering Progress*, 3, 171–197.

Danesh, P., Niaparast, P., Ghorbannezhad, P. & Ali, I. 2023. Biochar production: Recent developments, applications, and challenges. *Fuel*, 337, 126889.

Domingo, J. L., Marquès, M., Mari, M. & Schuhmacher, M. 2020. Adverse health effects for populations living near waste incinerators with special attention to hazardous waste incinerators. A review of the scientific literature. *Environmental Research*, 187, 109631.

Durak, H. 2023. Comprehensive assessment of thermochemical processes for sustainable waste management and resource recovery. *Processes*, 11, 2092.

Eke, J., Onwudili, J. A. & Bridgwater, A. V. 2020. Influence of moisture contents on the fast pyrolysis of trommel fines in a bubbling fluidized bed reactor. *Waste and Biomass Valorization*, 11, 3711–3722.

Environmental Protection Agency U.S. *Wastes-Non-Hazardous Waste-Municipal Solid Waste*. 2011. https://archive.epa.gov/epawaste/nonhaz/municipal/web/html/index.html

Ethaib, S., Omar, R., Kamal, S. M. M., Awangbiak, D. R. & Zubaidi, S. L. 2020. Microwave-assisted pyrolysis of biomass waste: A mini review. *Processes*, 8, 1190.

Fonseca, F. G., Funke, A., Niebel, A., Dias, A. P. S. & Dahmen, N. 2019. Moisture content as a design and operational parameter for fast pyrolysis. *Journal of Analytical and Applied Pyrolysis*, 139, 73–86.

Ghysels, S., Acosta, N., Estrada, A., Pala, M., DE Vrieze, J., Ronsse, F. & Rabaey, K. 2020. Integrating anaerobic digestion and slow pyrolysis improves the product portfolio of a cocoa waste biorefinery. *Sustainable Energy & Fuels*, 4, 3712–3725.

Giwa, A. S., Xu, H., Chang, F., Zhang, X., Ali, N., Yuan, J. & Wang, K. 2019. Pyrolysis coupled anaerobic digestion process for food waste and recalcitrant residues: Fundamentals, challenges, and considerations. *Energy Science & Engineering*, 7, 2250–2264.

González-Arias, J., Gil, M. V., Fernández, R. Á., Martínez, E. J., Fernández, C., Papaharalabos, G. & Gómez, X. 2020. Integrating anaerobic digestion and pyrolysis for treating digestates derived from s ewage sludge and fat wastes. *Environmental Science and Pollution Research*, 27, 32603–32614.

Grams, J., Ryczkowski, R., Sadek, R., Chałupka, K., Przybysz, K., Casale, S. & Dzwigaj, S. 2020. Hydrogen-rich gas production by upgrading of biomass pyrolysis vapors over NiBEA catalyst: Impact of dealumination and preparation method. *Energy & Fuels*, 34, 16936–16947.

Greco, G., Videgain, M., DI Stasi, C., Pires, E. & Manya, J. J. 2021. Importance of pyrolysis temperature and pressure in the concentration of polycyclic aromatic hydrocarbons in wood waste-derived biochars. *Journal of Analytical and Applied Pyrolysis*, 159, 105337.

Griffin, G., Ward, L., Madapusi, S., Shah, K. & Parthasarathy, R. 2022. A study of chemical pre-treatment and pyrolysis operating conditions to enhance biochar production from rice straw. *Journal of Analytical and Applied Pyrolysis*, 163, 105455.

Guedes, R. E., Luna, A. S. & Torres, A. R. 2018. Operating parameters for bio-oil production in biomass pyrolysis: A review. *Journal of Analytical and Applied Pyrolysis*, 129, 134–149.

Hamel, S., Funk, G., Krumm, W. & Mertens, C. 2001. Integrated pyrolysis and combustion of biomass. *Integrierte Pyrolyse und Verbrennung von Biomassen*. 3–10.

Haque, N. & Azad, A. K. 2023. Comparative study of hydrogen production from organic fraction of municipal solid waste and its challenges: A review. *Energies*, 16, 7853.

Hassan, Q., Algburi, S., Sameen, A. Z., Salman, H. M. & Jaszczur, M. 2023a. Green hydrogen: A pathway to a sustainable energy future. *International Journal of Hydrogen Energy*, 50, 310–333.

Hassan, Q., Azzawi, I. D., Sameen, A. Z. & Salman, H. M. 2023b. Hydrogen fuel cell vehicles: Opportunities and challenges. *Sustainability*, 15, 11501.

Hassan, Q., Sameen, A. Z., Salman, H. M., Jaszczur, M. & Al-Jiboory, A. K. 2023c. Hydrogen energy future: Advancements in storage technologies and implications for sustainability. *Journal of Energy Storage*, 72, 108404.

Hong, Z., Zhong, F., Niu, W., Zhang, K., Su, J., Liu, J., Li, L. & Wu, F. 2020. Effects of temperature and particle size on the compositions, energy conversions and structural characteristics of pyrolysis products from different crop residues. *Energy*, 190, 116413.

Hornung, A. 2012. Biomassbiomass pyrolysis biomass pyrolysis. *Encyclopedia of Sustainability Science and Technology*. Springer, New York, 1517–1531.

Hossain, S. S. & Roy, P. 2020. Sustainable ceramics derived from solid wastes: A review. *Journal of Asian Ceramic Societies*, 8, 984–1009.

Hosseini, S. E., Abdul Wahid, M., Jamil, M., Azli, A. A. & Misbah, M. F. 2015. A review on biomass-based hydrogen production for renewable energy supply. *International Journal of Energy Research*, 39, 1597–1615.

Hubble, A. H., Ryan, E. M. & Goldfarb, J. L. 2022. Enhancing pyrolysis gas and bio-oil formation through transition metals as in situ catalysts. *Fuel*, 308, 121900.

Hussain, M., Ali, O., Raza, N., Zabiri, H., Ahmed, A. & Ali, I. 2023. Recent advances in dynamic modeling and control studies of biomass gasification for production of hydrogen rich syngas. *RSC Advances*, 13, 23796–23811.

Igliński, B., Kujawski, W. & Kiełkowska, U. 2023. Pyrolysis of waste biomass: Technical and process achievements, and future development-A review. *Energies*, 16, 1829.

Im-Orb, K., Detchusananard, T., Ponpesh, P. & Arpornwichanop, A. 2017. Investigation of integrated biomass pyrolysis and gasification process for green fuel production. *Energy Procedia*, 142, 204–209.

Inglezakis, V. J. & Moustakas, K. 2015. Household hazardous waste management: A review. *Journal of Environmental Management*, 150, 310–321.

Ipiales, R., de la Rubia, M., Diaz, E., Mohedano, A. & Rodriguez, J. J. 2021. Integration of hydrothermal carbonization and anaerobic digestion for energy recovery of biomass waste: An overview. *Energy & Fuels*, 35, 17032–17050.

Ishaq, H., Dincer, I. & Crawford, C. 2022. A review on hydrogen production and utilization: Challenges and opportunities. *International Journal of Hydrogen Energy*, 47, 26238–26264.

Jouhara, H., Ahmad, D., van den Boogaert, I., Katsou, E., Simons, S. & Spencer, N. 2018. Pyrolysis of domestic based feedstock at temperatures up to 300°C. *Thermal Science and Engineering Progress*, 5, 117–143.

Kibria, M. G., Masuk, N. I., Safayet, R., Nguyen, H. Q. & Mourshed, M. 2023. Plastic waste: Challenges and opportunities to mitigate pollution and effective management. *International Journal of Environmental Research*, 17, 20.

Kulkarni, A., Mishra, G., Palla, S., Ramesh, P., Surya, D. V. & Basak, T. 2023. Advances in computational fluid dynamics modeling for biomass pyrolysis: A review. *Energies*, 16, 7839.

Kumar, A., Singh, E., Mishra, R. & Kumar, S. 2022. Solid waste to energy: Existing scenario in developing and developed countries. In *Handbook of Solid Waste Management: Sustainability through Circular* Economy(pp. 2023–2045). Singapore: Springer Nature Singapore.

Kumar, A., Thakur, A. K., Gaurav, G. K., Klemeš, J. J., Sandhwar, V. K., Pant, K. K. & Kumar, R. 2023. A critical review on sustainable hazardous waste management strategies: A step towards a circular economy. *Environmental Science and Pollution Research*, 30, 105030–105055.

Lachos-Perez, D., Martins-Vieira, J. C., Missau, J., Anshu, K., Siakpebru, O. K., Thengane, S. K., Morais, A. R. C., Tanabe, E. H. & Bertuol, D. A. 2023. Review on biomass pyrolysis with a focus on bio-oil upgrading techniques. *Analytica*, 4, 182–205.

Lam, C. H., Ip, A. W., Barford, J. P. & Mckay, G. 2010. Use of incineration MSW ash: A review. *Sustainability*, 2, 1943–1968.

Li, J., Yang, D., Yao, X., Zhou, H., Xu, K. & Geng, L. 2023a. Slow pyrolysis experimental investigation of biomass tar formation and hydrogen production by tar reforming. *International Journal of Hydrogen Energy*, 52, 74–87.

Li, S. & Skelly, S. 2023. Physicochemical properties and applications of biochars derived from municipal solid waste: A review. *Environmental Advances*, 13, 100395.

Li, T., Wang, J., Chen, H., Li, W., Pan, P., Wu, L., Xu, G. & Chen, H. 2023b. Performance analysis of an integrated biomass-to-energy system based on gasification and pyrolysis. *Energy Conversion and Management*, 287, 117085.

Liu, Y. & Liu, Y. 2005. Novel incineration technology integrated with drying, pyrolysis, gasification, and combustion of MSW and ashes vitrification. *Environmental Science & Technology*, 39, 3855–3863.

Lopez, G., Santamaria, L., Lemonidou, A., Zhang, S., Wu, C., Sipra, A. T. & Gao, N. 2022. Hydrogen generation from biomass by pyrolysis. *Nature Reviews Methods Primers*, 2, 20.

Louwes, A. C., Basile, L., Yukananto, R., Bhagwandas, J., Bramer, E. A. & Brem, G. 2017. Torrefied biomass as feed for fast pyrolysis: An experimental study and chain analysis. *Biomass and Bioenergy*, 105, 116–126.

Malik, F. R., Yuan, H.-B., Moran, J. C. & Tippayawong, N. 2023. Overview of hydrogen production technologies for fuel cell utilization. *Engineering Science and Technology, an International Journal*, 43, 101452.

Marouani, I., Guesmi, T., Alshammari, B. M., Alqunun, K., Alzamil, A., Alturki, M. & Hadj Abdallah, H. 2023. Integration of renewable-energy-based green hydrogen into the energy future. *Processes*, 11, 2685.

Márquez Negro, A., Patlaka, E., Sfakiotakis, S., Ortiz, I. & Sánchez-Hervas, J. M. 2023. Pyrolysis of municipal solid waste: A kinetic study through multi-step reaction models. *Waste Management*, 172, 171–181 Available at SSRN 4486213.

Masud, M. H., Mourshed, M., Hossain, M. S., Ahmed, N. U. & Dabnichki, P. 2023. Generation of waste: Problem to possible solution in developing and underdeveloped nations. In Pardeep Singh, Rishikesh Singh, André C.S. Batalhão, Pramit Verma, Arif Ahamad (eds), *Waste Management and Resource Recycling in the Developing World* (pp. 21–59). Elsevier.

Matamba, T., Tahmasebi, A., Yu, J., Keshavarz, A., Abid, H. & Iglauer, S. 2023. A review on biomass as a substitute energy source: Polygeneration influence and hydrogen rich gas formation via pyrolysis. *Journal of Analytical and Applied Pyrolysis*, 175, 106221.

Möslinger, M., Ulpiani, G. & Vetters, N. 2023. Circular economy and waste management to empower a climate-neutral urban future. *Journal of Cleaner Production*, 421, 138454.

Nandhini, R., Berslin, D., Sivaprakash, B., Rajamohan, N. & Vo, D.-V. N. 2022. Thermochemical conversion of municipal solid waste into energy and hydrogen: A review. *Environmental Chemistry Letters*, 20, 1645–1669.

Nguyen-Thi, T. X., Nguyen, P. Q. P., Tran, V. D., Ağbulut, Ü., Nguyen, L. H., Balasubramanian, D., Tarelko, W., Bandh, S. A. & Pham, N. D. K. 2023. Recent advances in hydrogen production from biomass waste with a focus on pyrolysis and gasification. *International Journal of Hydrogen Energy*, 54, 127–160.

Nhubu, T., Muzenda, E., Mbohwa, C. & Belaid, M. 2022. Biogas potential from the biomethanization of biodegradable municipal solid waste generated in Harare. In Chinnappan Baskar, Seeram Ramakrishna, Shikha Baskar, Rashmi Sharma, Amutha Chinnappan, Rashmi Sehrawat (eds), *Handbook of Solid Waste Management: Sustainability through Circular Economy* (pp. 2197–2227). Singapore: Springer Nature Singapore.

Ni, L., Feng, Z., Zhang, T., Gao, Q., Hou, Y., He, Y., Su, M., Ren, H., Hu, W. & Liu, Z. 2022. Effect of pyrolysis heating rates on fuel properties of molded charcoal: Imitating industrial pyrolysis process. *Renewable Energy*, 197, 257–267.

Ni, M., Leung, D. Y., Leung, M. K. & Sumathy, K. 2006. An overview of hydrogen production from biomass. *Fuel Processing Technology*, 87, 461–472.

Nnabuife, s. G., Oko, E., Kuang, B., Bello, A., Onwualu, A. P., Oyagha, S. & Whidborne, J. 2023. The prospects of hydrogen in achieving net zero emissions by 2050: A critical review. *Sustainable Chemistry for Climate Action*, 2, 100024.

GLANCE, A., 2014. ENERGY DEMAND DOMESTIC 11. *The Energy and Resources Institute Energy and Environment Data Directory and Yearbook, 2013/14*, p.263..

Opatokun, S. A., Lopez-Sabiron, A. M., Ferreira, G. & Strezov, V. 2017. Life cycle analysis of energy production from food waste through anaerobic digestion, pyrolysis and integrated energy system. *Sustainability*, 9, 1804.

Osman, A. I., Elgarahy, A. M., Eltaweil, A. S., Abd El-Monaem, E. M., El-Aqapa, H. G., Park, Y., Hwang, Y., Ayati, A., Farghali, M. & Ihara, I. 2023. Biofuel production, hydrogen production and water remediation by photocatalysis, biocatalysis and electrocatalysis. *Environmental Chemistry Letters*, 21, 1315–1379.

Osman, A. I., Mehta, N., Elgarahy, A. M., Hefny, M., AL- Hinai, A., AL- Muhtaseb, A. A. H. & Rooney, D. W. 2022. Hydrogen production, storage, utilisation and environmental impacts: A review. *Environmental Chemistry Letters*, 20, 153–188.

Pecchi, M. & Baratieri, M. 2019. Coupling anaerobic digestion with gasification, pyrolysis or hydrothermal carbonization: A review. *Renewable and Sustainable Energy Reviews*, 105, 462–475.

Pham, A.-Q. 2005. Co-production of hydrogen and electricity using pyrolysis and fuel cells. Google Patents.

Pheakdey, D. V., Quan, N. V., Khanh, T. D. & Xuan, T. D. 2022. Challenges and priorities of municipal solid waste management in Cambodia. *International Journal of Environmental Research and Public Health*, 19, 8458.

Potnuri, R., Suriapparao, D. V., Rao, C. S. & Kumar, T. H. 2022. Understanding the role of modeling and simulation in pyrolysis of biomass and waste plastics: A review. *Bioresource Technology Reports*, 20, 101221.

Purnell, P., Velenturf, A. & Marshall, R. 2019. New governance for circular economy: Policy, regulation and market contexts for resource recovery from waste. *Resource Recovery from Wastes*, 63, 395.

Rangel, M. D. C., Mayer, F. M., Carvalho, M. D. S., Saboia, G. & de Andrade, A. M. 2023. Selecting catalysts for pyrolysis of lignocellulosic biomass. *Biomass*, 3, 31–63.

Raza, M., Inayat, A., Ahmed, A., Jamil, F., Ghenai, C., Naqvi, S. R., Shanableh, A., Ayoub, M., Waris, A. & Park, Y.-K. 2021. Progress of the pyrolyzer reactors and advanced technologies for biomass pyrolysis processing. *Sustainability*, 13, 11061.

Rogdakis, E. D. & Bitsikas, P. I. 2018. Applications of bioenergy-modeling of anaerobic digestion. In Emmanuel D. Rogdakis, Irene P. Koronaki (eds), *Renewable Energy Engineering: Solar, Wind, Biomass, Hydrogen and Geothermal Energy System*, Bentham Science Publishers, Sharjah, UAE: Bentham Science Publishersm 166–184,

Sadeghi, A., Kermani Alghorayshi, S., Shamsi, M. & Mirjani, F. 2023. Multi-criteria evaluation of the extraction methods of rare earth elements from aqueous streams. *International Journal of Environmental Science and Technology*, 20, 9707–9716.

Safdari, M.-S., Amini, E., Weise, D. R. & Fletcher, T. H. 2019. Heating rate and temperature effects on pyrolysis products from live wildland fuels. *Fuel*, 242, 295–304.

Saha, N., Klinger, J., Islam, M. T. & Reza, T. 2023. Advanced biorefinery feedstock from non-recyclable municipal solid waste by mechanical preprocessing. *Frontiers in Fuels*, 1, 1105637.

Sarker, S., Lamb, J. J., Hjelme, D. R. & Lien, K. M. 2019. A review of the role of critical parameters in the design and operation of biogas production plants. *Applied Sciences*, 9, 1915.

Setiabudi, H., Aziz, M., Abdullah, S., Teh, L. & Jusoh, R. 2020. Hydrogen production from catalytic steam reforming of biomass pyrolysis oil or bio-oil derivatives: A review. *International Journal of Hydrogen Energy*, 45, 18376–18397.

Shah, H. H., Amin, M., Iqbal, A., Nadeem, I., Kalin, M., Soomar, A. M. & Galal, A. M. 2023. A review on gasification and pyrolysis of waste plastics. *Frontiers in Chemistry*, 10, 960894.

Sharma, A., Pareek, V. & Zhang, D. 2015. Biomass pyrolysis-A review of modelling, process parameters and catalytic studies. *Renewable and Sustainable Energy Reviews*, 50, 1081–1096.

Shen, J., Igathinathane, C., Yu, M. & Pothula, A. K. 2015. Biomass pyrolysis and combustion integral and differential reaction heats with temperatures using thermogravimetric analysis/differential scanning calorimetry. *Bioresource technology*, 185, 89–98.

Siddiqua, A., Hahladakis, J. N. & AL- Attiya, W. A. K. 2022. An overview of the environmental pollution and health effects associated with waste landfilling and open dumping. *Environmental Science and Pollution Research*, 29, 58514–58536.

Silva, R., DE Brito, J., LYE, C. & Dhir, R. 2017. The role of glass waste in the production of ceramic-based products and other applications: A review. *Journal of Cleaner Production*, 167, 346–364.

Singh, R., Paritosh, K., Pareek, N. & Vivekanand, V. 2022a. Integrated system of anaerobic digestion and pyrolysis for valorization of agricultural and food waste towards circular bioeconomy. *Bioresource Technology*, 360, 127596.

Singh, S., Jain, S., Venkateswaran, P., Tiwari, A. K., Nouni, M. R., Pandey, J. K. & Goel, S. 2015. Hydrogen: A sustainable fuel for future of the transport sector. *Renewable and Sustainable Energy Reviews*, 51, 623–633.

Singh, V. K., Solanki, P., Ghosh, A. & Pal, A. 2022b. Solid waste management and policies toward sustainable agriculture. In Chinnappan Baskar, Seeram Ramakrishna, Shikha Baskar, Rashmi Sharma, Amutha Chinnappan, Rashmi Sehrawat (eds), *Handbook of Solid Waste Management: Sustainability through Circular Economy*. Singapore: Springer Nature Singapore. pp. 523–544.

Solar, J., De Marco, I., Caballero, B., Lopez-Urionabarrenechea, A., Rodriguez, N., Agirre, I. & Adrados, A. 2016. Influence of temperature and residence time in the pyrolysis of woody biomass waste in a continuous screw reactor. *Biomass and Bioenergy*, 95, 416–423.

Squadrito, G., Maggio, G. & Nicita, A. 2023. The green hydrogen revolution. *Renewable Energy*, 216, 119041.

Sridevi, V., Surya, D. V., Reddy, B. R., Shah, M., Gautam, R., Kumar, T. H., Puppala, H., Pritam, K. S. & Basak, T. 2023. Challenges and opportunities in the production of sustainable hydrogen from lignocellulosic biomass using microwave-assisted pyrolysis: A review. *International Journal of Hydrogen Energy*, 52, 507–531.

Taylor, C. J., Pomberger, A., Felton, K. C., Grainger, R., Barecka, M., Chamberlain, T. W., Bourne, R. A., Johnson, C. N. & Lapkin, A. A. 2023. A brief introduction to chemical reaction optimization. *Chemical Reviews*, 123, 3089–3126.

Tihay, V. & Gillard, P. 2010. Pyrolysis gases released during the thermal decomposition of three Mediterranean species. *Journal of Analytical and Applied Pyrolysis*, 88, 168–174.

Tomczyk, A., Sokołowska, Z. & Boguta, P. 2020. Biochar physicochemical properties: Pyrolysis temperature and feedstock kind effects. *Reviews in Environmental Science and Bio/Technology*, 19, 191–215.

Traven, L. 2023. Sustainable energy generation from municipal solid waste: A brief overview of existing technologies. *Case Studies in Chemical and Environmental Engineering*, 8, 100491.

Wang, K. & Tester, J. W. 2023. Sustainable management of unavoidable biomass wastes. *Green Energy and Resources*, 1, 100005.

Williams, P. T. 2021. Hydrogen and carbon nanotubes from pyrolysis-catalysis of waste plastics: A review. *Waste and Biomass Valorization*, 12, 1–28.

Xie, R., Zhu, Y., Zhang, H., Zhang, P. & Han, L. 2021. Effects and mechanism of pyrolysis temperature on physicochemical properties of corn stalk pellet biochar based on combined characterization approach of microcomputed tomography and chemical analysis. *Bioresource Technology*, 329, 124907.

Xu, S., Chen, J., Peng, H., Leng, S., Li, H., Qu, W., Hu, Y., Li, H., Jiang, S. & Zhou, W. 2021. Effect of biomass type and pyrolysis temperature on nitrogen in biochar, and the comparison with hydrochar. *Fuel*, 291, 120128.

Yu, S., Fan, Y., Shi, Z., Li, J., Zhao, X., Zhang, T. & Chang, Z. 2023. Hydrogen-based combined heat and power systems: A review of technologies and challenges. *International Journal of Hydrogen Energy*, 48 (89), 34906–34929.

Zainal, B. S., Ker, P. J., Mohamed, H., Ong, H. C., Fattah, I., Rahman, S. A., Nghiem, L. D. & Mahlia, T. I. 2024. Recent advancement and assessment of green hydrogen production technologies. *Renewable and Sustainable Energy Reviews*, 189, 113941.

Zamri, M. F. M. A., Shamsuddin, A. H., Ali, S., Bahru, R., Milano, J., Tiong, S. K., Fattah, I. M. R. & Raja Shahruzzaman, R. M. H. 2023. Recent advances of triglyceride catalytic pyrolysis via heterogenous dolomite catalyst for upgrading biofuel quality: A review. *Nanomaterials*, 13, 1947.

Zhang, X., Liu, C., Chen, Y., Zheng, G. & Chen, Y. 2022. Source separation, transportation, pretreatment, and valorization of municipal solid waste: A critical review. *Environment, Development and Sustainability*, 24, 11471–11513.

Zhao, S., Su, J. & Wu, J. 2023. Release performance and kinetic behavior of volatile products from controlled pressure pyrolysis of oil shale in nitrogen atmosphere. *Scientific Reports*, 13, 10676.

Zhu, Y., Wang, Q., Yan, J., Cen, J., Fang, M. & Ye, C. 2022. Influence and action mechanism of pressure on pyrolysis process of a low rank Naomaohu coal at different temperatures. *Journal of Analytical and Applied Pyrolysis*, 167, 105682.

6 Gasification of Lignin for Hydrogen Production

Kang Kang, Sonil Nanda, Mohammad Latifi,
Gholamreza Roohollahi, Pedram Fatehi,
Ajay K. Dalai, Janusz Kozinski, and Yulin Hu

LIST OF ABBREVIATIONS

BFB	Bubbling fluidized bed gasifier
CCD	Central composite design
CFB	Circulating fluidized bed reactor
EFG	Entrained flow gasifier
ER	Equivalence ratio
LHV	lower heating value
SCW	Supercritical water
SCWG	Supercritical water gasification
UFB	Updraft fixed bed gasifier

6.1 IMPORTANCE OF LIGNIN UTILIZATION AND HYDROGEN PRODUCTION

6.1.1 IMPORTANCE OF LIGNIN AS A BIOMASS RESOURCE

The urgency of seeking renewable materials and energy sources has been triggering research activities into biomass conversion for decades. Lignocellulosic biomass is identified as a promising feedstock to produce various sustainable fuels, platform chemicals, and materials. In general, biomass can be categorized into three types such as raw biomass (lignocelluloses, crops, and vegetables), biomass waste (municipal solid waste and sewage), and agricultural wastes (K N et al., 2022). Lignocellulosic biomass refers to plant-based biomass which mainly consists of three major components cellulose, hemicelluloses, and the aromatic polymer lignin (Zoghlami and Paës, 2019). Out of the three major components, lignin is a unique biomass resource as it is the most abundant natural reservoir of aromatic compounds (Yoo et al., 2020). Specifically, Lignin is the second-most abundant component (15–40 wt.%) of lignocellulosic biomass, which acts as a binder consolidating hemicelluloses and cellulose in woody biomass (Vanholme et al., 2008). To effectively convert biomass into useful products, the first step is to reduce or overcome biomass recalcitrance, which refers to the natural resistance presented by plant cell walls as the result of their structural complexity and compositional heterogeneity.

DOI: 10.1201/9781003382270-8

In the conventional routes of biorefining, lignin is often separated from the other components of biomass via the delignification process, resulting in the production of different types of technical lignin. The lignin mentioned in this chapter refers to the "technical lignin" rather than "native lignin" in the plant body.

In industry, lignin is produced as waste from pulping and ethanol plants at the rate of ~110 million tons/year globally (Aro and Fatehi, 2017). As shown in Figure 6.1, there are many types of technical lignin produced including the delignification process of the lignocellulosic biomass, including Kraft lignin, lignosulfonates, soda lignin, hydrolysis lignin, organosolv lignin, ionic liquid lignin, etc. (Eraghi Kazzaz and Fatehi, 2020). Despite the growing interest in converting lignin into value-added products, most of the lignin (~95 wt.%) is produced from the pulping industry and is used as cheap combustion fuel (Sun, 2020). Interestingly, several technologies have been commercialized to produce lignin, including LignoForce™ and LignoBoost™ (Dessbesell et al., 2020). This commercial production of lignin enables the large-scale generation of lignin-derived materials, and research has progressed in converting lignin into adhesives, adsorbents, and surfactants (Zhang and Fatehi, 2019). Thanks to its high aromatic content, lignin has the attractive benefit of retaining high carbon yield and product homogeneity upon thermochemical conversion, which makes it a flexible raw material for different conversion products (Qu et al., 2021). However, on the other hand, lignin is also the difficult part of biomass conversion. Specifically, lignin is a complex and resistant phenolic macromolecule made up of the three methoxylated phenylpropane units, namely guaiacyl (G), syringyl (S), and

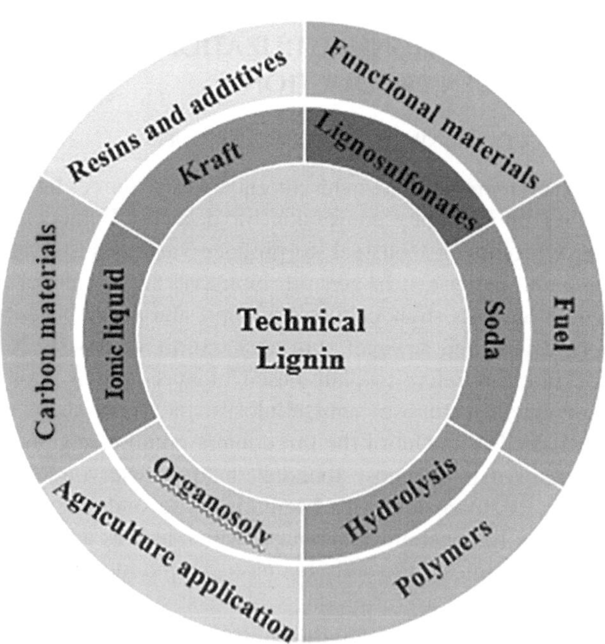

FIGURE 6.1   Sources, and potential applications of technical lignin.

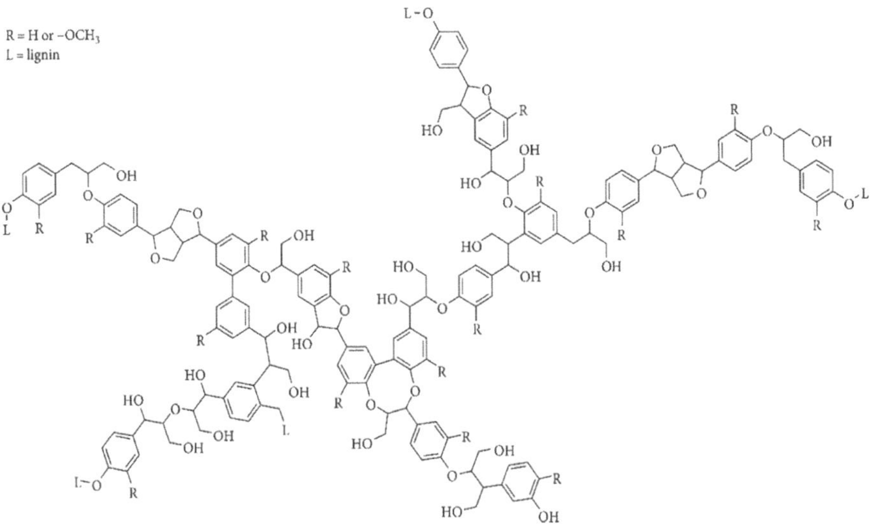

FIGURE 6.2   A typical chemical structure of lignin (Lu et al., 2017).

p-hydroxyphenyl (H) (Yuan et al., 2022). These units are connected by resinol β-β, β-1, and 4-O-5', β-5, β-O-4, 5-5', and α-O-4 linkages. A typical chemical structure of lignin is indicated in Figure 6.2. As a result of its highly erratic polymeric structure, lignin is immune to microbial attack and can stop water from damaging the polysaccharide-protein matrix of plant cells. Therefore, many research efforts have been published on the decoding of the lignin structure (Lai et al., 2022). It is known that biomass conversion technologies mainly have two pathways, including biological conversions (alcohol fermentation, anaerobic digestion, etc.) and thermochemical conversions (torrefaction, pyrolysis, gasification, etc.) (Kang et al., 2021). Without denying the importance of the biological routes, the higher content of lignin present in the biomass is often the cause of the inefficiency of the conversion due to the aforementioned biomass recalcitrance. Therefore, in recent years there have been more and more research efforts emerging on the conversion of lignin via thermochemical technologies (Margarida Martins et al., 2022). This chapter will focus on the gasification of lignin into hydrogen.

6.1.2   IMPORTANCE OF HYDROGEN PRODUCTION FROM LIGNIN

Hydrogen is widely accepted as the "energy vector" for constructing green and sustainable energy systems for the future. One of the most important benefits of hydrogen is that it only contains hydrogen elements so the combustion product is water, so when combusted as a fuel it will not release NOx, SOx, greenhouse gases, or any other particulate atmospheric pollutants. Another advantage is the flexibility of hydrogen as an energy carrier, e.g., hydrogen is the main component of syngas, which is an important industrial chemical source and could be converted into different hydrocarbons via the Fischer–Tropsch (FT) synthesis process (Rostrup-Nielsen, 2002).

As another example, hydrogen is the energy source of the hydrogen fuel cell, which has generated lots of research interest in recent years due to the need for sustainable and clean energy in the transportation sector and the boom in research into hydrogen-fueled vehicles (Camacho et al., 2022). Moreover, the energy density of hydrogen is approximately 120 MJ/kg, which is significantly higher than that of gasoline or diesel (~ 46 MJ/kg). Also, hydrogen is a critical industrial gas used in petroleum refining (hydrodesulfurization and hydrodeoxygenation) and ammonia production. Conventionally, the industrial production of hydrogen is mainly dependent on the steam reforming process of natural gas. Therefore, switching from fossil-based energy sources (i.e., natural gas) to biomass for hydrogen production can help to mitigate CO_2 and reduce the consumption of fossil resources. Although the safe storage of hydrogen is still a globally unresolved technical issue, the efforts of several countries to build hydrogen stations show their interest in building hydrogen-based energy systems (Kurtz et al., 2019).

Production of hydrogen from biomass can be achieved by many conversion routes. Specifically, the biological processes include water photolysis, photo-fermentation, dark-fermentation, and dark-photo co-fermentation, whereas the thermochemical processes include the pyrolysis, gasification, and catalytic-reforming of small organic molecules (Xu et al., 2022). Lignin can be viewed as a special biomass precursor for hydrogen production. However, due to the structural complexity of lignin than other biomass components (e.g., cellulose and hemicellulose), it is common to use lignin to evaluate the efficiency of a proposed process or a catalyst, with the assumption that technology that could effectively convert lignin into hydrogen could also be effective in converting other lignocellulosic materials. Compared to the colossal number of publications on biomass gasification, lignin gasification into hydrogen remains a significantly less explored research area. Therefore, this chapter will focus on the recent progress made in gasification technology, discussing different aspects including reactor technologies, process conditions optimizations, as well as catalyst designs.

6.2 REACTOR TECHNOLOGIES USED FOR LIGNIN GASIFICATION

Lignin gasification under similar operating conditions employed for woody biomass gasification is more challenging; this is basically due to lignin's physicochemical characteristics that make it recalcitrant to the gasification process. For instance, a relatively higher C/O molar ratio and an aromatic-based polymeric structure of lignin promote the production of tar during gasification. Tar is an undesired by-product of the gasification process. This product is constituted of a complex mixture of various condensable hydrocarbons so that its condensation at downstream processing units is detrimental. For example, it may result in fouling and blockage of piping and equipment. Additionally, it can cause filter and valve clogging while producing metallic corrosion. More importantly, the introduction of tar into the downstream catalytic reactors gives rise to the blockage, contamination, and/or early deactivation of the catalysts.

Various parameters affect tar production during lignin gasification. The most significant ones are the design of the gasification reactor (gasifier), bed geometry, and gasification conditions such as temperature, reaction time, gasifying agent, equivalence ratio (ER), and catalyst loading rate (Liakakou et al., 2019). Therefore, one should design a suitable gasification reactor with optimum operating conditions to suppress the tar formation and provide a sustainable process while achieving the desired syngas quality with a satisfactory carbon conversion. Considering lignin as a highly heterogeneous organic matter with widely varying properties, different gasification technologies and preparation steps have been studied in recent years. For instance, one or several pretreatment steps need to be carried out to feed the lignin into the gasifier efficiently because uniformity in size, moisture, and composition of a lignin feedstock is essential. Furthermore, lignin is the precursor of adhesives and may contribute to a failure in the feeding section of the gasifier. The pretreatment might involve lignin size reduction and sieving, partial moisture removal, and an increase in particle density through torrefaction, pelleting, and briquetting (Widjaya et al., 2018).

Gasifiers may be designed to operate either at atmospheric pressure or at a high pressure. In a high-pressure gasifier, the quality of the product gas, in terms of heat content is improved, the reactor size may be reduced, and the necessity to pressurize the gas before introducing it into a pipeline can be eliminated. On the other hand, introducing a feedstock into a high-pressure gasifier might cause some difficulties (Speight, 2014). The following factors distinguish gasifiers from one another:

I. The method of contact between the fuel and the gasifier —whether in the mode of countercurrent or cocurrent
II. The mode and the rate of heat transfer —whether in the type of allo-thermal or autothermal reactor, depending on the source of the heat supplied to the reactor (in the former, heat is supplied externally whereas in the latter the required heat is provided by exothermal reactions)
III. The residence time of the fuel within the reaction zone —in the order of hours in static gasifiers or rotary kilns, or within a minute in fluidized bed gasifiers

In terms of the design of lignin gasifiers, updraft fixed bed gasifiers (UFB), fluid bed gasifiers including bubbling fluidized bed gasifiers (BFG) and circulating fluidized bed gasifiers (CFB), entrained flow gasifiers (EFG), supercritical water gasifiers (SCWG) and indirect gasification have been employed in various research and development stages (Molino et al., 2016). A summary of lignin gasification reactors and associated operating conditions is presented in Table 6.1. A detailed review of different types of reactors is provided in the following sections.

6.2.1 Updraft Fixed Bed Gasifier

Updraft gasifier, (also called the countercurrent fixed bed gasifier or counterflow gasifier), is a moving-bed class of gasifiers. It is one of the most common and effective types of gasifiers with a relatively plain design. The feedstock with an approximate size of 3–25 mm diameter can be fed to the gasifier in such a way that it moves downward, while the gasifying streams flow upward at relatively low velocity and move through the voids of the moving bed (Speight, 2014). As the oxidants enter the bottom

TABLE 6.1

Reactors and Associated Operating Conditions for Gasification of Lignin and Lignin-Rich Residues

Catalyst	Gasifier Type	Feed Size, mm	Feed Rate	Oxidants	Steam Rate	EO (O_2)	Temperature (°C)	Pressure	Notes	Ref.
No catalyst	UFB	20–50 Two Samples A, B	27.9–28.8 kg/h	Air/ Steam	2.6–3.0 kg/h	0.17–0.18	687–776	Slightly above atmospheric	• RT: N/A • Pilot Scale • Autothermal • Calculated H_2 & CO yield: 0.442 & 0.422 Nm^3/kg_{feed} resp. (Sample A) • Highest Tar among other techniques • Sample A Tar formation: 80 g/Nm^3	Liakakou et al. (2019)
Olivine	BFB	2–10 Two Samples A, B	0.3–0.35 kg/h	Oxygen/ Steam	0–0.9 g/g daf of feedstock	0.13	750–900	Atmospheric	• RT: 90–120 min • Bench scale • Electrical Heated (Allo-thermal) • Silica Sand as fluidization bed • Calculated H_2 & CO yield: 0.198 & 0.252 Nm^3/kg_{feed} resp. (Sample A) • Sample A Tar Formation: 20 g/Nm^3	

(Continued)

TABLE 6.1 (*Continued*)

Reactors and Associated Operating Conditions for Gasification of Lignin and Lignin-Rich Residues

Catalyst	Gasifier Type	Feed Size, mm	Feed Rate	Oxidants	Steam Rate	EO (O_2)	Temperature (°C)	Pressure	Notes	Ref.
Olivine	Indirect	0.5–6 mm Two Samples A, B	1.8–2.9 kg/h	Steam	1.0–1.3 kg/h	-	780–870	Slightly above atmospheric	• RT: N/A • Lab Scale • Indirect Heating • Olivine as bed material • Calculated H_2 & CO yield: 0.086 & 0.139 Nm^3/kg_{feed} resp. (sample A) • Sample A Tar Formation: 30 g/Nm^3	
No catalyst	UFB	<250 mm	4 g/min	Steam	0–18 g/min	-	920–1,220	Slightly above atmospheric	• RT: 8 min • Lab Scale • Electrical Heating (Allo-thermal) • Hydrogen yield: 1.09 Nm^3/kg_{feed} at 1,020°C	Tian et al. (2017)
Sodium carbonate, Dolomite	EFG	<0.3 mm	4 g/min (cat: 0.2 g/min)	Air	–	0.25	1,000	N/A	• RT: N/A • Lab Scale • Electrical Heating (Allo-thermal) • H_2: 8% vol & CO: 13.5% vol & 46% C conversion for Na_2CO_3 • Na_2CO_3 increased CO yield and reduce H_2 yield	Yu et al. (2018)

(Continued)

TABLE 6.1 (*Continued*)

Reactors and Associated Operating Conditions for Gasification of Lignin and Lignin-Rich Residues

Catalyst	Gasifier Type	Feed Size, mm	Feed Rate	Oxidants	Steam Rate	EO (O_2)	Temperature (°C)	Pressure	Notes	Ref.
No catalyst	EFG	<0.3 mm	4 g/min	Oxygen	--	0.2–0.35	800–1,100	N/A	• RT: N/A • Lab Scale • Electrical Heating (Allo-thermal) • Tar yield decreased by raising the temperature (800°C to 1,100°C) • Tar yield of alkali lignin: 1.99 to 0.82 mg/g • Carbon conversion increased by raising the temperature (800°C–1100°C) • Carbon conversion efficiency from 59.85 to 83.2 wt%	Yu et al. (2014)
Lime, dolomite & olivine	BFB	Cylindrical Pellet Length: 5–23 D: 8	• Bench: 5 g daf/min • Pilot: N/A	Steam/oxygen	Steam/lignin ratio=0.7–1.2 Steam flowrate: 5 g/min	0–0.3	750–900	Atmospheric	• RT: N/A • Both bench Scale & Pilot Scale • Electrical Heating (Allo-thermal) • Catalysts demonstrated a positive impact on the cracking and steam reforming of tar and gaseous hydrocarbons • The presence of dolomite led to the highest H_2 production (35% v/v), lowest C_nH_m (2% v/v) and tar (4 g/Nm^3) contents • Residence time in the pilot-scale reactor was nearly 17% higher than a batch-scale reactor, and thus produces a lower tar formation	Ghosh et al. (2022)

(*Continued*)

TABLE 6.1 (*Continued*)

Reactors and Associated Operating Conditions for Gasification of Lignin and Lignin-Rich Residues

Catalyst	Gasifier Type	Feed Size, mm	Feed Rate	Oxidants	Steam Rate	EO (O_2)	Temperature (°C)	Pressure	Notes	Ref.
Olivine	CFB	6 mm cylindrical pellets	21.80–58.73 kg/h	$H_2O/CO_2/$ O_2	17.60–53.00 kg/h	12.08–26.49 kg/h	827.4–835.8	1.18–4.83 bar	• RT: N/A • Pilot Scale • Auto-thermal • Gasification agent: CO_2 flowrate: 13.80 & 38.61 kg/h, and H_2O/C close to 1 • H_2 & CO composition: 17.62 & 21.16 vol%, resp.	Szul et al. (2022)
No catalyst	Horizontal Fixed bed reactor	0.075–0.150 mm	12 g per each run	Air	–	Air EO = 0.1	500–900	0.95 ar	• RT: 20 min • Lab scale • Electrical Heating (Allo-thermal) • Maximum gas yield: 57% at 900 C • Tar concentration of lignin increases with temperature	Zhou et al. (2018)
No catalyst	EFG	< 0.75 mm	55 kg/h	70% O_2 & 30% N_2	–	ER (O2) = 0.45	1,200	1 bar	• RT: N/A • Pilot scale • Auto-thermal • Two samples • Sample A: H_2 & CO composition: 24 & 44.7 vol% resp. • Sample A H_2/CO=0.54 • Insignificant amount of tar	Öhrman et al. (2013)

(*Continued*)

TABLE 6.1 (*Continued*)

Reactors and Associated Operating Conditions for Gasification of Lignin and Lignin-Rich Residues

Catalyst	Gasifier Type	Feed Size, mm	Feed Rate	Oxidants	Steam Rate	EO (O_2)	Temperature (°C)	Pressure	Notes	Ref.
Olivine	UFB	~ 4 cm	14.7–18.0 kg/h	Steam/ Air Steam/O_2 O_2	0–8.5 kg/h	ER (O_2) = ER (H_2O) = 0.18–0.22 0–0.41	611–727	Slightly above atmospheric	• Lignin RT: 1.66–2.12 h • Pilot scale • Auto thermal • Tar production is significantly decreased below 50 g/kg$_{feedstock}$ using pretreated (torrefied) biomass	Cerone and Zimbardi (2021)

BFB: Bubbling fluidized bed gasifier; CFB: Circulating fluidized bed reactor, EFG: Entrained flow gasifier; EO: Equivalence ratio, N/A: Not available; UFB: Updraft fixed bed gasifier.

of the bed, they encounter the hot ash and chars moving downward, and due to the presence of a high concentration of oxidants (air or oxygen), complete combustion reactions begin to occur. This exothermic reaction consumes most of the oxidants and produces H_2O and CO_2 and a huge amount of heat. Subsequently, the released heat raises the temperature of the top parts of the bed and supplies the required heat for the endothermic thermal conversion reactions in the entire length of the reactor (Farzad et al., 2016).

The gasifying agents include steam, oxygen, and/or air. The integrated pretreatment, such as the drying of the feedstock is conducted once the fuel descends into the gasifier and is exposed to heat for the first time. In light of the mechanism of bed movement and the mode of contact between gas and moving bed, it is an essential requirement that the fuel must have high mechanical strength and must be non-caking to become a permeable bed. The ash is either removed dry or as a slag. Slagging gasifiers require a higher ratio of steam and oxygen to carbon to reach temperatures higher than the ash fusion temperature (Speight, 2014). There are many advantages associated with operating an updraft gasifier including ease of use, low-cost process, effective heat exchange in the reactor and energy efficiency, capability to handle feedstock with high moisture content (up to 50% w/w), and feedstock conversion efficiency. In addition, this gasifier is a proven technology and can be used to treat all types of biomass (Speight, 2014).

Despite a high residence time compared to other types of gasifiers, the exit gas temperature in an updraft gasifier is relatively low; therefore, there would be a high content of tar (10–20 wt%) in the effluent gas. As a result, this type of gasifier is not considered a suitable choice for most of the sophisticated applications (Liakakou et al., 2019, Farzad et al., 2016). Generally, UFB gasification is carried out on medium and small scales (10–15 MW) (Liakakou et al., 2019) and its throughput is relatively low (Speight, 2014).

6.2.2 ENTRAINED FLOW GASIFIER

In an EFG, a dry pulverized solid is gasified with oxygen in a cocurrent mode. During the gasification reaction, fine particles (i.e., less than 75 microns) are suspended in a dense cloud at significantly high pressures, which can be 20–70 bars. This type of gasifier operates at elevated temperatures (1,300–1,500°C). The residence time of feedstock in an entrained flow reactor is within several seconds and thus the operation should be carried out at high temperatures to allow a high carbon conversion. Therefore, such gasifiers work with oxygen instead of air to generate more heat and a higher temperature.

6.2.3 FLUIDIZED BED GASIFIER

The fluidized-bed gasification involves suspending the solid feedstock like lignin in an oxidative media including oxygen (or air) and steam. The feedstock to the gasifier needs to be crushed into small pieces and the gasifying agents are entered from a perforated plate installed at the bottom of the bed. A fluidized bed gasifier operates at a temperature ranging from 700 to 1,000°C to prevent agglomeration due to the presence of ash. Fluidized beds consist of an inert granular bed of materials (such as alumina or quartz sand) that is held in a condition of fluidization, which serves both as a heat carrier and a mixing medium, while the gasifying agents serve as the fluidizing medium. As soon as lignin particles encounter hot bed solids, the particles tend to reach the bed temperature where they are rapidly dried and pyrolyzed, resulting in the generation of char and gases. Upon encountering hot solids, pyrolysis products turn into non-condensable gases.

The most conventional types of fluidized bed gasifiers are the BFBs and CFBs (Lian et al., 2021, Speight, 2014). For the bed to begin bubbling, the gasification agent must be moving at a velocity exceeding the minimum fluidization velocity but the velocity should not be high enough to carry them outside the top of the vessel. BFB gasifiers are characterized by a high degree of solid mixing, which results in temperature uniformity. However, a BFB cannot provide a complete char conversion due to the back-mixing of solids. In addition, the presence of a high content of oxygen in the bubbling phase, which is due to the slow oxygen diffusion to the emulsion phase, gives rise to the combustion reaction and a drop in gasification efficiency. BFB gasifiers may be used for large to medium-scale applications. In a CFB gasifier, the effluent stream is sent to a cyclone where the solid bed materials effectively are separated from the gas stream and return to the first stage of the gasifier.

The flexibility in the use of a wide range of feedstocks, the ability to easily scale up for commercial utilization, the capability for the potential use of a catalyst along with an inert bed with the aim of reduction in tar formation, and enhanced heat and mass transfer make the fluidized bed as a most promising technology in lignin gasification. The absorption of the sulfur- and chlorine-containing components present in the fuel by inert bed materials eliminates the fouling hazard and reduces the maintenance costs. It is worth noting that the rate of feedstock inlet capacity for a fluidized bed is higher than for the updraft flow gasifier, but still not as high as for the EFG (Molino et al., 2016, Speight, 2014, Marcantonio et al., 2019). In addition, a fluidized bed gasifier requires a shorter residence time for feedstock particles than a moving bed gasifier, yet longer than an EFG. In addition to their ability to operate at a range of throughputs, fluidized bed systems also have the advantage of being able to operate with large-diameter units (Speight, 2014). In summary, the comparison of different types of reactors is presented in Table 6.2.

TABLE 6.2

Advantages of Different Types of Conventional Gasifiers

Parameter	Fixed Bed Gasifier	Fluidized Bed Gasifier	Entrained Flow Gasifier
Feedstock flexibility	No	Yes	Yes
Bed temperature uniformity	No	Yes	Yes
Solid-gas contact	Good	Good	Good
Tar formation	High	Low	Low
Residence time	Hours	Minutes	Seconds
Scale-up capability	Limited	Yes	Yes

6.3 NON-CATALYTIC GASIFICATION OF LIGNIN FOR HYDROGEN PRODUCTION

Without the presence of any catalyst, effectively decomposing the lignin into smaller molecules is a tough task. Thanks to the relatively high temperature of the gasification processes, the research into non-catalytic lignin gasification for hydrogen production is considered a potentially viable route and is still in progress (Table 6.3). In 2014, Yu et al. conducted a study of biomass precursors' gasification using cellulose, hemicellulose, and lignin gasification (Yu et al., 2014). Aside from the gas phase, the characteristics of tar obtained from different feedstocks were investigated in this work. The results indicated that the tars produced from lignin had a higher tar yield, and the tar was more chemically & thermally stable and toxic. The researchers also suggested that for better gasification efficiency, it is important to remove the lignin-derived tar for tar control. Another study published in 2019 compared the performances of different gasification technologies on the gasification of steam explosion and enzymatic hydrolysis lignin (Liakakou et al., 2019). The study shows that lignin

TABLE 6.3

Papers Published in Non-Catalytic Gasification of Lignin for Hydrogen Production

Lignin Type	Reactor	Temperature	Gasification Agent, Flow Rate	Residence Time	Maximum Hydrogen Yield Obtained	Reference
Alkali lignin	Entrained-flow gasifier	800°C–1,100°C	O_2	1.5–2.2 s	n.s.	Yu et al. (2014)
n.s.	Batch reactor	350°C	SCW	22 min	5.33 mmol/g	Yoshida and Matsumura (2001)
Organosolv lignin	Quartz reactor	365°C–725°C	SCW	2.5–75 min	6.5 mmol/g	Resende et al. (2008)
Alkali lignin	Batch reactor	399°C–651°C	SCW	50 min	1.60 mmol/g	Kang et al. (2015)
Simulated lignin molecule	Simulated cubic box reactor	1,927°C–2,527°C	SCW	3,000 ps	~ 14 mmol/g	Li et al. (2019)
Soda lignin	Autoclave	750°C	SCW	5–50 min	n.s.	Cao et al. (2020)
Simulated guaiacyl dimer lignin molecule	Simulated reactor	1,727°C–5,727°C	SCW	2,000 ps	n.s.	Liu et al. (2020)

n.s. represents not specified.

could be efficiently gasified with all three technologies including air gasification in an updraft gasifier, oxygen/steam gasification in a BFB gasifier, and steam gasification in an indirect gasifier, and the composition of the lignin feedstock will significantly impact the composition of the produced gas.

It is well known that the harvested biomass usually contains a considerable amount of moisture and drying is one of the most cost-intensive steps of biorefining. To avoid the cost of biomass drying and to enhance the lignin decomposition, SCWG is a relatively new technology studied to gasify biomass in a supercritical water (SCW) medium. SCW is defined as the conditions of water are above the critical point of water (i.e., 374°C, 22 MPa), which possesses unique properties and could serve as an effective reaction medium for biomass conversion. Other important advantages of SCWG include: (i) the process occurs in SWG so is CO_2 neutral, and during the life cycle the growing biomass utilizes the same amount of CO_2 as released during the reaction; (ii) the temperature for complete gasification reaction is relatively lower than the conventional gasification processes at approximately 600°C, and the process suffers less from tar accumulation; (iii) with an eased separation of H_2 and CO_2 as CO_2 has a high solubility in pressurized water under room temperature; (iv) the effluent can be potentially used as fertilizer (Chen et al., 2010). Early in 2001, Yoshida and Matsumura (2001) gasified different mixtures of cellulose, xylan, and lignin in SCW. It was found that the increase in the content of lignin impacted the gas phase product yield, whereas cellulose and xylan tend to function as hydrogen donors to facilitate lignin decomposition. The first non-catalytic SCWG study of lignin was reported by Resende et al. (2008), in the complete absence of metal catalysts, where the reaction was conducted in quartz reactors to avoid any catalytic effect coming from the reactor material, which are often metals. The main conclusion is that high temperature and low biomass loading will favor the production of hydrogen from lignin. Later, the same research group published results on the construction and validation of kinetic models for the non-catalytic SCWG of cellulose and lignin (Resende and Savage, 2010). In this study, it is found that the critical reaction and stage for the formation of hydrogen is steam reforming at a short reaction time, while with the increase in reaction time, the water–gas shift reaction (WGS) will gradually become the dominant reaction for the production of hydrogen.

A later study of non-catalytic lignin SCWG was published by Kang et al. (2015), where the central composite design (CCD) methodology was first used. This study identified and discussed the interaction effect between reaction temperature and the water-to-biomass ratio, i.e., at temperatures above 600°C, an excessive amount of water present in the reactor can significantly reduce the hydrogen yield. Cao et al. (2020) found the gasification of soda lignin in the presence of different plastic wastes. The results suggest that adding plastic waste can enhance the gasification efficiency when compared to the gasification of pure lignin, and the beneficial effect of adding different plastics follows the order of PE (polyethylene) > PC (polycarbonate) ≈ PP (polypropylene) > ABS (acrylonitrile butadiene styrene). Li et al used the simulation method to study the molecular dynamics during the lignin gasification in SCW at very high temperatures 1,927–2,527°C, which is difficult to achieve using existing reactors (Li et al., 2019). The simulations help to understand the reaction mechanism of lignin under extreme conditions like this, and the researchers found that if the target product is H_2 then raising the reaction temperature will be beneficial since the impact of temperature on the generation of H ion is much more significant than the breakage of C-C bonds. Another simulation work is published by Liu et al to discuss the cleavage mechanism of lignin SCWG (Liu et al., 2020). The simulation results reveal that the H_2 and CO are produced via the combination of free radicals and the reaction between water and the molecular fragment produced by the decomposed lignin.

At the time of this review, the number of works dedicated to the non-catalytic gasification of lignin is still quite low. Out of the existing studies, it is clear that without catalysts the highly efficient lignin to hydrogen conversion with a reasonable cost is very difficult to achieve. Clearly, it can be concluded that the knowledge of the gasification chemistry of certain types of lignin is lacking; however, such knowledge could be valuable reference data for designing an industrial conversion process. Moreover, more research effort should be given to elaborate on the reaction mechanism of lignin during non-catalytic gasification, which will help the design of a better catalytic gasification process. Here, one point worth mentioning is that even when detected in low concentrations, the extracted lignin samples might contain certain contents of alkaline earth metals, which can act as catalysts during the gasification. Therefore, without the addition of external catalysts, the internal catalysis reaction might happen and would be an interesting topic to investigate. In terms of simulation studies, many recent works have been published suggesting that simulation is a powerful tool to decode the non-catalytic gasification mechanism of lignin, even at very high temperatures that existing reactors cannot stand. However, no matter how high the temperature is, the gasification rate of lignin continues to increase, suggesting that to build a feasible process, a balance between energy consumption and hydrogen production rate should be made (Liu et al., 2020). The catalysis studies on lignin gasification will be discussed in the following section.

6.4 CATALYTIC GASIFICATION OF LIGNIN FOR HYDROGEN PRODUCTION

In general, a catalyst is not need to be added to gasification for hydrogen production; however, adding a catalyst can: (i) increase the reaction rate, (ii) reduce the activation energy, (iii) reduce the required reaction temperature and residence time, (iv) improve carbon conversion, and (v) hinder tar formation (Alptekin and Celiktas, 2022). In the gasification, the catalyst can be: (i) directly added to the feedstock by impregnation, (ii) mixed with feedstock before loading to the gasifier, and (iii) utilized as the bed material for a bubbling/circulating fluidized bed gasifier (Narnaware and Panwar, 2021). Zhang et al. (2023) reported a 91.05% tar-cracking efficiency in biomass gasification over Fe-Mo/biochar, and Slatter et al. (2022) applied CaO as a tar-cracking catalyst in biomass gasification for syngas production.

To date, few investigations have been carried out to catalytically gasify lignin for producing hydrogen, and the relevant studies are summarized in Table 6.4. Alkali and alkaline earth metals (AAEM) are one type of catalyst that can be applied in the gasification of lignin. Yu et al. (2018) compared the catalytic performance of Na_2CO_3 and dolomite in the gasification of lignin, cellulose, and hemicellulose in an EFG. It was

TABLE 6.4

The Summary of the Most Recent Investigations Regarding Catalytic Supercritical Water Gasification of Lignin for Hydrogen Production

Lignin Type	Catalysts	Catalyst Synthesis Methods	Main Conclusion	Reference
n.s	CuO-ZnO	Sol-gel; co-precipitation; deposition precipitation	• Sol-gel was the best catalyst providing the smallest crystallite size, largest surface area, and high metal dispersion. • During sol-gel synthesis, the addition of ethanol and PEG helped improve the catalytic performance.	Cao et al. (2022)
Dealkaline lignin	Ni/zeolites	Impregnation	• The order for catalytic activity: Ni/4A (Na-A type) > Ni/3A (K-A type) > Ni/5A (Ca-A type) > Ni/13X (Na-X type) > Ni/ZSM-5 • Using Ni/4A resulted in an increase in the hydrogen selectivity (~27% → ~43%) and hydrogen yield (3.14 mol/kg → 5.63 mol/kg).	Yu et al. (2022)
Soda lignin and plastic waste	NaOH		• NaOH promoted H_2 production and deterred CH_4 and C_2H_6 formation by catalyzing steam reforming reactions. • The positive synergistic interaction between soda lignin and plastics was observed in gasification efficiency.	Cao et al. (2020)
n.s	Na_2CO_3; K_2CO_3; Ni-Ce/Al_2O_3; Ni/Al_2O_3; Ru/Al_2O_3	Impregnation and co-precipitation	• Using K_2CO_3 produced the highest hydrogen yield. • Hydrogen production was independent of the specific surface area and metal dispersion was not	Kang et al. (2016a)
n.s	Ni-Co/Al-Mg	Impregnation and co-precipitation	• 2.6Ni-5.2Co/2.6Mg-Al prepared by co-precipitation was the best catalyst for hydrogen production with the highest hydrogen yield of 2.36 mml/g. • A better resistance to coke formation was found in this catalyst.	Kang et al. (2017)
Alkaline lignin	Ru/C; Pt/C; Pd/C; Ni/Al_2O_3-SiO_2		• Ru/C led to the biggest increase in the hydrogen yield from 2.5 to 12 mmol/g. • The presence of sulfur improved the stability and durability of the catalyst in the SCWG.	Guan et al. (2014)

(Continued)

TABLE 6.4 (*Continued*)
The Summary of the Most Recent Investigations Regarding Catalytic Supercritical Water Gasification of Lignin for Hydrogen Production

Lignin Type	Catalysts	Catalyst Synthesis Methods	Main Conclusion	Reference
Dealkaline lignin	CeO_2-ZrO_2	Co-precipitation	• CeO_2-ZrO_2 showed a better gasification performance and hydrogen yield at a lower temperature ($500°C >$ $600°C$). • Adding ZrO_2 increased the dispersion of CeO_2 and hydrogen adsorption on the catalyst surface.	Cao et al. (2021b)
Dealkaline lignin	CuO-ZnO; Fe_2O_3-Cr_2O_3	Co-precipitation	• CuO-ZnO exhibited a better gasification performance than Fe_2O_3-Cr_2O_3 in terms of hydrogen yield.	Cao et al. (2021a)

*n.s. represents not specified.

observed that dolomite demonstrated a positive impact on lignin gasification but a negative influence of Na_2CO_3 was found in the gasification of lignin with respect to carbon conversion, cold gasification efficiency, and lower heating value (LHV). Soomro et al. (2018) found that the hydrogen vol.% increased when adding CaO to the steam gasification of lignin, and a maximum vol.% of hydrogen reached 85.40 vol.% at a weight ratio of CaO/lignin of 1.5. While the trend for hydrogen yield with CaO loading was different. The reason could be due to the formation of $Ca(OH)_2$ through the reaction between CaO and CO_2, which leads to an increase in Ca atoms, and one Ca atom is capable of binding up to seven hydrogen molecules in the molecular form (Nguyen et al., 2009). In short, the literature summarizes that the presence of Na and Ca facilitates H_2 production from lignin gasification by resulting in higher gas yield gasification reactivity and removing tar (Yu et al., 2021). In another study, the authors doped CaO with FeO, ZrO_2, and CeO_2 to enhance its mechanical and cyclic stability in lignin gasification, especially caused by tar formation (Soomro et al., 2020). The catalytic activity of the synthesized catalysts for lignin conversion and hydrogen production was as follows: Zr-Fe/CaO > Fe/CaO > Ce-Fe/CaO, and Zr-Fe/CaO also produced high stability in the catalyst recyclability test. In addition, a Ni-based catalyst has been explored for several industrial applications due to its low cost, great stability and activity balance, and ability to be easily modified with different supports and promoters (Galadima et al., 2022). For instance, Ru-promoted Ni/Al_2O_3 was synthesized and its catalytic performance in lignin gasification was evaluated in a fluidized bed reactor (Calzada Hernandez et al., 2020). The authors reported that the addition of Ru was effective for deterring coke formation by promoting char gasification ($C + H_2O \rightarrow CO + H_2$) and Boudouard reactions, meanwhile, an increase in the hydrogen yield was observed.

In recent years, SCWG has attracted attention as an alternative approach to conventional gasification for hydrogen production. Unlike gasification, the use of a catalyst in SCWG is more common like when gasifying lignin in SCW. Another main difference between conventional gasification is related to the use of water in SCWG, during which water not only acts as a reaction medium but also as a hydrogen donor (Hu et al., 2020). A Ni-based catalyst has been widely applied in the SCWG of lignin for producing hydrogen, such as Yu et al. (2022) using Ni supported over different zeolites, and Kang et al. (2016b) synthesizing 29 Ni-based catalysts with different supports (Al_2O_3, activated carbon, TiO_2, ZrO_2, and MgO) and promoters (Co, Cu, and Ce). Fe is another transition metal that has been utilized for hydrogen production from SCWG of lignin. Zhang et al. (2021) simulated the catalytic performance of various Fe-based catalysts (i.e., Fe, FeO, Fe_3O_4, and Fe_2O_3) in SCW, and it was observed that the zero valent Fe catalyst produced a significantly higher yield of H_2 yield than other iron oxide catalysts with different valence, and such differences in valence also led to a difference in carbon gasification efficiency. Besides, noble metals like Pt, Rh, and Ru have been tested in SCW for producing hydrogen lignin. Wang et al. (2015) investigated SCWG of alkali lignin over Ru/C for enhancing hydrogen yield, and it was found that the optimized catalyst dosage was 0.5 wt.%/wt.% (catalyst/feedstock). In addition to different metals and supports that have been investigated in SCWG of lignin for producing hydrogen, the effect of the catalyst preparation method has also been evaluated. Cao et al. (2022) synthesized a bimetallic catalyst, CuO-ZnO, using sol-gel, co-precipitation, and deposition methods, among which sol-gel was found to be the best synthesis method in terms of the highest catalytic performance and gasification efficiency. In the SCWG of lignin, the homogenous catalyst is another major category of catalyst that has been broadly employed to improve gasification efficiency and hydrogen production, and the investigated homogenous catalysts include Na_2CO_3, NaOH, KOH, K_2CO_3, and $Ca(OH)_2$ and most of them have demonstrated promotion of WGS reaction to form H_2. The comparative investigation between homogenous and heterogeneous catalysts in the SCWG of lignin for producing hydrogen was also carried out, and the authors reported that the use of K_2CO_3 resulted in the highest hydrogen yield from lignin (2.86 mmol/g) when compared to other catalysts including Ni/Al_2O_3 of 0.43 mmol/g, $Ni-Ce/Al_2O_3$ of 0.37 mmol/g, and Ru/Al_2O_3 of 1.13 mmol/g. To further improve the hydrogen production, co-solvents including tetrahydrofuran (THF), N-methyl-2-pyrrolidone (NMP), acetone, methanol, xylene, and tetralin have been added to the SCWG of lignin, and it was found that the addition of a co-solvent was able to produce hydrogen at a relatively lower temperature, i.e., 350°C (Seçer et al., 2023).

In short, the catalyst plays an important role in conventional gasification (without water used as the reaction medium) and SCWG of lignin for producing hydrogen by promoting carbon conversion and hydrogen yield; however, the catalyst's lifetime is limited due to the side reactions and/or the structure change occurring to the catalyst. Catalyst deactivation might be caused by the CH_4 cracking and the Boudouard reaction (equation (6.1)), leading to coke deposition on the active site of the catalyst, and hence catalyst regeneration or replacement is a need after a certain period (Alptekin and Celiktas, 2022).

$$C + CO_2 \leftrightarrow 2CO \tag{6.1}$$

As a result, there is a great need to develop approaches to prevent coke deposition and extend the catalyst's lifetime in hydrogen production from either conventional gasification or SCWG of lignin. Some relevant reviewer papers have been published regarding catalyst development, the mechanism of catalyst deactivation, and catalyst regeneration, like Wu et al. (2022), Ren et al. (2019), Lee et al. 2021), and Abdpour and Santos (2021).

6.5 CO-GASIFICATION OF LIGNIN WITH OTHER WASTES

Apart from independent waste-to-energy processes such as pyrolysis, liquefaction, firing, and gasification, co-processing technologies are recently gaining attention. Co-processing technologies including co-pyrolysis, co-liquefaction, co-firing, and co-gasification have several advantages over their counterparts. Co-processing can be defined as a technology or process that involves two or more waste raw materials to recover energy and material more efficiently and economically. Some advantages of co-processing waste-to-energy technologies are as follows (Kamińska-Pietrzak and Smoliński, 2013, Nanda et al., 2022):

    i. conversion of complex and heterogeneous feedstocks
   ii. synergistic interactions between feedstocks leading to efficient conversion
  iii. ability to accommodate mixed feedstocks within the existing infrastructure without intensive retrofitting
  iv. ability to study the interactive effects of feedstocks and process parameters on the process efficiency and product yields
   v. improvement in product yields and quality
  vi. reduction in greenhouse gas emissions
 vii. relatively lower energy requirement
viii. reduction in waste generation
  ix. ability to recycle wastes within the integrated process
   x. lower operating costs due to maximized feedstock, low energy requirements and utilization, and reduced wastes

Based on these advantages, several municipalities, legislations, and resource recovery facilities encourage the co-processing of complex feedstocks (e.g., coal, plastics, sewage sludge, and municipal solid wastes) with lignocellulosic biomass to produce clean alternative fuels. There are several studies available on the co-gasification of lignin with other waste materials leading to cleaner products with improved fuel properties. During co-gasification, the mineral composition of one feedstock can pose synergistic impacts on the conversion of other feedstocks. For example, the kinetics of co-gasification of recalcitrant feedstocks such as coal and plastics is improved by the ash or mineral matter of lignocellulosic biomass, which primarily contains AAEM contents (Mallick et al., 2017). The AAEMs can have catalytic effects on co-gasification by reducing the activation energy and boosting the reaction kinetics. However, the presence of alumina and silica in coal and sewage sludge during co-gasification can have an antagonistic effect by impeding the catalytic activities of alkali metals by generating aluminosilicate complexes during co-gasification. This largely impacts the process efficiency and reduces gas yields. Alkali metals can also lose their catalytic effects during co-gasification at higher process temperatures. On the other hand, the silica present in coal can also act as an in-situ catalyst leading to the reforming and cracking of tar produced during co-gasification (Mallick et al., 2017).

As mentioned earlier, lignin is a phenyl-propane polymer (Nanda et al., 2014). Hence, each monomer or precursor of lignin behaves differently when co-gasified with other feedstocks. Su et al. (2015) studied the interactive effects of lignin and proteins to improve SCW gasification of sewage sludge. The mixtures of model compounds of lignin i.e., phenol and alanine were co-gasified at 400°C under 22–26 MPa pressure for 30 min. The synergistic effects of the phenol and alanine (at 60 wt%) led to a twofold rise in H_2 yield than its average value. The levels of total organic carbon and nitrogenous compounds in the aqueous phase also reduced as a result of co-gasification. The authors suggested that phenols favored the decomposition of alanine leading to a rise in gas and oil yields, whereas the condensation of phenol was promoted by alanine.

The study by Seçer et al. (2018) compared co-gasification efficiencies of lignocellulosic biomass (sorghum and kenaf), biomass hydrolysate, and lignite coal to produce H_2. Biomass hydrolysate was prone to hydrothermal cracking leading to higher H_2-rich gas products compared to lignocellulosic biomass. The temperature was a governing parameter leading to high gas yields. While high temperature increased gas production from co-gasification, high feed flow rate led to low gas yields due to short reaction time and less interaction between the co-feedstocks. The synergistic effects of coal were more pronounced during its co-gasification with biomass hydrolysate compared to that of coal and biomass. The application of CaO was found to reduce the yields of CO_2 in the gas products while enhancing H_2 levels. Cao et al. (2020) co-gasified the black liquor-obtained soda lignin and waste plastics in SCW. The authors studied the influence of the lignin/plastic ratio (20:80, 40:60, 50:50, 60:40, and 80:20), temperature (500°C–750°C), reaction time (5–50 min), feed concentration (50–25 wt.%), and plastic types were studied. Different types of plastics that were co-gasified with soda lignin included polyethylene, polypropylene, polycarbonate, and acrylonitrile butadiene styrene. At higher temperatures and reaction times along with low feed concentration and the plastic/lignin ratio of 50:50, synergistic effects in co-gasification were noticed. A five-fold increase in H_2 yield of 57 mol/kg was achieved with an increase in temperature from 500°C to 750°C. When the feed concentration was reduced to 5 wt%, a maximum H_2 yield of 63 mol/kg was obtained at 700°C. The alkaline nature of soda lignin also enhanced gasification efficiency by depolymerizing plastics resulting in the yields of H_2, CH_4, and C_2H_6. Polyethylene showed maximum co-gasification efficiency with lignin followed by polycarbonate, polypropylene, and acrylonitrile butadiene styrene.

Wang et al. (2022a) implemented reactive molecular dynamics simulations to model the co-gasification of waste plastics (i.e., polyethylene) with β-O-4 lignin in SCW. The synergistic interactions between lignin and polyethylene resulted in their breakdown in SCW to generate long carbon chain (C_5-C_{19}) compounds, which further degraded into short carbon chain molecular fragments. Co-gasification temperature of 1,725°C led to the depolymerization of lignin without much effect on polyethylene. On the other hand, higher temperatures of 2,725°C–4,725°C resulted in the breakdown of polyethylene along with lignin to produce short carbon chain compounds indicating enhanced hydrothermal cracking. In addition, at higher co-gasification temperatures, gas production increased in the order of H_2 followed by CO, CH_4, and CO_2. In a subsequent study, Wang et al. (2022b) compared two supercritical fluids i.e., SW and supercritical carbon dioxide ($SCCO_2$) for co-gasification of waste lignin and plastics (i.e., polypropylene and polyethylene). The co-gasification experiments were designed using reactive force field molecular dynamics simulation modeling to study the decomposition and depolymerization reactions. The studies indicated that the plastics played a synergistic role in increasing H_2 production from co-gasification in both SCW and $SCCO_2$ systems. However, the synergistic effects of plastic addition on lignin decomposition were more evident in SCW co-gasification compared to that of $SCCO_2$ leading to higher yields of H_2 and CH_4. Plastic addition also resulted in the occurrence of more unsaturated short carbon chain compounds and CO in the $SCCO_2$ system. The reaction time for lignin decomposition into monomers and cleavage of aromatic rings increased at a lower blending ratio of plastic/lignin.

6.6 CONCLUSION

As the largest natural reservoir of aromatics, lignin is an important natural resource. With a complicated structure, the directional conversion of lignin into value-added fuels and chemicals is still in its infant stage. Gasification technology provides a unique option to convert lignocellulosic biomass into gaseous fuels. Based on the current scenario, it is critical to understand the conversion chemistry of lignin during gasification. Since converting lignin into hydrogen requires almost "complete" cleavage of lignin, it becomes particularly challenging using conventional technologies and without catalysts. As suggested by this review, the recommendations of future research directions are: (i) continue to advance in the non-catalytic and catalytic gasification of lignin, from different sources (attention should be given to different lignin sources other than only Kraft lignin), to understand the hydrogen product mechanism, (ii) invent innovative catalysts that could be used to reduce the high reaction temperature required by lignin gasification, be resistant to coke formation, and can provide high selectivity to hydrogen, (iii) continue the improvement and find novel reactors that could handle lignin and lignin-contained feedstock at high temperatures, and (iv) closely investigate the synergistic effects between lignin and other possible feedstocks (preferably waste material), which might lead to a co-gasification route that converts the mixture containing lignin into hydrogen with promoted efficiency. Despite all the challenges, it is foreseeable that because of the commercialized industrial production of lignin products, the need for green conversion of lignin and lignin-containing residues will increase, which will turn into more fundamental research on this topic.

REFERENCES

Abdpour, S. & Santos, R. M. 2021. Recent advances in heterogeneous catalysis for supercritical water oxidation/gasification processes: Insight into catalyst development. *Process Safety and Environmental Protection*, 149, 169–184.

Alptekin, F. M. & Celiktas, M. S. 2022. Review on catalytic biomass gasification for hydrogen production as a sustainable energy form and social, technological, economic, environmental, and political analysis of catalysts. *ACS Omega*, 7, 24918–24941.

Aro, T. & Fatehi, P. 2017. Production and application of lignosulfonates and sulfonated lignin. *ChemSusChem*, 10, 1861–1877.

Calzada Hernandez, A. R., Gibran González Castañeda, D., Sánchez Enriquez, A., De Lasa, H. & Serrano Rosales, B. 2020. Ru-promoted Ni/Al$_2$O$_3$ fluidized catalyst for biomass gasification. *Catalysts*, 10, 316. doi:10.3390/catal10030316.

Camacho, M. D. L. N., Jurburg, D. & Tanco, M. 2022. Hydrogen fuel cell heavy-duty trucks: Review of main research topics. *International Journal of Hydrogen Energy*, 47, 29505–29525.

Cao, C., Bian, C., Wang, G., Bai, B., Xie, Y. & Jin, H. 2020. Co-gasification of plastic wastes and soda lignin in supercritical water. *Chemical Engineering Journal*, 388, 124277.

Cao, C., Xie, Y., Chen, Y., Lu, J., Shi, J., Jin, H., Wang, S. & Zhang, L. 2021a. Hydrogen production from supercritical water gasification of lignin and cellulose with coprecipitated CuO-ZnO and Fe2O3-Cr2O3. *Industrial & Engineering Chemistry Research*, 60, 7033–7042.

Cao, C., Xie, Y., Li, L., Wei, W., Jin, H., Wang, S. & Li, W. 2021b. Supercritical water gasification of lignin and cellulose catalyzed with co-precipitated CeO2-ZrO2. *Energy & Fuels*, 35, 6030–6039.

Cao, C., Yu, L., Xie, Y., Wei, W. & Jin, H. 2022. Hydrogen production by supercritical water gasification of lignin over CuO-ZnO catalyst synthesized with different methods. *International Journal of Hydrogen Energy*, 47, 8716–8728.

Cerone, N. & Zimbardi, F. 2021. Effects of oxygen and steam equivalence ratios on updraft gasification of biomass. *Energies*, 14, 2675.

Chen, J., Lu, Y., Guo, L., Zhang, X. & Xiao, P. 2010. Hydrogen production by biomass gasification in supercritical water using concentrated solar energy: System development and proof of concept. *International Journal of Hydrogen Energy*, 35, 7134–7141.

Dessbesell, L., Paleologou, M., Leitch, M., Pulkki, R. & Xu, C. 2020. Global lignin supply overview and kraft lignin potential as an alternative for petroleum-based polymers. *Renewable and Sustainable Energy Reviews*, 123, 109768.

Eraghi Kazzaz, A. & Fatehi, P. 2020. Technical lignin and its potential modification routes: A mini-review. *Industrial Crops and Products*, 154, 112732.

Farzad, S., Mandegari, M. A. & Görgens, J. F. 2016. A critical review on biomass gasification, co-gasification, and their environmental assessments. *Biofuel Research Journal*, 3, 483–495.

Galadima, A., Masudi, A. & Muraza, O. 2022. Catalyst development for tar reduction in biomass gasification: Recent progress and the way forward. *Journal of Environmental Management*, 305, 114274.

Ghosh, S., Timsina, R. & Moldestad, B. 2022. Study of gasification behavior for a biorefinery lignin waste in a fluidized bed gasification reactor. *Proceedings of the 63rd International Conference of Scandinavian Simulation Society,* SIMS 2022, Trondheim, Norway.

Guan, Q., Mao, T., Zhang, Q., Miao, R., Ning, P., Gu, J., Tian, S., Chen, Q. & Chai, X.-S. 2014. Catalytic gasification of lignin with Ni/Al2O3-SiO2 in sub/supercritical water. *The Journal of Supercritical Fluids*, 95, 413–421.

Hu, Y., Gong, M., Xing, X., Wang, H., Zeng, Y. & Xu, C. C. 2020. Supercritical water gasification of biomass model compounds: A review. *Renewable and Sustainable Energy Reviews*, 118, 109529.

Kamińska-Pietrzak, N. & Smoliński, A. 2013. Selected environmental aspects of gasification and co-gasification of various types of waste. *Journal of Sustainable Mining*, 12, 6–13.

Kang, K., Azargohar, R., Dalai, A. K. & Wang, H. 2015. Noncatalytic gasification of lignin in supercritical water using a batch reactor for hydrogen production: An experimental and modeling study. *Energy & Fuels*, 29, 1776–1784.

Kang, K., Azargohar, R., Dalai, A. K. & Wang, H. 2016a. Hydrogen production from lignin, cellulose and waste biomass via supercritical water gasification: Catalyst activity and process optimization study. *Energy Conversion and Management*, 117, 528–537.

Kang, K., Azargohar, R., Dalai, A. K. & Wang, H. 2016b. Systematic screening and modification of Ni based catalysts for hydrogen generation from supercritical water gasification of lignin. *Chemical Engineering Journal*, 283, 1019–1032.

Kang, K., Azargohar, R., Dalai, A. K. & Wang, H. 2017. Hydrogen generation via supercritical water gasification of lignin using Ni-Co/Mg-Al catalysts. *International Journal of Energy Research*, 41, 1835–1846.

Kang, K., Klinghoffer, N. B., Elghamrawy, I. & Berruti, F. 2021. Thermochemical conversion of agroforestry biomass and solid waste using decentralized and mobile systems for renewable energy and products. *Renewable and Sustainable Energy Reviews*, 149, 111372.

Kurtz, J., Sprik, S. & Bradley, T. H. 2019. Review of transportation hydrogen infrastructure performance and reliability. *International Journal of Hydrogen Energy*, 44, 12010–12023.

Lai, C., Yang, C., Jia, Y., Xu, X., Wang, K. & Yong, Q. 2022. Lignin fractionation to realize the comprehensive elucidation of structure-inhibition relationship of lignins in enzymatic hydrolysis. *Bioresource Technology*, 355, 127255.

Lee, C. S., Conradie, A. V. & Lester, E. 2021. Review of supercritical water gasification with lignocellulosic real biomass as the feedstocks: Process parameters, biomass composition, catalyst development, reactor design and its challenges. *Chemical Engineering Journal*, 415, 128837.

Li, H., Xu, B., Jin, H., Luo, K. & Fan, J. 2019. Molecular dynamics investigation on the lignin gasification in supercritical water. *Fuel Processing Technology*, 192, 203–209.

Liakakou, E. T., Vreugdenhil, B. J., Cerone, N., Zimbardi, F., Pinto, F., André, R., Marques, P., Mata, R. & Girio, F. 2019. Gasification of lignin-rich residues for the production of biofuels via syngas fermentation: Comparison of gasification technologies. *Fuel*, 251, 580–592.

Lian, Z., Wang, Y., Zhang, X., Yusuf, A., Famiyeh, L., Murindababisha, D., Jin, H., Liu, Y., He, J., Wang, Y., Yang, G. & Sun, Y. 2021. Hydrogen production by fluidized bed reactors: A quantitative perspective using the supervised machine learning approach. *J*, 4, 266–287. doi:10.3390/j4030022

Liu, X., Wang, T., Chu, J., He, M., Li, Q. & Zhang, Y. 2020. Understanding lignin gasification in supercritical water using reactive molecular dynamics simulations. *Renewable Energy*, 161, 858–866.

Lu, Y., Lu, Y.-C., Hu, H.-Q., Xie, F.-J., Wei, X.-Y. & Fan, X. 2017. Structural characterization of lignin and its degradation products with spectroscopic methods. *Journal of Spectroscopy*, 2017, 1–15.

Mallick, D., Mahanta, P. & Moholkar, V. S. 2017. Co-gasification of coal and biomass blends: Chemistry and engineering. *Fuel*, 204, 106–128.

Marcantonio, V., DE Falco, M., Capocelli, M., Bocci, E., Colantoni, A. & Villarini, M. 2019. Process analysis of hydrogen production from biomass gasification in fluidized bed reactor with different separation systems. *International Journal of Hydrogen Energy*, 44, 10350–10360.

Margarida Martins, M., Carvalheiro, F. & Gírio, F. 2022. An overview of lignin pathways of valorization: From isolation to refining and conversion into value-added products. *Biomass Conversion and Biorefinery*, 14, 3183–3207.

Molino, A., Chianese, S. & Musmarra, D. 2016. Biomass gasification technology: The state of the art overview. *Journal of Energy Chemistry*, 25, 10–25.

Nanda, S., Mohammad, J., Reddy, S. N., Kozinski, J. A. & Dalai, A. K. 2014. Pathways of lignocellulosic biomass conversion to renewable fuels. *Biomass Conversion and Biorefinery*, 4, 157–191.

Nanda, S., Okolie, J. A., Patel, R., Pattnaik, F., Fang, Z., Dalai, A. K., Kozinski, J. A. & Naik, S. 2022. Catalytic hydrothermal co-gasification of canola meal and low-density polyethylene using mixed metal oxides for hydrogen production. *International Journal of Hydrogen Energy*, 47, 42084–42098.

Narnaware, S. L. & Panwar, N. L. 2021. Catalysts and their role in biomass gasification and tar abetment: A review. *Biomass Conversion and Biorefinery*. doi:10.1007/s13399-021-01981-1

Nguyen, M. C., Cha, M.-H., Choi, K., Lee, Y. & Ihm, J. 2009. Calcium-hydroxyl group complex for potential hydrogen storage media: A density functional theory study. *Physical Review B*, 79, 233408.

Öhrman, O. G. W., Weiland, F., Pettersson, E., Johansson, A.-C., Hedman, H. & Pedersen, M. 2013. Pressurized oxygen blown entrained flow gasification of a biorefinery lignin residue. *Fuel Processing Technology*, 115, 130–138.

Qu, W., Yang, J., Sun, X., Bai, X., Jin, H. & Zhang, M. 2021. Towards producing high-quality lignin-based carbon fibers: A review of crucial factors affecting lignin properties and conversion techniques. *International Journal of Biological Macromolecules*, 189, 768–784.

Ren, J., Cao, J.-P., Zhao, X.-Y., Yang, F.-L. & Wei, X.-Y. 2019. Recent advances in syngas production from biomass catalytic gasification: A critical review on reactors, catalysts, catalytic mechanisms and mathematical models. *Renewable and Sustainable Energy Reviews*, 116, 109426.

Resende, F. L. P., Fraley, S. A., Berger, M. J. & Savage, P. E. 2008. Noncatalytic gasification of lignin in supercritical water. *Energy & Fuels*, 22, 1328–1334.

Resende, F. L. P. & Savage, P. E. 2010. Kinetic model for noncatalytic supercritical water gasification of cellulose and lignin. *AIChE Journal*, 56, 2412–2420.

Rostrup-Nielsen, J. R. 2002. Syngas in perspective. *Catalysis Today*, 71, 243–247.

Seçer, A., Küçet, N., Fakı, E. & Hasanoğlu, A. 2018. Comparison of co-gasification efficiencies of coal, lignocellulosic biomass and biomass hydrolysate for high yield hydrogen production. *International Journal of Hydrogen Energy*, 43, 21269–21278.

Seçer, A., Şayan, E., Üzden, Ş. T. & Hasanoğlu, A. 2023. Low temperature catalytic co-solvent gasification of biomass for hydrogen rich gas. *Journal of the Energy Institute*, 107, 101183.

Slatter, N. L., Vichanpol, B., Natakaranakul, J., Wattanavichien, K., Suchamalawong, P., Hashimoto, K., Tsubaki, N., Vitidsant, T. & Charusiri, W. 2022. Syngas production for fischer-tropsch synthesis from rubber wood pellets and eucalyptus wood chips in a pilot horizontal gasifier with CaO as a tar removal catalyst. *ACS Omega*, 7, 44951–44961.

Soomro, A., Chen, S., Ma, S., Sun, Z., Xu, C. & Xiang, W. 2020. Promoting effect of ZrO2/CeO2 addition on Fe/CaO catalyst for hydrogen gas production in the gasification process. *Biomass and Bioenergy*, 142, 105712.

Soomro, A., Chen, S., Ma, S., Xu, C., Sun, Z. & Xiang, W. 2018. Elucidation of syngas composition from catalytic steam gasification of lignin, cellulose, actual and simulated biomasses. *Biomass and Bioenergy*, 115, 210–222.

Speight, J. G. 2014. Chapter 3- Gasifier types. In: Speight, J. G. (ed.) *Gasification of Unconventional Feedstocks*, 54–90. Boston: Gulf Professional Publishing.

Su, Y., Zhu, W., Gong, M., Zhou, H., Fan, Y. & Amuzu-Sefordzi, B. 2015. Interaction between sewage sludge components lignin (phenol) and proteins (alanine) in supercritical water gasification. *International Journal of Hydrogen Energy*, 40, 9125–9136.

Sun, R.-C. 2020. Lignin source and structural characterization. *ChemSusChem*, 13, 4385–4393.

Szul, M., Iluk, T. & Zuwała, J. 2022. Use of CO2 in pressurized, fluidized bed gasification of waste biomasses. *Energies*, 15, 1395.

Tian, T., Li, Q., He, R., Tan, Z. & Zhang, Y. 2017. Effects of biochemical composition on hydrogen production by biomass gasification. *International Journal of Hydrogen Energy*, 42, 19723–19732.

Vanholme, R., Morreel, K., Ralph, J. & Boerjan, W. 2008. Lignin engineering. *Current Opinion in Plant Biology*, 11, 278–285.

Wang, H.-M., Miao, R.-R., Yang, Y., Qiao, Y.-H., Zhang, Q.-F., Li, C.-S. & Huang, J.-P. 2015. Study on the catalytic gasification of alkali lignin over Ru/C nanotubes in supercritical water. *Journal of Fuel Chemistry and Technology*, 43, 1195–1201.

Wang, T., Liu, X., Huang, S., Waheed, A. & He, M. 2022a. Modelling co-gasification of plastic waste and lignin in supercritical water using reactive molecular dynamics simulations. *International Journal of Hydrogen Energy*, 47, 21060–21066.

Wang, T., Xu, J., Liu, X. & He, M. 2022b. Co-gasification of waste lignin and plastics in super-critical liquids: Comparison of water and carbon dioxide. *Journal of CO$_2$ Utilization*, 66, 102248.

Widjaya, E. R., Chen, G., Bowtell, L. & Hills, C. 2018. Gasification of non-woody biomass: A literature review. *Renewable and Sustainable Energy Reviews*, 89, 184–193.

Wu, Y., Wang, H., Li, H., Han, X., Zhang, M., Sun, Y., Fan, X., Tu, R., Zeng, Y., Xu, C. C. & Xu, X. 2022. Applications of catalysts in thermochemical conversion of biomass (pyrolysis, hydrothermal liquefaction and gasification): A critical review. *Renewable Energy*, 196, 462–481.

Xu, X., Zhou, Q. & Yu, D. 2022. The future of hydrogen energy: Bio-hydrogen production technology. *International Journal of Hydrogen Energy*, 47, 33677–33698.

Yogalakshmi, K. N., Poornima Devi, T., Sivashanmugam, P., Kavitha, S., Yukesh Kannah, R., Varjani, S., Adishkumar, S., Kumar, G. & J, R. B. 2022. Lignocellulosic biomass-based pyrolysis: A comprehensive review. *Chemosphere*, 286, 131824.

Yoo, C. G., Meng, X., Pu, Y. & Ragauskas, A. J. 2020. The critical role of lignin in lignocel-lulosic biomass conversion and recent pretreatment strategies: A comprehensive review. *Bioresource Technology*, 301, 122784.

Yoshida, T. & Matsumura, Y. 2001. Gasification of cellulose, xylan, and lignin mixtures in supercritical water. *Industrial & Engineering Chemistry Research*, 40, 5469–5474.

Yu, H., Wu, Z. & Chen, G. 2018. Catalytic gasification characteristics of cellulose, hemicel-lulose and lignin. *Renewable Energy*, 121, 559–567.

Yu, H., Zhang, Z., Li, Z. & Chen, D. 2014. Characteristics of tar formation during cellulose, hemicellulose and lignin gasification. *Fuel*, 118, 250–256.

Yu, J., Guo, Q., Gong, Y., Ding, L., Wang, J. & Yu, G. 2021. A review of the effects of alkali and alkaline earth metal species on biomass gasification. *Fuel Processing Technology*, 214, 106723.

Yu, L., Zhang, R., Cao, C., Liu, L., Fang, J. & Jin, H. 2022. Hydrogen production from super-critical water gasification of lignin catalyzed by Ni supported on various zeolites. *Fuel*, 319, 123744.

Yuan, J.-M., Li, H., Xiao, L.-P., Wang, T.-P., Ren, W.-F., Lu, Q. & Sun, R.-C. 2022. Valorization of lignin into phenolic compounds via fast pyrolysis: Impact of lignin structure. *Fuel*, 319, 123758.

Zhang, H., Chen, F., Zhang, J. & Han, Y. 2021. Supercritical water gasification of fuel gas production from waste lignin: The effect mechanism of different oxidized iron-based catalysts. *International Journal of Hydrogen Energy*, 46, 30288–30299.

Zhang, S., Wang, J., Ye, L., Li, S., Su, Y. & Zhang, H. 2023. Investigation into biochar sup-ported Fe-Mo carbides catalysts for efficient biomass gasification tar cracking. *Chemical Engineering Journal*, 454, 140072.

Zhang, Y. & Fatehi, P. 2019. Periodate oxidation of carbohydrate-enriched hydrolysis lignin and its application as coagulant for aluminum oxide suspension. *Industrial Crops and Products*, 130, 81–95.

Zhou, B., Dichiara, A., Zhang, Y., Zhang, Q. & Zhou, J. 2018. Tar formation and evolution during biomass gasification: An experimental and theoretical study. *Fuel*, 234, 944–953.

Zoghlami, A. & Paës, G. 2019. Lignocellulosic biomass: Understanding recalcitrance and pre-dicting hydrolysis. *Frontiers in Chemistry*, 7, 874.

7 Gasification of Sewage Sludge for Hydrogen Production

Abhishek N. Srivastava, Michal Jeremias, and Vineet Singh Sikarwar

7.1 INTRODUCTION

Over the decades, advancements in municipal wastewater handling and treatment planning have resulted in perseverant efforts to improve effluent quality by upgrading existing treatment plants. These efforts concurrently proceeded with the implementation of industrial and household measures to diminish the discharge of toxic contaminants into the sewers (Liu et al. 2019). In line with the advancement of effluent quality, the expanding awareness of the problems associated with sewage sludge while treating municipal wastewater is observed. Such problems are mainly the ever-increasing production of sewage sludge, the high inventory cost required for sludge treatment, and the plausible risk imposed by sewage sludge on human health and the environment. In general, the treatment cost of sewage sludge most often represents more significant than 50% of total wastewater treatment costs. Sewage sludge, the deposited waste from the wastewater treatment process employed for municipalities, is quantitatively increasing due to the increasing population, consumerism, and lifestyle. The worldwide volume of municipal wastewater was predicted to be 360–380 km^3 in the year 2020, which was further estimated to increase by 24% by 2030 and 51% by 2050 (Di Giacomo and Romano 2022). Only a little more than half of the total wastewater is being treated in more than 60,000 treatment plants worldwide, while the leftover is expelled into the environment (Thipkhunthod et al. 2005). Accordingly, sewage sludge production is estimated to be more than 45 million tons worldwide, with more than 70 g of dry sludge per capita per day (Ehalt Macedo et al. 2022). As far as sludge is concerned, that is, wet organic and inorganic matter, the quantity varies from 5 to 10 times higher than that of dry sludge (Di Giacomo and Romano 2022). The ever-increasing sewage sludge volume at a global scale is reinforcing authorities to implement advanced technologies to utilize sewage sludge for energy generation.

Municipal wastewater is treated through floatation, sedimentation, and biological filtration and clubbed in a primary, secondary, and tertiary system. The basic objectives of these treatment procedures are removing contaminants, providing convenience in the management of produced products, and meeting the legislative standards regarding the discharged water quality (Srivastava and Chakma 2020). The

DOI: 10.1201/9781003382270-9"

activated sludge or sewage sludge generated from a tertiary unit of the treatment plant consists of 59%–88% degradable organic matter, containing 50%–55% of carbon, 25%–30% of oxygen, 10%–15% of nitrogen, and 6%–10% of hydrogen with a slight quantity of phosphorus and sulfur (Raheem et al. 2018). The sludge from the treatment plant could be aerobic or anaerobic; the structural composition, chemical characterization, and thermodynamic behavior could vary. The sewage sludge is a complex mixture of valuable materials such as organic matter, inorganic elements, degrading microorganisms, and dissolved and suspended solids (Di Giacomo and Romano 2022). Furthermore, it also contains valuable micro- and macronutrients, including N, P, K, Ca, Na, Mg, and S, which could participate in various biochemical reactions during the biological conversion of sewage sludge (Xue et al. 2018). Multiple microbiological constituents, including proteins, carbohydrates, lipids, and their biochemical by-products clubbed with inorganic matter and lignocellulose components, are also inherent elements of sewage sludge (Sikarwar, Mašláni et al. 2022).

A range of treatment and valorization techniques are currently present for the resourceful conversion of sewage sludge. The most commonly used technologies are based on biological and thermochemical conversion methods. Here, anaerobic digestion and composting are considered biological conversion techniques; however, incineration, pyrolysis, and gasification are thermal conversion technologies used for sewage sludge conversion into energy and valuable products. In biological techniques, anaerobic digestion is a promising bioenergy production technique for such wastes, which promotes biomethane generation for energy production. Anaerobic digestion is a viable pathway for sewage sludge processing, keeping its simplicity in mind for treating a wide range of organic sludges varying from municipalities to industries (Srivastava and Chakma 2021a). The produced biogas has a composition of CH4 (55%–75%) and CO_2 (25%–45%) with nuanced amounts of trace gases such as H_2S, NH_3, and H_2 (Sikarwar, Pfeifer, et al. 2022). These gases are usually implemented for real-time application for cogeneration in internal combustion engines or can be supplied to the natural gas grid after its upgrading (i.e., removal of CO_2 and impurities) (Sikarwar, Reichert, et al. 2021). Nevertheless, such an anaerobic digestion process seldom experiences skilled supervision, long reaction times, stability issues, digestates of lesser standards, underperformance of bioconversion efficacy due to lignocellulose content, etc. (Srivastava et al. 2022; Srivastava and Chakma 2022). In general, sewage sludge contains high moisture content (~98% by wt), but for disposal in landfills and agricultural applications, the required moisture must be lower than 60% by weight (Zhang et al. 2018). However, for thermal processing, the maximum limit of moisture should be 15% by weight (Zhang et al. 2018). Various preprocessing techniques could help remove and condition excess moisture from sewage sludge for thermal processing. Various drying processes, including natural, mechanical, thermal, and biodrying, are the major processes to accomplish it (Gao et al. 2020). Solar drying could be an example of natural sewage sludge drying, which can economically remove excess moisture. However, mechanical drying could be an underhand process compared to thermal drying. In addition, the biodrying process is comparable to the composting process, which can take several days to reduce the moisture content (Gao et al. 2020). Sewage sludge can also possess a high content of heavy metals, which raises contamination levels upon raw application.

Based on moisture, contamination levels, and socioeconomic factors, many developed and developing economies have adopted potential sewage sludge applications for building materials, bioconversion, and thermal conversion (Liu et al. 2019). As a significant amount of sewage sludge is still being dumped openly, there is an urgent need for sustainable technology that can utilize sewage sludge as a substrate and produce energy to meet the energy requirement. Therefore, to get the most conversion efficiency, the implementation of thermal technologies is inevitable.

Currently, gasification, incineration, pyrolysis, and hydrothermal carbonization (HTC) are major thermochemical approaches that are efficient waste treatment technologies available, which produce various products along with the generation of other valuable gases such as H_2- and CO-rich gases, polymer oil, and multifunctional charcoal (Sikarwar et al. 2021). Apart from this, a controlled processing environment, ecological compatibility, and energy recuperation are the additional advantages of thermal methods. The organic matter in the waste is converted into useful gaseous products (H_2, CO, and CO_2) for cogeneration; oil produced from pyrolysis can be used as a fuel substitute (Huang et al. 2017). Moreover, the biochar produced from the thermal process is unique in surface quality, morphology, surface chemistry, and hydrophobicity, which make it utilizable as a soil amendment and biosorbent (Domínguez et al. 2006). Sewage sludge, consisting of high-quality organic matter and microbiological cultures, could be one of the best substrates for thermochemical technologies. Among other available biological, mechanical, and thermochemical technologies, pyrolysis and gasification are recommended as replacements for energy recovery from sewage sludge (Jayaraman and Gökalp 2015). Thermochemical processes rely on corresponding technological advancements and the socioeconomic growth of a country. Despite being cost-intensive and requiring high upfront costs, thermochemical processes exhibit better long-term economic performance, improved efficacy, and higher volume consumption (Sikarwar, Peela, et al. 2022). Mulchandani and Westerhoff (2016) have comprehensively discussed the advantages and disadvantages of biological and thermochemical methods and concluded that the bioconversion method or anaerobic digestion requires several days of solid retention time, specific food-to-microorganism ratio, large footprint area requirement for units, low variation in influent sludge characteristics, and additional capital investment compared to thermochemical processes. The current interests in sewage sludge management and its resourceful conversion deal with process intensification, resource recovery, advanced treatment processes, improved valorization techniques, and catalyst inventions to intensify the process (Gherghel et al. 2019).

Hydrogen, the lightest of all the elements worldwide, is also one of the most abundantly available elements, constituting around 75% of the earth's elemental mass (Midilli et al. 2021). Being lighter than air, hydrogen is freely available in the earth's atmosphere. Hence, natural occurrences are almost rare on earth. Although rare and nearly nonexistent, it can still be produced from various sources, including natural gas, coal, oil, and water, by utilizing multiple process technologies (Ehalt Macedo et al. 2022). Compared to all other fuels, hydrogen possesses the highest energy density; the energy yield of hydrogen is 122 kJ/kg, which is nearly 2.75 times higher than that of other hydrocarbons (Liu et al. 2019). Moreover, hydrogen is considered an efficient alternative to gasoline, as 9.5 kg of hydrogen can easily replace 25 kg of gasoline (Mašláni et al. 2021). Hydrogen, upon combustion, gives clear energy without any emissions, and its by-product is only water. As far as end use is concerned, hydrogen is considered utilizable as an intermediate storage fuel. It has an extensive implementation potential in combustion engines and fuel cell technologies. Hydrogen, being versatile enough, can be used as gas or liquid fuel in the automobile industry to generate gasoline and lubricants. Hydrogen is the main synthesis element for ammonia in the fertilizer industry; furthermore, its implementation in the food industry for hydrogenation for vegetable oil making is exemplary (Parthasarathy and Narayanan 2014). Hydrogen is also one of the protruding elements in the electronics industry for the production of silicon, along with its usage as a refining agent in the metallurgical industry (Sikarwar, Pfeifer, et al. 2022). Although hydrogen is seeking applications in various sectors, it is assumed to impact the energy sector significantly. It is forecasted that if hydrogen is utilized in the transportation sector, more than 40 million tons of hydrogen per year could make transportation greener (Parthasarathy and Narayanan 2014). With its versatility in mind and its greener contribution to the environment, hydrogen production needs to be promoted in the future. Therefore, the domain of this chapter revolves around the comprehensive exploration of thermochemical technologies available for sewage sludge conversion into energy. This chapter covers explicitly the mechanism of gasification, its kinetics, and a detailed review of this technology employed for hydrogen production from sewage sludge.

7.2 SEWAGE SLUDGE AS A SUBSTRATE FOR THERMOCHEMICAL CONVERSION

7.2.1 HIGH MOISTURE CONTENT

Various types of sewage sludges, including primary, activated, and digested sludge, have conversion potential into energy. In municipal wastewater treatment units, the clear water is recovered, separating sewage sludge resembling slurry containing high water content. Generally, a typical sewage sludge contains more than 95% moisture content, reinforcing unlikely gasification conditions. Therefore, before using sewage sludge in the gasification process, drying is recommended to reduce moisture. Thermal dryers such as trays, pneumatic, rotary, tunnel, and cyclone dryers are the most recommended for drying before sewage sludge gasification (Moško et al. 2022). Various filters and hydraulic presses (belt and vacuum type) are also most common for mechanical sewage sludge dewatering (Wu et al. 2018). To balance the moisture, the co-processing of sewage sludge with other organic wastes such as straws, cotton, wood, and industrial organic wastes is also recommended (Srivastava and Chakma 2021b). Gasification requires amendment in moisture so that the additional cost of reducing moisture can be eliminated. However, time dilation has several disadvantages, which is one of the prime issues when simply drying sewage sludge. Therefore, thermal drying could benefit sewage sludge drying before gasification. During thermal drying operation, nitrogen gas is liberated at a temperature range of 160°C–200°C, which can be recovered for utilization in the fertilizer industry (Kan et al. 2016). Mechanical dewatering involves propane hydrate formation for sewage

TABLE 7.1

Comparison of Various Drying Processes Implemented for Sewage Sludge before Thermochemical Conversion

S. No.	Drying Process	Remarks	Outcomes	References
1	Thermal	Removes pathogens with catalytic conversion	Drum-drying and fry-drying have removed significant drying efficacy with only survival of hepatitis virus A	Romdhana et al. (2009)
		Electromagnetic induction with the amendment of CaO and sawdust was used	The moisture content reduced from 84% to 7% in 30 minutes	Xue et al. (2018)
2	Biodrying	Understood airflow effect with a moisture content of 59%, 68%, and 78%	It was observed that initial moisture content has more impact than air flow rate	Huiliñir and Villegas (2015)
		Sewage sludge: biodried product and sawdust ratio 3:2:1	Biodrying led to moisture removal from 66.1 to 54.7% in 20 days	Cai et al. (2016)
3	Free dewatering	Water conversion to its hydrate formation efficacy was investigated in 14 units	Hydrate-based dewatering was half that of the thermal drying process	Wu et al. (2018)
4	Electro-osmotic dewatering combined with biodrying	Sewage sludge was examined with and without inoculum	Reduction in moisture from 85% to 60% in 7.5 minutes with energy consumption of 0.065 kWh	Li et al. (2018)

sludge drying, which requires less inventory than thermal drying and can remove ~45% of moisture (Huiliñir and Villegas 2015). The comparative assessment of various drying processes for sewage sludge gasification is mentioned in Table 7.1. The cost of drying is a vital factor to consider before gasification. Before using sewage sludge, conditioning is also required, which requires incorporating chemical agents such as calcium oxide (Gao et al. 2020). For gasification, the moisture content in sewage sludge creates the constituent's difference in CO, CO_2, and H_2. With increased moisture content, incomplete combustion promotes CO production, which hampers H_2 yield (Huiliñir and Villegas 2015).

7.2.2 HEAVY METAL CONTENT AND ORGANIC STRENGTH

The heavy metal content and other organic pollutants are vital challenges while utilizing sewage sludge. Despite treatment technologies employed at municipal and industrial levels, heavy metals in sewage sludge persist as a pivotal challenge to eradicate. The characterization of primary and secondary sewage provides insights about its

TABLE 7.2

Comparative Insights of Primary and Secondary Sewage Sludge Characterizations

Parameters	Tyagi and Lo (2013)		Bora et al. (2020)	
	Primary Sludge	Secondary Sludge	Primary Sludge	Secondary Sludge
pH	5.0–8.0	6.5–8.0	5.61–6.42	6.2–7.9
Alkalinity (mg/L as $CaCO_3$)	500–1,500	580–1,100	NA	NA
Volatile solids (wt.%)	60–80	59–68	60–80	59–68
Total dry solids (wt.%)	5–9	0.8–1.2	27.58	25.36
Nitrogen (wt.%)	1.5–4	2.4–5.0	33.82	49.91
Phosphorous (wt.%)	0.8–2.8	0.5–0.71	NA	NA
Potash (wt.%)	0–1	0.5–0.71	NA	NA
Cellulose (wt.%)	8–15	7–9.7	NA	NA
Protein (wt.%)	20–30	33–41	2–30	4–32
Fats (wt.%)	7–35	5–12	NA	NA
Organic acids (mg/L as acetate)	200–2,000	1,100–1,700	NA	200–1,700
Energy content (kJ/kg TS)	23,000–29,000	19,000–23,000	23,000–29,000	19,000–23,000
Iron (g/kg)	2–4	NA	NA	NA
Silica (wt.%)	15–20	NA	NA	15–20

wt.%, weight (dry basis); NA, not available.

physicochemical properties, as shown in Table 7.2. The existence of heavy metals and other strong organic pollutants makes sewage sludge a rigid substrate for utilization (Bora et al. 2020). Sewage sludge is a heterogeneous mixture of microbes, leftover organic matter, fecal matter, inorganic substances, and water content. However, the influent characterization of heavy metals and their distribution in sewage sludge does not impact thermal methods; nevertheless, the posttreatment distribution of heavy metals changes (Srivastava and Chakma 2021a). Table 7.3 depicts the proximate and ultimate sewage sludge, which also provides insights into characterization. Sewage sludge is a good source of organic matter and volatile solid content. Moreover, the metals considered toxic (e.g., Hg, Cd, Cu, Zn, Se, Ni, As, and Cr) are also present along with emerging contaminants, which recommends that sewage sludge disposal must not be performed in open landfills or agricultural lands. Thermal sewage sludge treatment at low temperatures transforms heavy metals like Cd, Cr, Cu, Pb, and Zn into their respective stable forms and can decrease leaching (Xue et al. 2018).

7.3 THERMOCHEMICAL CONVERSION OF SEWAGE SLUDGE FOR HYDROGEN PRODUCTION

Sewage sludge is a versatile substrate for energy recovery as it contains a considerable energy potential but with the shortcoming of being rich in moisture, ash content, heavy metals, and organic pollutants. There are biological and chemical pathways

TABLE 7.3

Proximate and Ultimate Analyses of Sewage Sludge by Various Researchers

Proximate Analysis (wt%)	Thipkhunthod et al. (2005)	Kan et al. (2016)	Wu et al. (2019)	Gerasimov et al. (2019)	Liu et al. (2015)
Moisture content	6.1	7.65	7.33	10.84	1.05
Volatile solids	53.0	51.66	62.97	48.31	47.92
Ash	38.4	35.02	16.33	33.88	45.51
Fixed carbon	8.6	5.67	13.37	6.97	5.52
Heating value	13.9	13.16	15.2	11.79	NA
		Ultimate Analysis			
C (%)	31.1	58.5	38.28	27.38	25.93
H (%)	4.2	5.8	5.92	3.92	4.21
O (%)	24.3	33.74	31.06	13.64	22.02
N (%)	3.3	0.53	1.00	9.90	4.78
S (%)	1.1	1.43	0.09	0.45	1.03

available for energy recovery from sewage sludge. As mentioned, biological methods such as anaerobic digestion could be financially advantageous but can produce detrimental environmental effects through pollutant release and inefficient process dynamics (Sikarwar, Reichert, et al. 2021; Sikarwar, Pohořelý, et al. 2021). Therefore, selecting suitable pathways must involve the consideration of distinct technologies incorporating socioeconomic and environmental advantages in terms of resourceful recovery from sewage sludge. In this regard, thermochemical conversion could be very beneficial for the efficient and mass conversion of sewage sludge. Notably, the technology readiness level (TRL) for the thermochemical conversion of sewage sludge for its treatment and recovery proved to be commercially viable. Specifically, the TRL for gasification ranges between 6 and 7, whereas for pyrolysis and hydrothermal treatment, the TRL is 5 and 6, respectively (Gherghel et al. 2019). It can be inferred that thermochemical treatment through pyrolysis, gasification, and combustion is commercially viable technology in the long term. Catalysis is the current research domain for producing energy from biohydrogen from sewage sludge while using thermochemical technologies (Gao et al. 2020). Most of the procedures are still in the demonstration phase. However, few have been incorporated successfully. Considering the merits of thermochemical processes, that is, the elimination of destructive bacteria and significant volume reduction, using catalysts and additives to improve the quality and quantity of end products is attracting broader interest. A comprehensive discussion of thermochemical conversion processes and the utilization of catalysts for process intensification have been incorporated further. The consolidated diagram (Figure 7.1) overviews the plausible thermochemical processes for sewage sludge conversion.

7.3.1 Pyrolysis of Sewage Sludge

Pyrolysis is considered a favorable sludge treatment method as it produces bio-oil, char, and light gases, including H_2 and CO, from sewage sludge treated under low to medium thermal range in an inert atmosphere (Sikarwar et al. 2020).

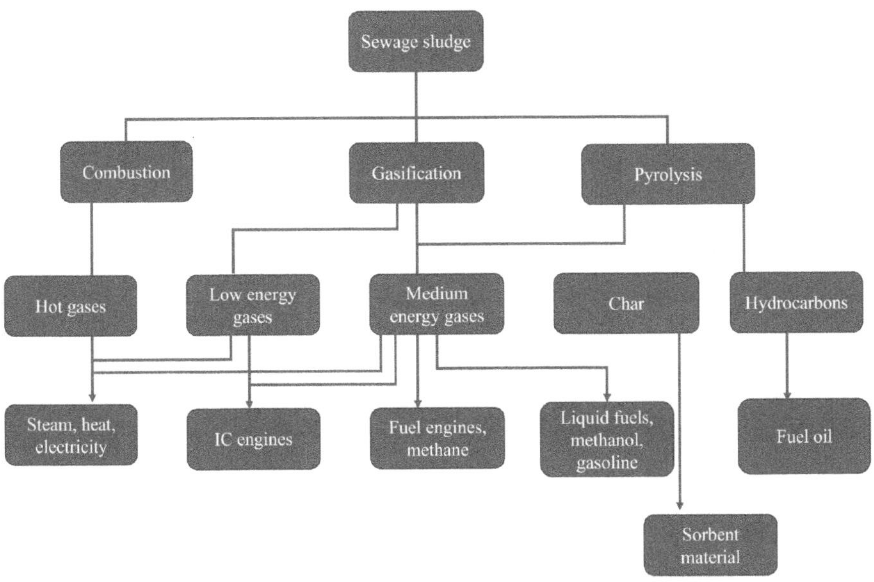

FIGURE 7.1   Thermochemical conversion processes for sewage sludge (Werle and Dudziak 2019).

During pyrolysis, the thermal disintegration of sewage sludge structure takes place at temperature ranges from 400°C to 600°C, in the absence of oxygen or the presence of inert gases such as N_2 and CO_2. The residence reaction time, temperature, reaction environment, feed characteristics, and heating rate are the vital parameters influencing the pyrolysis rate. In particular, there are several types of pyrolysis: slow pyrolysis, fast pyrolysis, vacuum pyrolysis, flash pyrolysis, and hydro-pyrolysis (Mašláni et al. 2021). During the pyrolysis process, CH_4 and water vapors are further treated through steam reforming and water–gas shift (WGS) reactions responsible for higher H_2 production. The major chemical reactions (equations (7.1)–(7.3)) are responsible during this treatment process. Figure 7.2 shows the fundamental pyrolysis mechanism for sewage sludge treatment with the aim of hydrogen production.

$$\text{Sewage sludge pyrolysis} \rightarrow H_2 + CO_2 + CO + \text{hydrocarbons} + \text{oil} + \text{biochar} \tag{7.1}$$

$$C_aH_b + aH_2O \rightarrow aCO + \left(a + \frac{b}{2}\right)H_2 \tag{7.2}$$

$$CO + H_2O \rightarrow CO_2 + H_2 \tag{7.3}$$

Rising moisture, temperature, and heating rates higher than 5°C/min create favorable conditions for hydrogen production (Zhang et al. 2018). The literature confirmed that the advantages of higher temperatures (~900°C) and longer residence time also contribute to the production of other gaseous products; however, catalytic steam reforming of the product gas at temperature optimal for the catalyst and WGS reaction at temperatures < 600°C are usually beneficial for the maximization of hydrogen

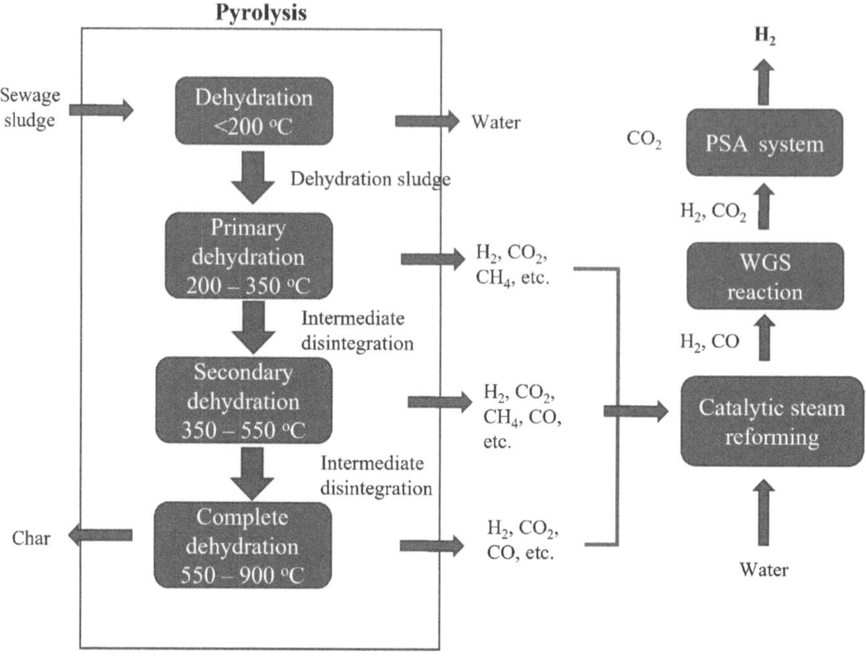

FIGURE 7.2 Pyrolysis mechanism of sewage sludge to produce hydrogen. Reproduced from Liu et al. (2019).

production. Using catalysts during pyrolysis of sewage sludge could accelerate the decomposition of raw sewage and act as a flocculent (Romdhana et al. 2009).

The slow pyrolysis, which requires low heating rates and more considerable reaction time, is known as carbonization or torrefaction. In general, this process involves comparatively lower temperatures (<400°C) to reduce the moisture content and increase the density and handling ability of produced char (Gao et al. 2020). The torrefaction process leads to less formation of biofuel and gases, rather than charcoal. The carbonization process not only reduces the volume of sewage sludge but also liberates oxides of nitrogen, sulfur, and other gaseous pollutants, improving the heating value of the produced char (Domínguez et al. 2006). Most studies have reported that higher heating values of torrefied char increased to a temperature of 270°C and finally decreased while considering energy and mass yields as process parameters (Mašláni et al. 2021). For the treatment of sewage sludge with high moisture content (75%–80%), another thermochemical process is beneficial, known as HTC, which produces hydro-char, liquid, and nominal CO_2 under a temperature range of 180°C–250°C in a closed pressurized environment (Zhang et al. 2018).

On the contrary, fast pyrolysis rapidly converts sewage sludge into bio-oil. The main difference between fast and slow pyrolysis is the higher heating rate and relatively shorter reaction/residence time with temperatures ranging from 400°C to 600°C. In this process, higher production of bio-oil is attained rather than char and gases (Gao et al. 2020). The gases generated from the fast pyrolysis of sewage sludge (~500°C) are mostly light gases, including H_2, CO, CH_4, C_2H_4, CO_2, C_3H_8, and C_4H_{10} (Magdziarz and Werle 2014). The incorporation of other feedstocks is practiced to enhance the quality of the end product from sewage sludge pyrolysis. Various researchers have explored distinct organic wastes to improve the thermal treatability of sewage sludge. For example, lignocellulose word waste, microalgae, sawdust, rice straw, bamboo waste, fruit waste, cotton stalk, and bagasse have been proven to be excellent co-substrates for sewage sludge (Huang et al. 2017; Zhu et al. 2018). Incorporation of such sludges improves the surface properties of biochar and liberation of oxygenated compounds, bio-oils, and water, which further increases the efficacy of co-pyrolysis assay.

Catalysts are employed in a fast pyrolysis process to produce liquid fuel with high energy content. This emerging methodology is environmentally sustainable, concerning the versatility and efficacy of the pyrolysis system. However, compared to co-pyrolysis, catalysis in pyrolysis is challenging due to coke formation and regeneration risks (Huang et al. 2017). Nevertheless, many researchers have reported its benefits over conventional techniques, which can improve the values of end products. The catalysts can be incorporated at three stages: before, during, and post-pyrolysis. By their incorporation and type of catalyst, the characteristics of end products vary. For example, the pyrolytic kinetics of sewage sludge with SiO_2, Al_2O_3, and Fe_2O_3 in the red mud indicated that hydrogen production could be significantly enhanced through the selective incorporation of catalysts (Yang et al. 2018). In this study, Fe_2O_3 and Al_2O_3 were involved as in situ catalysts, which promoted the cracking of C–C and C–H bonds to yield more gases. Specifically, these catalysts increased hydrogen production by 268%, 50.7%, 56.0%, 10.9%, and 10.3% through sewage sludge pyrolysis at 700°C, 800°C, and 900°C, respectively (Yang et al. 2018). The incorporation of metal oxides (CaO, Al_2O_3, CaO, Fe_2O_3, TiO_2, and ZnO) can promote the decomposition of organic matter in demineralized sewage sludge, which can produce a high quantity of solid residues and liquids (Gao et al. 2020). Metal oxides also promote the preliminary breakdown of the sewage sludge structure, reducing the pyrolysis reaction time.

While using sewage sludge as a substrate for the pyrolysis process for hydrogen production, there are specific challenges that very often arise: (i) product quality could be compromised as moisture content is very high in sewage sludge; (ii) high energy input is required for moisture elimination and conditioning of sludge, which affects the overall efficacy of the pyrolysis system; (iii) drying technologies applied to increase the end product quality often experience challenges of deoxygenation and denitrogenation; (iv) a limited number of catalysts have been explored for process intensification of sewage sludge pyrolysis; and (v) finally, the most vital challenge for successful pyrolysis or catalytic pyrolysis is to make end products commercially viable. For the production of green fuel, including syngas and H_2, pyrolysis is not preferred due to the thermal destruction of the hydrocarbon chain in an inert environment. Gasification is advantageous for syngas production as it utilizes steam or air to oxidize the hydrocarbon chain to H_2 and CO partially.

7.3.2 Gasification of Sewage Sludge

Gasification is a thermochemical conversion method that generates syngas or product gas, ideally a hydrogen and carbon monoxide composition. Still, a content of CO_2, CH_4, tar water vapor, and light hydrocarbons is also possible (Sikarwar et al. 2017).

The gasification process is performed at temperatures higher than 800°C with partial application of oxygen as a gasifying agent. Moreover, a mixture of airstream, O_2–steam, and steam–CO_2 has also been utilized as gasifying agents in the sewage sludge gasification process, which imparts a significant role in syngas composition. In recent research scenarios, the application of torrefied sludge, carbonized sludge, hydro-char, pyrolytic char, and raw sewage sludge and its co-gasification and catalytic gasification are being studied. The reactions and mechanisms involved must be understood to understand the gasification process properly. The gasification can be divided into four stages, viz., (i) drying, (ii) pyrolysis, (iii) oxidation, and (iv) reduction (Jayaraman and Gökalp 2015). In the drying phase, essential moisture reduction occurs from sewage sludge; the prevailing nominal temperature is usually 70°C–200°C. Further, the pyrolysis phase reinforces the thermal decomposition of sewage sludge structure at the temperature range of 350°C–600°C. In the oxidation phase, the leftover char and volatile matter are oxidized due to a series of exothermic reactions; here, the temperature rapidly touches >1,100°C. Ultimately, the final reduction zone experiences the chemical reduction of char into CO and H_2 through partial oxidation, Boudouard reaction, steam gasification, and hydro-gasification. It was observed during the gasification process that with the incorporation of steam, the temperatures of oxidation and reduction zones decrease, and correspondingly, the hydrogen production rate increases (Jayaraman and Gökalp 2015). The detailed reactions involved in gasification are mentioned in Table 7.4; for sewage sludge gasification, the specific mechanism is depicted in Figure 7.3. Under high temperature (800°C–1,400°C) and pressure (atmospheric to 33 bars), a combination of H_2, CO, N_2, and CO_2 is generated; such temperature-raised phase can be combined with pyrolysis in the oxygen-absent environment (Figure 7.2). As depicted in Table 7.4, produced syngas can be further treated through steam reformation and WGS reactions, which would enhance hydrogen production similar to the case of pyrolysis (Jayaraman and Gökalp 2015). During the gasification process, various pollutant gases are also produced, along with syngas, at higher temperatures. Also, the ash and heavy metal content could affect the syngas quality. To fetch the clean syngas for further utilization, tar formation has to be prevented through optimum operating conditions or through the incorporation of co-gasification, catalyst addition, and efficient reactor systems (Sikarwar and Zhao 2017), or the tar content has to be minimized by downstream gas cleaning techniques.

Various studies have been conducted on sewage sludge gasification with a target for hydrogen production. Manara and Zabaniotou (2012) mentioned several findings that characterized gaseous products from distinct reactors and chemical agents. For hydrogen production, downdraft gasification is one of the prime methods, which promotes the substitution of steam air as a gasifying agent to enhance hydrogen yield. Similarly, Midilli et al. (2021) identified a cluster of operations conditions such as influent rate, air–fuel ratio, gasification rate, and a turndown ratio of a downdraft gasifier for good results in the form of qualitative gases. Further, Manara and Zabaniotou (2012) also compared the hydrogen production from sewage sludge pyrolysis and steam gasification. They concluded that increasing temperature promotes hydrogen yield and a good portion of hydrogen composition in gasification end products. They mentioned that the ideal temperature for augmented H_2 production from sewage sludge gasification is 700°C–1,100°C. Another study on the type of gasifier clarifies that H_2 production

TABLE 7.4

Fundamental Reactions for Gasification of Sewage Sludge (Nowicki and Markowski 2015; Sikarwar et al. 2017; Xie et al. 2010)

Reaction Name	Reaction	Endothermic/ Exothermic	Heat of Reaction	Temp. (°C)
	Drying/De-volatilization			
	Sewage sludge + heat → biochar + volatile solids + water vapor + light gases + primary coal tar	Endothermic reaction	NA	<400
	Secondary Tar Cracking			
	$Tar + H_2O \rightarrow H_2 + CO$	Endothermic reactions	NA	>500
	$Tar + H_2 \rightarrow$ light hydrocarbons + gases		NA	>700
	$Tar + aH_2O \rightarrow bCO_2 + cH_2$		NA	>700
	$Tar \rightarrow CH_4 + H_2 + H_2O + C_xH_y$		NA	>700
	Oxidation of Carbon			
Boudouard reaction	$C + CO_2 \leftrightarrow 2CO$	Endothermic reactions	173 kJ/kmol	>700
Water gas primary reaction	$C + H_2O \leftrightarrow CO + H_2$		131 kJ/kmol	>700
	$C + 2H_2O \leftrightarrow CO_2 + 2H_2$		90 kJ/mol	>700
Carbon combustion	$C + O_2 \leftrightarrow CO_2$	Exothermic reaction	−393 kJ/mol	>300
Methane steam reforming	$CH_4 + H_2O \leftrightarrow CO + 3H_2$	Endothermic reaction	206 kJ/mol	>500
WGS reaction	$CO + H_2O \leftrightarrow CO_2 + H_2$	Exothermic reactions	−41 kJ/mol	>200
Steam reforming reaction	$C_xH_y + xH_2O \leftrightarrow (x + (y/2))H_2 + xCO$	Exothermic reactions		>300

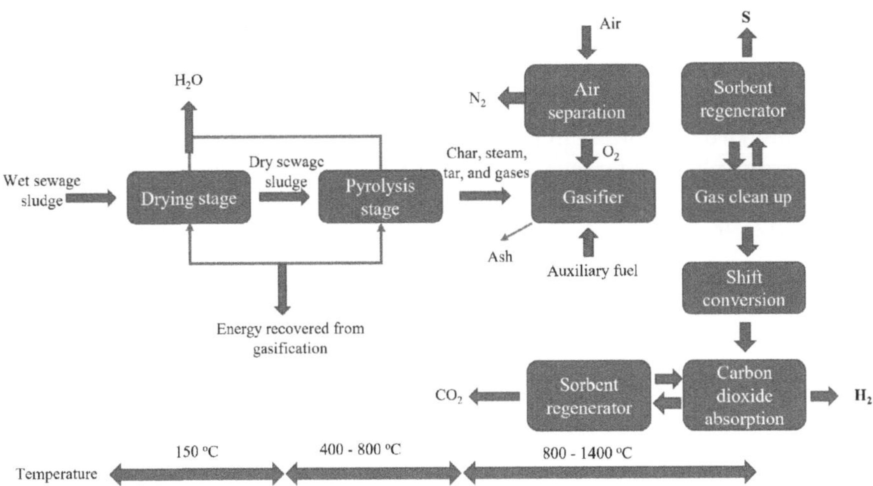

FIGURE 7.3 Gasification mechanism of sewage sludge. Reproduced from Liu et al. (2019).

is higher in fluidized bed gasifiers than in fixed-bed reactors (Schnell et al. 2020). Moreover, steam gasification can further enhance hydrogen production. Manara and Zabaniotou (2012) have concluded that combustible gas proportion by sewage sludge gasification is usually between 8.9% and 11% by volume. Similarly, a downdraft gasification study by Midilli et al. (2021) showcased that a fixed-bed gasifier could yield 10%–11% by vol. of hydrogen, among other gaseous products. Dogru et al. (2002) also employed a throated downdraft gasifier for sewage sludge and attained a hydrogen composition of 8%–12%, among other clusters of gases. Gai et al. (2016) verified that incorporating Fe, Ni, alkali and alkaline earth metals, and alkali catalysts can promote higher hydrogen production. In this study, the hydrogen composition of steam gasification from sewage sludge was 15%–30% (by vol.).

Efficient hydrogen production and its excellent yield can be obtained through sewage sludge gasification. However, prerequisites for its higher output than pyrolysis are higher temperatures and efficient reaction rates, which can lead to higher energy ingesting and stern equipment requirements. Moreover, the low content of hydrogen elements in sludge could also impact the composition of hydrogen in gasification end products. Therefore, future studies have enough scope for concentrating hydrogen gas due to sewage sludge gasification.

7.3.2.1 Impact of Reactor Configuration in Gasification

Reactor configuration has vital importance in the gasification process. Various reactors, including fixed-bed reactors, rotary reactors, circulating fluidized bed reactors, simple fluidized bed reactors, and auger and plasma reactors, are central gasification units used (Magdziarz and Werle 2014). Moreover, the fixed-bed reactors for gasification utilize downdraft and updraft as the methodology for injecting gasifying agents, as shown in Figure 7.4. In the updraft type, the feedstock or raw sewage sludge is applied from the top, and gasifying agents such as air, steam, oxygen, and the mixture of these are injected from the bottom of the reactor. The syngas and mixture of gases are released from the side. Various studies have identified that updraft gasifier produces gas with high tar content but has higher thermal efficiency since the gas leaves at a low temperature (<300°C). However, sewage sludge is applied from the top of the downdraft gasifier. Gasifying agents are entered from the oxidation zone and pass the reduction area with high temperature at the bottom. In this case, the produced gases are released from the side bottom of the downdraft reactor (Figure 7.4). The downdraft gasifier comparatively produces good-quality syngas with a lesser volume of tar; however, because of the high temperature of leaving gas, the thermal efficiency is lower (Dogru et al. 2002). Some shortcomings have been recorded, such as slag accumulation and pressure drop inside the fixed-bed reactors.

7.3.2.2 Role of Operating Parameters in Gasification of Sewage Sludge

Apart from moisture content in sewage sludge, as discussed before, several other factors significantly impact the gasification of sewage sludge and the composition of end products. Feed size, feeding rate, temperature, pressure, equivalence ratio, catalyst addition, and co-gasification are the most vital factors.

To procure a high degree of hydrogen, feed size and rate of feeding impact considerably. The sensitivity of particle sizes is in preference of increasing particle size, that

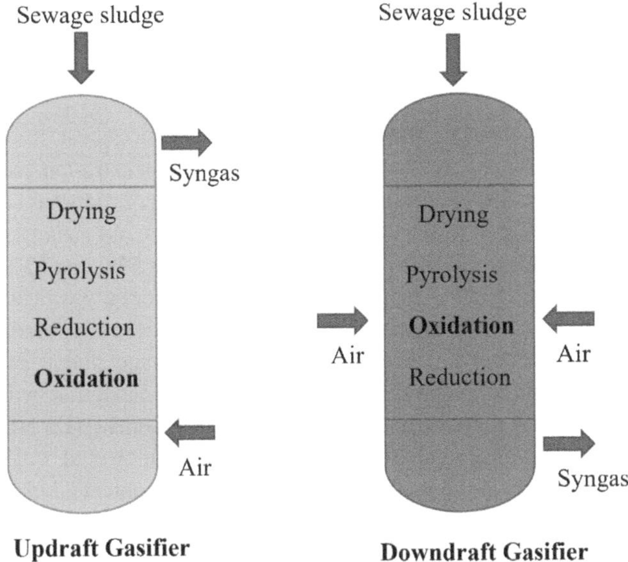

FIGURE 7.4 Comparison of stages in updraft and downdraft gasifiers. Reproduced from Gao et al. (2020).

is, circulating fluidized bed reactor followed by bubbling fluidized bed reactor and finally fixed-bed reactor (Gao et al. 2020). The evenly distributed particles facilitate the efficient transfer of heat, which improves the gasification rate, thereby ultimately increasing syngas yield and reducing tar content. However, a high feeding rate might reduce reaction time in the gasifier, impacting the end composition of gases in quality and quantity. As far as the equivalence ratio is considered, the injecting air-to-sewage sludge ratio is required for stoichiometric combustion. A lower equivalence ratio fetches higher H_2 and CO volume, but at a lower ratio, the other gas production would also be hampered, similar to the beginning of pyrolysis. On the other hand, a higher equivalence ratio would reflect higher CO_2 production and lower H_2 and CO production, resulting in a lower heating value of syngas. Similarly, a much lower or much higher equivalence ratio can also create adverse conditions for the gasification of sewage sludge. Hence, 0.2–0.35 is the optimum suggested ratio for good results (Schnell et al. 2020).

Catalytic gasification is performed to enhance the degree of thermal treatment and degradation of tar contaminants, which will ultimately increase the adequate volume of syngas. Tar, the complex and high-end hydrocarbon, is degraded by catalyst incorporation during gasification to obtain more H_2, CO, and CH_4. Such catalysts also enable heat and mass transfer, benefit overall degrading reactions, and produce a higher yield of gases (Magdziarz and Werle 2014). Nickel-based catalysts, dolomite, olivine, and metal oxides have been extensively investigated as beneficial catalysts for promoting the gasification efficiency of sewage sludge. Apart from that, biochar and activated carbons produced from pyrolysis and gasification also have the potential to be utilized as catalysts. Biochar being produced in lower quantity than that of pyrolysis, its carbon content and surface area make it excellent catalysts as it can impart additional macro- and micronutrients (Na, K, Ca, P, Mg, and Al) and minerals which can assist catalytic reaction during gasification (Gao et al. 2020).

7.3.2.3   Co-Gasification of Sewage Sludge

For increasing end product quality, reducing pollutant concentration, ash, and heavy metals, and co-gasification of sewage sludge with other organic biomass could be beneficial using air and steam (Sikarwar, Peela, et al. 2022). Cold gas efficiency (CGE) for co-gasification with wood chips and paper mill waste is 59% and 62% – more than control sewage sludge at 55%. The production of dry gas for co-gasification with coal was much higher (2.5–2.7 times) than the sole gasification of sewage sludge (Sikarwar, Peela, et al. 2022). Moreover, incorporating coal in sewage sludge gasifiers might reduce the CO_2 content in the producer gas. It can be speculated that hydrogen sulfide and ammonia contents could be higher than other constituents (HCl, HF, and PAHs), owing to higher S and N content in sewage sludge (Gerasimov et al. 2019). Moreover, co-gasification can reduce the ash and the contaminant's contents. For example, biomass co-gasification with sewage sludge can reduce ammonia and sulfur dioxide ammonia (Midilli et al. 2021). Therefore, there is enough scope to incorporate such organic wastes to complement sewage sludge gasification and provide good results.

7.3.3   SUPERCRITICAL WATER GASIFICATION (SCWG)

The SCWG process refers to the application of supercritical water, which is attained beyond the critical point (i.e., temperature > 374.15°C and pressure > 220.64 bar) and applied as a gasifying agent (Domínguez et al. 2006). Using water under normal circumstances, that is, below critical conditions, does not dissolve constituents of sewage sludge. However, the supercritical water application proceeds with miscibility with organic matter. Such conditions make the gasification system a rapid and tar-free system with a high content of hydrogen, carbon monoxide, and methane (Dogru et al. 2002). During the SCWG reaction phase, the molecules of supercritical water impose distinctive roles under different reaction phases, as either reactants or catalysts. These reactions are generally classified into three categories, as depicted in equations (7.4)–(7.6).

Reaction 1: Steam reforming

$$\text{Sewage sludge} + H_2O \rightarrow CO + H_2 \tag{7.4}$$

Reaction 2: WGS reaction

$$CO + H_2O \rightarrow CO_2 + H_2 \tag{7.5}$$

Reaction 3: Methanation reaction

$$CO + 3H_2 \rightarrow CH_4 + H_2O \tag{7.6}$$

Operational conditions and feedstock characteristics are significant factors that influence SCWG reaction routes and their performance (Correa and Kruse 2018).

The operating conditions mainly refer to feed content, temperature, pressure, detention time, and catalysis. The incorporation of catalysts intensifies the process and is economical as it reduces the cost of investment by increasing the system's overall efficacy. Mostly, alkali metal compounds (Na_2CO_3, KOH, and NaOH) and transition meals (Bi, Ru, Cu, and Co) are proven to provide positive impacts on the efficacy of the SCWG process (Correa and Kruse 2018).

A comprehensive discussion of present sewage sludge thermochemical treatment methods helps understand these technologies' fundamentals. However, experimental and analytical data for catalytic thermos-conversion must be explored extensively. More state-of-the-art research is expected to explore that this area for augmented hydrogen production is expected shortly.

7.4 CONCLUSION

Hydrogen production through sewage sludge gasification is an effective method for clean and green energy production with large sludge consumption and minor waste production. This chapter summarized the treatment methodologies available for sewage sludge, especially comparing biological and thermochemical technologies. Biological conversion methods for sewage sludge are critical yet beneficial for biomethane production. However, it does not offer hydrogen production. Moreover, anaerobic digestion as a biological method for sewage sludge treatment is often disadvantageous due to its critical operational parameters and process supervision. However, thermochemical conversion techniques are more reliable for clean energy production. The chapter comprehensively covered the mechanisms of pyrolysis and gasification and reported the advantage of gasification as it has a better potential of producing hydrogen with minor pollutants. It is inferred from the review that sewage sludge, besides having higher moisture content, ash, and detrimental heavy metal levels, thermochemical routes are promising for its resourceful conversion. The pollutant emissions in the case of gasification of sewage sludge are comparatively less while enabling hydrogen and bio-oil production. Moreover, torrefied char, pyrolytic char, and biochar have excellent properties for efficient gasification and can also act as catalysts for improving the efficacy of gasification. Apart from simple gasification, co-gasification and catalytic gasification of sewage sludge have a high potential for producing high hydrogen content in syngas. Catalytic gasification can improve sewage sludge's kinetic degradation, improving the quality of end gases. Utilizing metal oxides, biochar, dolomites, and zeolites as catalysts has further expanded the domain of developing new catalysts to enhance gasification technology's efficiency. Supercritical water gasification technology is a new paradigm in gasification technology that has significantly improved the hydrogen proportion in end gas. Given the socioeconomic application of thermal technologies, incineration and combustion of sewage sludge are not recommended because of high contamination and moisture content. These methods are responsible for producing more flue gas pollutants and ash, thereby requiring higher operational costs than gasification.

Developing and advancing reactor technology for efficient gasification are vital in hydrogen production from sewage sludge. Updraft and downdraft technologies are the most popular ones, whose mechanisms reflect that operational parameters such as injecting and feed conditions have significant roles in the end product of gasification. Based on this fact, an extensive opportunity exists for developing an efficient gasifier that can further enhance the hydrogen yield.

REFERENCES

Bora, A. P., Gupta, D. P., & Durbha, K. S. (2020). Sewage sludge to bio-fuel: A review on the sustainable approach of transforming sewage waste to alternative fuel. *Fuel, 259,* 116262.

Cai, L., Chen, T.-B., Gao, D., & Yu, J. (2016). Bacterial communities and their association with the bio-drying of sewage sludge. *Water Research, 90,* 44–51.

Correa, C. R., & Kruse, A. (2018). Supercritical water gasification of biomass for hydrogen production-Review. *The Journal of Supercritical Fluids, 133,* 573–590.

Di Giacomo, G., & Romano, P. (2022). Evolution and prospects in managing sewage sludge resulting from municipal wastewater purification. *Energies, 15*(15), 5633.

Dogru, M., Midilli, A., & Howarth, C. R. (2002). Gasification of sewage sludge using a throated downdraft gasifier and uncertainty analysis. *Fuel Processing Technology, 75*(1), 55–82.

Domínguez, A., Menéndez, J. A., & Pis, J. J. (2006). Hydrogen rich fuel gas production from the pyrolysis of wet sewage sludge at high temperature. *Journal of Analytical and Applied Pyrolysis, 77*(2), 127–132.

Ehalt Macedo, H., Lehner, B., Nicell, J., Grill, G., Li, J., Limtong, A., & Shakya, R. (2022). Distribution and characteristics of wastewater treatment plants within the global river network. *Earth System Science Data, 14*(2), 559–577.

Gai, C., Guo, Y., Liu, T., Peng, N., & Liu, Z. (2016). Hydrogen-rich gas production by steam gasification of hydrochar derived from sewage sludge. *International Journal of Hydrogen Energy, 41*(5), 3363–3372.

Gao, N., Kamran, K., Quan, C., & Williams, P. T. (2020). Thermochemical conversion of sewage sludge: A critical review. *Progress in Energy and Combustion Science, 79,* 100843.

Gerasimov, G., Khaskhachikh, V., Potapov, O., Dvoskin, G., Kornileva, V., & Dudkina, L. (2019). Pyrolysis of sewage sludge by solid heat carrier. *Waste Management, 87,* 218–227.

Gherghel, A., Teodosiu, C., & De Gisi, S. (2019). A review on wastewater sludge valorisation and its challenges in the context of circular economy. *Journal of cleaner production, 228,* 244–263.

Huang, H., Yang, T., Lai, F., & Wu, G. (2017). Co-pyrolysis of sewage sludge and sawdust/rice straw for the production of biochar. *Journal of Analytical and Applied Pyrolysis, 125,* 61–68.

Huiliñir, C., & Villegas, M. (2015). Simultaneous effect of initial moisture content and airflow rate on biodrying of sewage sludge. *Water research, 82,* 118–128.

Jayaraman, K., & Gökalp, I. (2015). Pyrolysis, combustion and gasification characteristics of miscanthus and sewage sludge. *Energy Conversion and Management, 89,* 83–91.

Kan, T., Strezov, V., & Evans, T. (2016). Effect of the heating rate on the thermochemical behavior and biofuel properties of sewage sludge pyrolysis. *Energy & Fuels, 30*(3), 1564–1570.

Li, Q., Lu, X., Guo, H., Yang, Z., Li, Y., Zhi, S., & Zhang, K. (2018). Sewage sludge drying method combining pressurized electro-osmotic dewatering with subsequent bio-drying. *Bioresource technology, 263,* 94–102.

Liu, H., Zhang, Q., Hu, H., Liu, P., Hu, X., Li, A., & Yao, H. (2015). Catalytic role of conditioner CaO in nitrogen transformation during sewage sludge pyrolysis. *Proceedings of the Combustion Institute, 35*(3), 2759–2766.

Liu, Y., Lin, R., Man, Y., & Ren, J. (2019). Recent developments of hydrogen production from sewage sludge by biological and thermochemical process. *International Journal of Hydrogen Energy*, *44*(36), 19676–19697.

Magdziarz, A., & Werle, S. (2014). Analysis of the combustion and pyrolysis of dried sewage sludge by TGA and MS. *Waste Management*, *34*(1), 174–179.

Manara, P., & Zabaniotou, A. (2012). Towards sewage sludge based biofuels via thermochemical conversion-a review. *Renewable and Sustainable Energy Reviews*, *16*(5), 2566–2582.

Mašláni, A., Hrabovský, M., Křenek, P., Hlina, M., Raman, S., Sikarwar, V. S., & Jeremiáš, M. (2021). Pyrolysis of methane via thermal steam plasma for the production of hydrogen and carbon black. *International Journal of Hydrogen Energy*, *46*(2), 1605–1614.

Midilli, A., Kucuk, H., Topal, M. E., Akbulut, U., & Dincer, I. (2021). A comprehensive review on hydrogen production from coal gasification: Challenges and opportunities. *International Journal of Hydrogen Energy*, *46*(50), 25385–25412.

Moško, J., Jeremiáš, M., Skoblia, S., Beňo, Z., Sikarwar, V. S., Hušek, M., et al. (2022). Residual moisture in the sewage sludge feed significantly affects the pyrolysis process: Simulation of continuous process in a batch reactor. *Journal of Analytical and Applied Pyrolysis*, *161*, 105387.

Mulchandani, A., & Westerhoff, P. (2016). Recovery opportunities for metals and energy from sewage sludges. *Bioresource Technology*, *215*, 215–226.

Nowicki, L., & Markowski, M. (2015). Gasification of pyrolysis chars from sewage sludge. *Fuel*, *143*, 476–483.

Parthasarathy, P., & Narayanan, K. S. (2014). Hydrogen production from steam gasification of biomass: Influence of process parameters on hydrogen yield-a review. *Renewable energy*, *66*, 570–579.

Raheem, A., Sikarwar, V. S., He, J., Dastyar, W., Dionysiou, D. D., Wang, W., & Zhao, M. (2018). Opportunities and challenges in sustainable treatment and resource reuse of sewage sludge: A review. *Chemical Engineering Journal*, *337*, 616–641.

Romdhana, M. H., Lecomte, D., Ladevie, B., & Sablayrolles, C. (2009). Monitoring of pathogenic microorganisms contamination during heat drying process of sewage sludge. *Process Safety and Environmental Protection*, *87*(6), 377–386.

Schnell, M., Horst, T., & Quicker, P. (2020). Thermal treatment of sewage sludge in Germany: A review. *Journal of environmental management*, *263*, 110367.

Sikarwar, V. S., Ji, G., Zhao, M., & Wang, Y. (2017). Equilibrium modeling of sorption-enhanced cogasification of sewage sludge and wood for hydrogen-rich gas production with in situ carbon dioxide capture. *Industrial & Engineering Chemistry Research*, *56*(20), 5993–6001.

Sikarwar, V. S., Hrabovský, M., Van Oost, G., Pohořelý, M., & Jeremiáš, M. (2020). Progress in waste utilization via thermal plasma. *Progress in Energy and Combustion Science*, *81*, 100873.

Sikarwar, V. S., Mašláni, A., Hlína, M., Fathi, J., Mates, T., Pohořelý, M., et al. (2022). Thermal plasma assisted pyrolysis and gasification of RDF by utilizing sequestered CO2 as gasifying agent. *Journal of CO_2 Utilization*, *66*, 102275.

Sikarwar, V. S., Peela, N. R., Vuppaladadiyam, A. K., Ferreira, N. L., Mašláni, A., Tomar, R., et al. (2022). Thermal plasma gasification of organic waste stream coupled with CO 2-sorption enhanced reforming employing different sorbents for enhanced hydrogen production. *RSC advances*, *12*(10), 6122–6132.

Sikarwar, V. S., Pfeifer, C., Ronsse, F., Pohořelý, M., Meers, E., Kaviti, A. K., & Jeremiáš, M. (2022). Progress in in-situ CO2-sorption for enhanced hydrogen production. *Progress in Energy and Combustion Science*, *91*, 101008.

Sikarwar, V. S., Pohořelý, M., Meers, E., Skoblia, S., Moško, J., & Jeremiáš, M. (2021). Potential of coupling anaerobic digestion with thermochemical technologies for waste valorization. *Fuel*, *294*, 120533.

Sikarwar, V. S., Reichert, A., Pohorely, M., Meers, E., Ferreira, N. L., & Jeremias, M. (2021). Equilibrium modeling of thermal plasma assisted co-valorization of difficult waste streams for syngas production. *Sustainable Energy & Fuels, 5*(18), 4650–4660.

Sikarwar, V.S., Mašláni, A., Van Oost, G., Fathi, J., Hlína, M., Mates, T., Pohořelý, M. and Jeremiáš, M., 2024. Integration of thermal plasma with CCUS to valorize sewage sludge. *Energy, 288*, p.129896.

Srivastava, A. N., & Chakma, S. (2020). Quantification of landfill gas generation and energy recovery estimation from the municipal solid waste landfill sites of Delhi, India. *Energy Sources, Part A: Recovery, Utilization, and Environmental Effects*, 1–14. https://doi.org/10.1080/15567036.2020.1754970

Srivastava, A. N., & Chakma, S. (2021a). Dry tomb-bioreactor landfilling approach for enhanced biodegradation and biomethane generation from municipal solid waste co-disposed with sugar mill pressmud. *Bioresource Technology*, 125895.

Srivastava, A. N., & Chakma, S. (2021b). Investigating leachate decontamination and biomethane augmentation through Co-disposal of paper mill sludge with municipal solid waste in simulated anaerobic landfill bioreactors. *Bioresource Technology, 329*, 124889.

Srivastava, A. N., & Chakma, S. (2022). Landfill gas utilization. In Eduardo Jacob-Lopes, Leila Queiroz Zepka, Mariany Costa Deprá (eds), *Handbook of Waste Biorefinery* (pp. 807–811). Springer: Switzerland.

Srivastava, A. N., Singh, R., Chakma, S., & Birke, V. (2022). Advancements in operations of bioreactor landfills for enhanced biodegradation of municipal solid waste. In Pankaj Pathak, Sankar Ganesh Palani (eds), *Circular Economy in Municipal Solid Waste Landfilling: Biomining & Leachate Treatment* (pp. 153–166). Springer: Switzerland.

Thipkhunthod, P., Meeyoo, V., Rangsunvigit, P., Kitiyanan, B., Siemanond, K., & Rirksomboon, T. (2005). Predicting the heating value of sewage sludges in Thailand from proximate and ultimate analyses. *Fuel, 84*(7–8), 849–857.

Tyagi, V. K., & Lo, S.-L. (2013). Sludge: A waste or renewable source for energy and resources recovery? *Renewable and Sustainable Energy Reviews, 25*, 708–728.

Werle, S., & Dudziak, M. (2019). Gasification of sewage sludge. In Majeti Narasimha Vara Prasad, Paulo Jorge de Campos Favas, Meththika Vithanage, S. Venkata Mohan (eds), *Industrial and Municipal Sludge* (pp. 575–593). Elsevier: The Netherlands.

Wu, B., Horvat, K., Mahajan, D., Chai, X., Yang, D., & Dai, X. (2018). Free-conditioning dewatering of sewage sludge through in situ propane hydrate formation. *Water Research, 145*, 464–472.

Wu, J., Liao, Y., Lin, Y., Tian, Y., & Ma, X. (2019). Study on thermal decomposition kinetics model of sewage sludge and wheat based on multi distributed activation energy. *Energy, 185*, 795–803.

Xie, L., Tao, L. I., Gao, J., Fei, X., Xia, W. U., & Jiang, Y. (2010). Effect of moisture content in sewage sludge on air gasification. *Journal of Fuel Chemistry and Technology, 38*(5), 615–620.

Xue, Y., Wang, C., Hu, Z., Zhou, Y., Liu, G., Hou, H., et al. (2018). Thermal treatment on sewage sludge by electromagnetic induction heating: Methodology and drying characterization. *Waste Management, 78*, 917–928.

Yang, J., Xu, X., Liang, S., Guan, R., Li, H., Chen, Y., et al. (2018). Enhanced hydrogen production in catalytic pyrolysis of sewage sludge by red mud: Thermogravimetric kinetic analysis and pyrolysis characteristics. *International Journal of Hydrogen Energy, 43*(16), 7795–7807.

Zhang, X.-P., Zhang, C., Li, X., Yu, S.-H., Tan, P., Fang, Q.-Y., & Chen, G. (2018). A two-step process for sewage sludge treatment: Hydrothermal treatment of sludge and catalytic hydrothermal gasification of its derived liquid. *Fuel Processing Technology, 180*, 67–74.

Zhu, J., Yang, Y., Yang, L., & Zhu, Y. (2018). High quality syngas produced from the co-pyrolysis of wet sewage sludge with sawdust. *International Journal of Hydrogen Energy, 43*(11), 5463–5472.

8 Liquefaction of Algal Material for Hydrogen Production

Yang Jia[1], Prince Ochonma[2], Akanksh Mamidala[1], and Greeshma Gadikota[1,2]

8.1 INTRODUCTION

The dual need to decarbonize industries and meet our growing energy demand motivates advances in various approaches to produce H_2 as a low-carbon fuel. To compete with low-cost conventional approaches to produce H_2 such as steam methane reforming (SMR), it is essential to develop alternative pathways that coproduce high-value resources in addition to H_2. Algal biomass is a promising green bioenergy source that coproduces several types of biofuels including biodiesel, bioethanol, biochar, and biohydrogen under various operating conditions. Algal biomass is considered an attractive feedstock due to its high energy density, high areal yields, and higher photosynthetic efficiency, and its usage does not significantly interfere with food supply (Zhu et al., 2019).

This chapter delves into a multifaceted exploration of harnessing the potential of algal resources for hydrogen (H_2) production along with the generation of other high-value products. Initially, we present a discussion of the feasibility and challenges of utilizing algae as feedstocks for H_2 production. Subsequently, the processes of algal biofuel production, encompassing algal cultivation, harvesting, drying, extraction, and conversion techniques are discussed. Among the algal biofuel production methods discussed here, biohydrogen emerges as a promising pathway for the utilization of algal resources. Commonly employed methods of converting algal biomass through pyrolysis, gasification, and liquefaction are discussed in terms of the conversion efficiency, energy consumption, and environmental impacts. This chapter places particular emphasis on the research progress, pivotal technologies, and prospects of liquefaction due to its inherent potential. A key aspect of this is the upcycling of low-value aqueous biomass oxygenates via hydrothermal treatment, offering a carbon-negative H_2 production pathway that holds significant potential for mitigating the environmental impacts of conventional H_2 production pathways. In summary, this chapter provides a comprehensive exploration of the diverse pathways for harnessing algal resources to produce H_2 and other high-value products, with a particular focus on thermochemical liquefaction approaches. These advances hold the promise of contributing to a greener and more sustainable energy landscape while efficiently utilizing the abundant resources provided by our aquatic ecosystems (Figure 8.1).

DOI: 10.1201/9781003382270-10

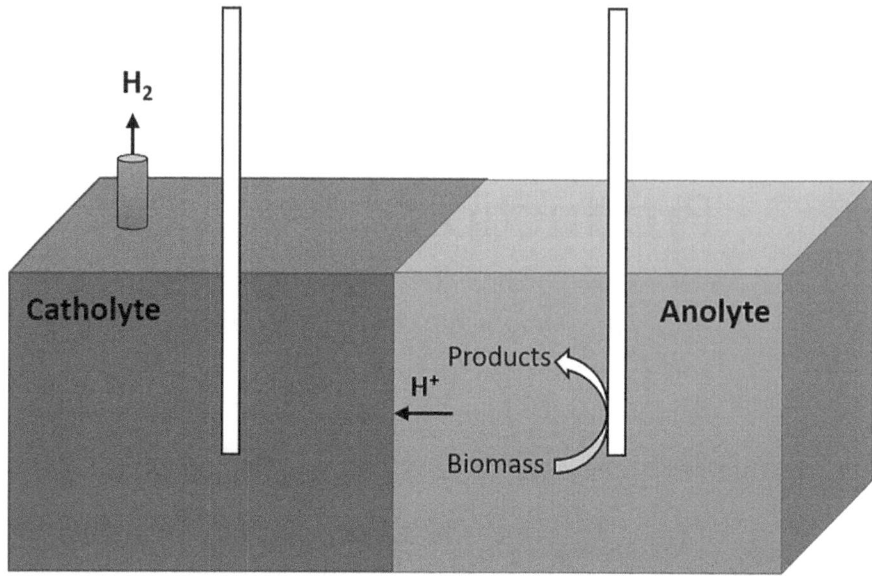

$$\text{Anolyte-} C_xH_yO_z \rightarrow \text{Products} + ne^-$$
$$\text{Catholyte-} nH^+ + ne^- \rightarrow (n/2)\ H_2$$

FIGURE 8.1 Schematic representation of biomass oxygenate electrolysis to produce H_2.

8.2 ALGAE AS FEEDSTOCKS FOR BIOHYDROGEN PRODUCTION

The feasibility of harnessing algae as a source for producing H_2 to address the increasing demand for energy resources while reducing detrimental environmental impacts (Chen, 2022; Melis & Melnicki, 2006; Tamburic et al., 2011) has been investigated for decades (Bothe et al., 1978; Saifuddin & Priatharsini, 2016; Stevens et al., 1973; Weaver et al., 1980). The scientific study of algae has a long history of more than 2000 years (Wang et al., 2020). Ancient civilizations such as the Chinese and the Aztecs used algae for medicinal and nutritional purposes (Barott et al., 2012). In the 19th and 20th centuries, major advances were made in our understanding of the biology and ecology of algal species (Borowitzka, 2013). Algae are a diverse group of photosynthetic organisms that range in size from single-celled microalgae to large multicellular seaweeds (Hoek et al., 1995). Algal resources are abundant, with an estimated number of algal species of more than one million (Guiry, 2012), existing in diverse habitats, including the ocean (Behrenfeld, 2011; Ding et al., 2022), brackish water (Lim et al., 2012; Ahmed et al., 2014), freshwater (Sheath & Wehr, 2015), snow (Hoham, 1975), ice (Lizotte, 2001), wetland (Francoeur et al., 2013), and even soil (Starks et al., 1981), owing to their superior adaptability (Cui et al., 2015; Thompson, 1996). On a global scale, algae generate around half of the atmospheric oxygen (Arumugam, 2022) and remove a significant portion of CO_2 in the air simultaneously (Singh & Ahluwalia, 2013). Algal biohydrogen can be generated via photobiological process (Hwang et al., 2018), dark fermentation (Sarangi and Nanda 2020), and indirect methods like liquefaction (Kavitha et al., 2023). Algal biohydrogen not only generates a clean, high-energy fuel but also effectively sequesters carbon dioxide, making it a remarkable contributor to carbon neutrality (Show et al., 2019). Additionally, the efficiency of algal biohydrogen production eliminates the need for expensive pretreatment processes, making it a front-runner among the portfolio of energy conversion solutions for a sustainable future.

8.3 ALGAL BIOFUEL PRODUCTION

Algal biofuels represent a promising frontier in sustainable energy production, offering diverse options to meet our growing energy demand while mitigating detrimental environmental concerns. These biofuels primarily encompass biodiesel, biogas, and biohydrogen. Biodiesel, mainly derived from algal fatty acids and triacylglycerols (Leite et al., 2013), is regarded a viable means to lower our reliance on fossil fuels (Smith et al., 2010). The production of biodiesel is challenged by the significant variation in the lipid content among different algae species (El Maghraby and Fakhry 2015), as well as the influence of experimental conditions like nitrogen supply and salinity (Peccia et al., 2013; Lawton et al., 2015). Biogas, generated from the fermentation process of algal biomass (Dębowski et al., 2013), serves as a versatile renewable energy source. Algal biogas production suffers from challenges in its use as a substrate, including factors like resistant cell walls, toxic compounds produced by certain algae strains, and unfavorable C:N ratios (Dębowski et al., 2013). Moody and coworkers (2014) estimated the global biofuel potential from microalgae and concluded that 13 quadrillion British thermal units (Btu) of energy can be generated annually if all regions suitable for microalgae cultivation are utilized (Moody et al., 2014).

8.3.1 Algal Cultivation, Harvesting, Drying, Extraction

Developing more efficient methods of generating algal biomass, as well as commercially cultivating algae in wastewater (Church et al., 2017; Jiang et al., 2011; Lavrinovičs et al., 2021), salt water (Barahoei et al., 2021; Figler et al., 2019), and toxic water (L. Wang et al., 2018), is of great interest given the potential to increase algal biofuel production while decreasing detrimental environmental impacts (Lam et al., 2012; Onyeaka et al., 2021; Rezvani et al., 2016). Furthermore, each unit mass of biomass fixes 1.83 times of CO_2 (Chisti, 2008). Therefore, there is considerable interest in harnessing wastewater resources and CO_2 emissions for producing algae (Church et al., 2017; Kang et al., 2017; Lavrinovičs et al., 2021). Since CO_2 supply can be a limiting factor in increasing algae yields, various approaches were developed to increase the use of anthropogenic CO_2 emissions for algae growth. In the context of CO_2 capture and conversion, innovative selective membrane-based approaches were developed to deliver CO_2 from flue gas streams to enhance the yield of microalgae species, *Chlorella sp.* (Zheng et al., 2016). This approach resulted in a maximum biomass productivity of 0.38 g/L/d and regenerated the absorbents, with an energy recovery rate of up to 53%. Another approach to enhance algal yields is by coupling the supply of nutrients for *Chlorella vulgaris* with the removal of ammonium and phosphorus from pretreated pig urine (Zou et al., 2020). Using this approach, the biomass yield increased to 1.72g/m^2/d in a light-receiving plate (LRP)-enhanced raceway pond and 98.2% of NH$_4^+$-N and 68.48% of total P were removed. Coupling the use of *Chlorella vulgaris* to the bioremediation of textile wastewater removed 41.8%–50% of color in high-rate algae ponds (HRAPs) (Lim et al., 2010).

While the source of algae is abundant, and the cultivation is environmentally friendly (Onyeaka et al., 2021), the low biomass concentration of raw algal materials has been limiting the development and commercialization of algal biohydrogen production (Elliott et al., 2013; Maddi et al., 2016; Ravichandran et al., 2022; Tian et al., 2014). The biomass concentration of algae in open ponds is typically about 0.05% (dry basis) when reaching the stationary phase (Gross et al., 2015; Johnson & Wen, 2010). The downstream processes of flocculation, flotation, or centrifugation are responsible for more than 20% of energy consumption in algal biohydrogen production (Gross et al., 2015; Maddi et al., 2016). While extensive efforts have been made to accelerate biomass production of algae, more cost-effective and energy-efficient approaches for preparing highly-concentrated raw algal materials are available (Lam et al., 2012; Ugya et al., 2021; Xie et al., 2017). For example, buoyant beads encapsulating carbonic anhydrase for enhancing CO_2 capture and delivery from air doubled the biomass density rate (Xu et al., 2021). This approach showed the possibility of enhancing CO_2 uptake to increase algae yields sustainably in open ponds. Selective membranes for supplying CO_2 to microalgae from flue gas increased the algae density to 0.2% (Zheng et al., 2016).

Another approach to increase algal yields is by harnessing attached biofilm systems, resulting in 10%–20% cell concentration (Johnson & Wen, 2010), which is suitable as raw material for liquefaction in terms of biomass concentration (Gross et al., 2015). One of the main reasons for the large difference in algal concentrations in the biofilm system and other cultivation methods is that, in the biofilm system, the algae are obtained by directly separating them from the film surface instead of harvesting the culture media (Gross and co-workers, 2015; Mantzorou & Ververidis, 2019). The challenge of applying biofilms in algal biohydrogen production is that other microorganisms (Mantzorou & Ververidis, 2019) in the harvested algae may influence the downstream processes (Alalayah et al., 2017). These challenges and opportunities motivate further investigation into approaches to increase the efficiency of algae cultivation for sustainable H$_2$ production.

8.3.2 Algal Biomass Conversion Techniques (Biochemical, Thermochemical, Electrochemistry)

Biochemical conversion involves alcoholic fermentation, dark fermentation, anaerobic digestion, and biophoton synthesis and generates bioethanol, biomethane, and biohydrogen as products (Kröger and Müller-Langer 2012). Living microorganisms can produce H$_2$ through bio-photolysis (Alalayah et al., 2017; Kapdan & Kargi, 2006), photofermentation (Chandra & Venkata Mohan, 2011; Liu et al., 2020, 2021; Winkler, 2002), and dark fermentation (Cheng et al., 2012; Nagarajan et al., 2016; Xia et al., 2015). These processes occur largely at room temperature and atmospheric pressure (Saifuddin & Priatharsini, 2016), thus they normally do not require additional resource input. The requirement for pretreatment to ensure efficient conversion results in increased costs for biochemical conversion. Thermochemical approaches produce algal bioenergy from algae substrates, generating gaseous products, biocrude, and aqueous products (Gu et al., 2019), which will be discussed in detail in the following section.

While gaseous products and biocrude can be used directly for generating energy, an indirect approach would be to further upgrade the aqueous product into value-added H_2 and other by-products. Electrochemical approaches are some of the promising techniques to produce H_2 on a large scale with simultaneous generation of high-value chemicals with less energy intensity. Conventional electrolysis primarily focuses on water splitting, but the key limitations are the thermodynamic energy penalty and slow kinetics. Energy- and material-efficient recovery of H_2 requires the determination of the chemical composition of aqueous products from the liquefaction process, as this composition determines the electrochemical process parameters, such as: (i) the selection of catalyst, (ii) oxidation and reduction reactions and associated potentials, and (iii) oxygenate decomposition mechanisms, including competing reactions. While the electrochemical conversion of more complex components has not been well studied, C_1-C_3 precursors such as methanol, ethanol, and glycerol have been well studied.

Methanol electrolysis has been studied extensively by changing the electrocatalyst from simple metal electrodes to complex metal–organic frameworks or nanoparticles, which can enhance the kinetics of the reaction. Methanol is a commonly available oxygenate and is one of the most widely used organic compounds obtained from the liquefaction of algae. Although the use of diverse electrocatalysts can result in the formation of varying intermediate species, the overall reactions remain the same (Ruiz-López et al., 2020) as shown below.

$$\text{Reaction at the Anode: } CH_3OH + 6OH^- \rightarrow CO_2 + 5H_2O + 6e^- \qquad (8.1)$$

$$\text{Reaction at the Cathode: } 6H_2O + 6e^- \rightarrow 3H_2 + 6OH^- \qquad (8.2)$$

$$\text{Overall reaction: } CH_3OH + H_2O \rightarrow CO_2 + 3H_2 \qquad (8.3)$$

Ethanol has a wide range of uses due to its simple preparation process and low toxicity. Although ethanol, after liquefaction, can only be used directly as a precursor in various processes, solutions with low concentrations of ethanol that have limited value can be electrochemically upgraded to value-added products and H_2 to extend potential uses. The general representations of reactions at the anode and the cathode are shown below (Liu et al., 2022):

$$\text{Reaction at the Anode: } C_2H_5OH + 6O_2 \rightarrow 2\,CO_2 + 3H_2O + 12e^- \qquad (8.4)$$

$$\text{Reaction at the Cathode: } 6H_2O + 12e^- \rightarrow 6H_2 + 6O_2 \qquad (8.5)$$

$$\text{Overall reaction: } C_2H_5OH + 3H_2O \rightarrow 2CO_2 + 6H_2 \qquad (8.6)$$

Glycerol is an important by-product from the liquefaction of biomass and is also obtained from biocrude. It is widely used in the industry as food additives, pharmaceuticals, personal care products, and cosmetics (Xiao et al., 2015). The energy density of glycerol is 5 kWh/kg, and this makes it a promising resource for H_2 generation and other value-added products (Marshall and Haverkamp, 2008). A dual oxidation process of glycerol is possible based on the composition of the electrolyte. If the glycerol–water reforming is performed in acidic media, then the reactions are as follows (Xiao et al., 2015):

$$\text{Reaction at the Anode: } CH_2OH - CHOH - CH_2OH + 3H_2O \rightarrow 3CO_2 + 14H^+ + 14e^- \tag{8.7}$$

$$\text{Reaction at the Cathode: } 14\,H^+ + 14e^- \rightarrow 7H_2 \tag{8.8}$$

$$\text{Overall Reaction : } CH_2OH - CHOH - CH_2OH + 3H_2O \rightarrow 3CO_2 + 7H_2 \tag{8.9}$$

The oxidation of glycerol in an alkali medium is much more complicated. However, if the complete oxidation of the glycerol–water mixture is assumed, then the reactions are as follows (Xiao et al., 2015):

$$\text{Reaction at the Anode: } CH_2OH - CHOH - CH_2OH + 3H_2O \rightarrow 3CO_2 + 14H^+ + 14e^- \tag{8.10}$$

$$\text{Reaction at the Cathode: } 14H_2O + 14e^- \rightarrow 7H_2 + 14OH^- \tag{8.11}$$

$$\text{Overall Reaction: } CH_2OH - CHOH - CH_2OH + 3H_2O \rightarrow 3CO_2 + 7H_2 \tag{8.12}$$

These advances reported in the electrochemical oxidation of C_1-C_3 molecules such as methanol, ethanol, and glycerol provide the foundation for investigating the use of more complex oxygenates resulting from hydrothermal liquefaction (HTL) of algal biomass.

8.4 THERMOCHEMICAL CONVERSION OF ALGAL BIOMASS TO BIOHYDROGEN (PYROLYSIS, LIQUEFACTION, GASIFICATION)

Thermochemical conversion of algal biomass includes pyrolysis (20%DM; 160°C–220°C; <20 bar; 1–12 h), liquefaction (20%DM; 280°C–370°C; 100–250 bar; 1–20 minutes), and gasification (20%DM; 400°C–750°C; >230 bar; 30 seconds to 5 minutes) (Kröger and Müller-Langer 2012). Pyrolysis operates within an oxygen-deprived environment and produces biochar as the main product. Biochar is widely used in adsorption, soil fertility restoration, and as catalysts (Sekar et al., 2021). Liquefaction approaches are highly versatile since they can use both wet and dry biomass and have high energy efficiencies by rendering the drying process unnecessary. Liquefaction generates a variety of products ranging from biochar and bio-oil to syngas and aqueous fertilizer (Gollakota et al., 2017). The process of gasification depends on several factors including biomass resources, oxidizers, and temperature (Felix et al., 2022). Main compositions of algal biomass gasification are H_2, methane, and carbon monoxide. In contrast, gasification provides fast kinetics but is limited by high energy consumption (Hognon et al., 2014), compared to other pyrolysis and liquefaction processes.

8.4.1 LIQUEFACTION OF ALGAE FOR HYDROGEN PRODUCTION

H_2 has the potential to help decarbonize the hardest to abate sectors such as industry, long-haul aviation, and shipping. However, conventional methods for H_2 production still rely on SMR which emits about 7–12 tons of CO_2 per ton of H_2 produced (Davis et al., 2018; Sun et al., 2019). The aqueous products obtained from HTL of algae has been reported to contain bio-derived oxygenates in quantities uneconomical for direct utilization. For example, alcohols such as methanol, ethanol, and glycerol have been reported as coproducts during algae HTL of various wastewater streams (Mincer et al., 2016; Ellis et al., 2012; Cui et al., 2020). Carbon-negative H_2 can be realized by upcycling low-value aqueous biomass oxygenates for H_2 production via hydrothermal treatment, with *in situ* CO_2 capture as shown in Figure 8.2 (Ochonma et al., 2021, 2023). These studies show that high H_2 yields can be obtained with a catalyst and an alkaline sorbent at $<250°C$ in pressurized N_2 environments. The low temperatures at which these reactions occur facilitate its integration with solar thermal energy technologies, thus enabling the distributed conversion of aqueous biomass oxygenates to H_2 using renewable energy (Martínez-Rodríguez, 2022; Sharma et al., 2017). In contrast to green electrolysis, economics and scalability issues could be resolved via process integration with already existing biorefineries such as HTL of algae. Upcycling these oxygenates to produce H_2 is a unique approach that would enable further valorization of the aqueous phase by-products and enhance resource utilization as opposed to treating them as wastes in need of remediation.

Although the hydrothermal treatment of aqueous bio-oxygenates with in situ CO_2 capture has not been well explored, hydrothermal treatment of aqueous bio-oxygenates without in situ CO_2 capture has been widely studied using different heterogeneous catalysts and more complex substrates including longer chain alcohols and polyols (Alvear et al., 2020; Murzin et al., 2017; Kirilin et al., 2014). Aqueous feed concentrations as high as 10–60 wt.% of glucose and glycols have been utilized as feedstocks, demonstrating that this process is also effective for converting nonvolatile biomass feedstocks into H_2 (Cortright, 2007). These results show promise for adapting hydrothermal upcycling of complex aqueous bio-oxygenates from algae HTL for H_2 production with *in situ* CO_2 capture via carbon mineralization.

8.4.1.1 Hydrothermal Liquefaction (HTL)

The two primary algal biohydrogen production approaches are direct biohydrogen generation from algae and thermochemical conversion (see Figure 8.3). The main challenge of direct biohydrogen generation from algae is the coevolution of oxygen, which can reduce the energy conversion efficiency to less than 1.5% (Show et al., 2018), thereby having a tremendous impact on H_2 production (Laurinavichene et al., 2008).

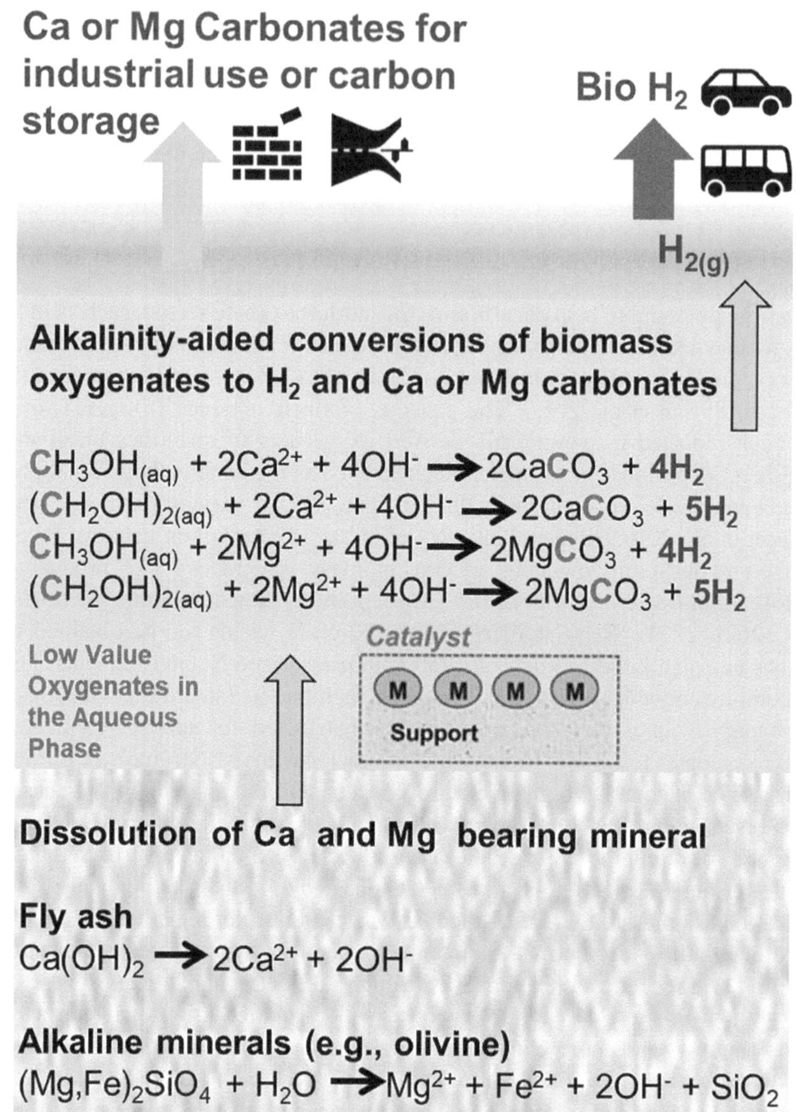

FIGURE 8.2 Schematic representation of the multiphase chemical interactions involved in the cogeneration of biohydrogen and inorganic carbonates from biomass and alkaline industrial residues or minerals (Ochonma et al., 2021, 2023).

In comparison, HTL (Cherad et al., 2016; Guo et al., 2015; Ibrahim et al., 2020; Ravichandran et al., 2022) results in higher energy conversion rates (Saifuddin & Priatharsini, 2016; Show et al., 2018; Yang et al., 2004) and generates high-added-value products such as bio-oil (Galadima & Muraza, 2018; Guo et al., 2015), biochar (Ibrahim et al., 2020; Karthik et al., 2021), and high-value gases and aqueous by-products (Biddy et al., 2013; Cherad et al., 2016; Galadima & Muraza, 2018; Tian et al., 2014).

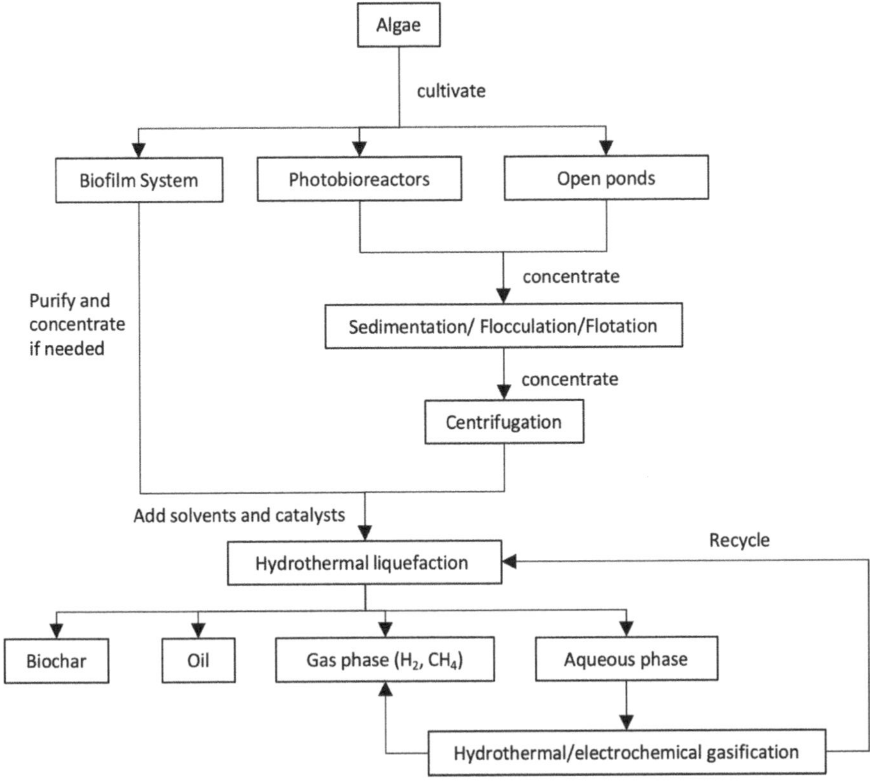

FIGURE 8.3   Algae cultivation and biohydrogen production from algae.

The HTL process involves the depolymerization of lipids, hydrocarbons, proteins, and other algal organisms (AOM) in the algae precursors and repolymerization of the intermediates into bioproducts (Swetha et al., 2021) (Figure 8.4). The reactions are typically carried out in continuous (Castello et al., 2018), batch (Elliott et al., 2015), or semi-batch reactors (Sunphorka et al., 2015), with a pressure of 8–28 MPa (Elliott et al., 2013; Guo et al., 2015; Tian et al., 2014) and a temperature around 220°C–440°C (Biddy et al., 2013; Ravichandran et al., 2022; Swetha et al., 2021).

The components of the bioproducts from hydrothermal liquefication of algae vary based on the algal strains as shown in Table 8.1. These observed differences in the compositions of bio-oil, biochar, aqueous, and gas phases are attributed to the varying content of carbohydrates, protein, and lipids in these materials (Table 8.2). Generally, lipids are the preferred component in microalgae for H_2 production through HTL because they have a high energy content and are easily converted into H_2-rich gas phases. Previous research shows that nitrogen deprivation and increasing salinity of the culture media facilitate the accumulation of lipids in microalgae (Table 8.3). Despite the enhancement of initial lipid content in algae, about 24%–49.4% of the recoverable energy resource remains in the aqueous phase (Maddi et al., 2016). Therefore, the effective utilization of the aqueous phase is essential for sustainable

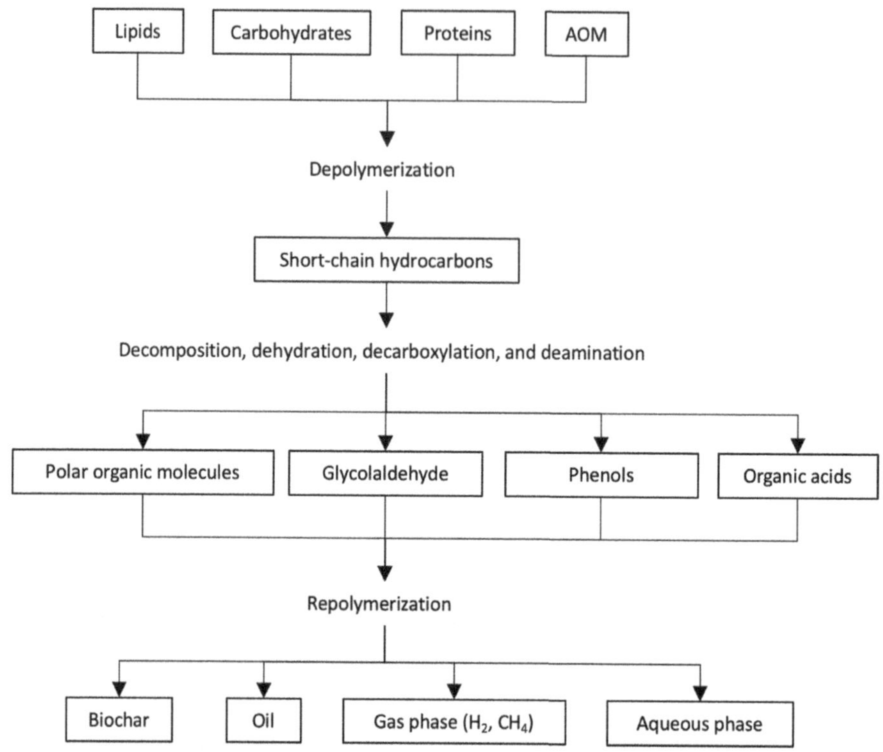

FIGURE 8.4 Mechanism of converting the primary components in algal materials into bio-products during the HTL process.

TABLE 8.1

Summary of Various Phases Resulting from HTL of Algae

Algae	Experiment Condition	Bio-Oil	Aqueous Phase	Gaseous Phase	Biochar	Reference
Microcystis viridis	30 minutes holding time, 300°C	29.40%	48.70%	1.10%	2.30%	Yang et al. (2004)
	30 minutes holding time, 340°C	27.60%	41.60%	1.10%	1.30%	
	30 minutes holding time, 5 wt% Na₂CO₃ dosage 300°C	37.60%	40.50%	1.40%	2%	
	30 minutes holding time, 5 wt% Na₂CO₃ dosage 340°C	39.50%	47.80%	1.50%	1.10%	
	60 minutes holding time, 300°C	30.50%	49.40%	1.70%	5%	
	60 minutes holding time, 340°C	31.60%	43.60%	1.50%	5.20%	
	60 minutes holding time, 5 wt% Na₂CO₃ dosage 300°C	29.90%	43.90%	1.50%	2.80%	
	60 minutes holding time, 5 wt% Na₂CO₃ dosage 340°C	37.80%	46.30%	1.80%	0.80%	

(Continued)

TABLE 8.1 (*Continued*)
Summary of Various Phases Resulting from HTL of Algae

Algae	Experiment Condition	Bio-Oil	Aqueous Phase	Gaseous Phase	Biochar	Reference
Pomace	15 minutes holding time, 310°C	48%	43%	–	–	Déniel et al. (2016)
	30 minutes holding time, 310°C	57%	37%	–	–	
–	Conceptual process, two-stage sequential hydrothermal liquefaction (SEQHTL), first stage 160°C, second stage 240°C	44.40%	43.90%	–	4.60%	Gu et al. (2019)
Ulva fasciata	15 minutes holding time, 300°C	11%	24%	9%	–	Singh et al. (2015)
	15 minutes holding time, 300°C, in CH_3OH	44%	32%	6%		
	15 minutes holding time, 300°C, in C_2H_5OH	40%	41%	8%		

TABLE 8.2
Carbohydrate, Protein, and Lipid Contents of Different Algae Strains

Algae	Carbohydrate	Protein	Lipids	References
Nannochloropsis sp.	12.4% ± 0.9%	36.4% ± 4.6%	27.8% ± 0.9%	Wang et al. (2017)
Nannochloropsis	8.92%	62.79%	18.12%	Shakya et al. (2015)
Pavlova	28%	46.94%	13.88%	
Isochrysis	25.46%	44.36%	18.98%	
Chlorella vulgaris	35% ± 0.1%	20%	25.10%	Alavijeh et al. (2020)
Botryococcus braunii	15.5% ± 1%	34.5% ± 1.1%	39.7% ± 0.6%	Hussain et al. (2020)
Chlamydomonas sp.	29.4% ± 1.4%	39.3% ± 1.4%	15.3% ± 0.8%	
Chlorella vulgaris	18.8% ± 0.8%	45.6% ± 0.7%	21.4% ± 0.7%	
Chlorococcum sp.	24.3% ± 0.8%	42.4% ± 0.7%	18.4% ± 0.6%	
Coelastrum microporum	33.6% ± 0.6%	34.1% ± 1.0%	18.6% ± 0.5%	
Scenedesmus obliquus	20.4% ± 0.7%	47.3% ± 1.3%	18.2% ± 0.8%	
Staurastrum sp.	36.5% ± 0.8%	40.4% ± 0.8%	7.5% ± 1.1%	
Synechocystis aquatilis	31.8% ± 1.4%	41.1% ± 0.5%	11.2% ± 0.5%	
Chlorella vulgaris	9%	55%	25%	Biller and Ross (2011)
Nannochloropsis oculata	8%	57%	32%	
Porphyridium cruentum	40%	43%	8%	
Spirulina	20%	65%	5%	

TABLE 8.3

Comparison of Lipid Contents of Same Algae Strains before and after Lipid Enhancement Treatments

Algae Strain	Experimental Method	Lipid Content before Treatment	Lipid Content after Treatment	References
Chlorella vulgaris	Adding sodium acetate, with sufficient N supply	$22.25\% \pm 0.22\%$	$25.05\% \pm 0.21\%$	Abedini Najafabadi et al. (2015)
	Adding sodium bicarbonate, sufficient N supply		$24.45\% \pm 0.32\%$	
	Two-stage production, the second stage is under N deprivation		$25.86\% \pm 2.84\%$	
	Two-stage production, adding sodium acetate		$42.5\% \pm 1.93\%$	
	Two-stage production, adding sodium bicarbonate		$36.32\% \pm 2\%$	
Dunaliella	Adding 0.5 M NaCl	26%	60%	Takagi et al. (2006)
	Adding 1 M NaCl		67%	
Isochrysis sp.	Increasing the salinity from 10 to 20 ppt	25%	25%	Renaud and Parry (1994)
	Increasing the salinity from 10 to 30 ppt		29%	
Nannochloropsis oculata	Increasing the salinity from 10 to 20 ppt	28%	30%	
	Increasing the salinity from 10 to 30 ppt		33%	

H_2 conversion. In addition to recycling the aqueous phase for liquefaction, hydrothermal and electrochemical pathways are less explored but highly transformative pathways for generating H_2 from the aqueous phase (Cherad et al., 2016; Ravichandran et al., 2022; Tian et al., 2014).

8.4.1.2 Catalytic Liquefaction

H_2 production via low-temperature hydrothermal treatment of bio-oxygenates, at high pressures (>25 bars), in the presence of a metal catalyst, **without** *in situ* CO_2 capture has been extensively studied (Cortright et al., 2010; Rahman et al., 2014; Hoang et al., 2016; Davda et al., 2004; Vaidya & Lopez-Sanchez, 2017; Boga et al., 2016; Shabaker et al., 2004; Coronado et al., 2017; Shabaker et al., 2003; Chen et al., 2015). This process is also known as aqueous phase reforming (APR) and has several advantages over conventional gasification, including relatively lower energy requirements, and low CO selectivity as the water–gas shift (WGS) is favored at low temperatures and high pressures. However, there are still significant limitations associated with H_2 selectivity because of competing methanation reactions and slow kinetics due to the endothermic nature of bio-oxygenates breakdown pathways, leading to lower H_2 production efficiencies (Hoang et al., 2016). Also, significant quantities of CO_2 are typically coproduced, resulting in the need for additional units for separation and purification of gas streams to produce low-carbon H_2. The capture and removal of CO_2 from mixed gases have been extensively investigated (Rochelle 2009). In particular, carbon mineralization, a thermodynamically downhill pathway for converting gaseous CO_2 into stable solid carbonates has gained a significant attention in recent years (Liu and Gadikota, 2020; Gadikota, 2021; Gadikota and Park, 2015; Yin et al., 2022). As an alternative to post-combustion CO_2 capture and removal, coupling carbon mineralization pathways with uphill energy and resource conversions has the net impact of lowering the overall energy needs while also removing CO_2. For example, coupling hydrothermal treatment of bio-oxygenates such as ethylene glycol with carbon mineralization of CaO resulted in lowering the standard heat of reaction from 195.5 to −153 kJ/mol, at 227°C (Ochonma et al., 2021).

8.4.1.3 Other Liquefaction Methods

In addition to lowering the thermodynamic energy barrier associated with H_2 production from aqueous bio-oxygenates, higher H_2 yields and selectivity can be achieved with in situ CO_2 capture via carbon mineralization. The in situ removal of CO_2 favors WGS reactions over CO and CO_2 methanation reactions, which limits H_2 production. According to Figure 8.5, the significant enhancement in H_2 yields is achievable from a thermodynamic standpoint. An increase in H_2 yields of methanol, ethylene glycol, and glycerol from <1% to 75%, 92%, and 93%, respectively, was reported

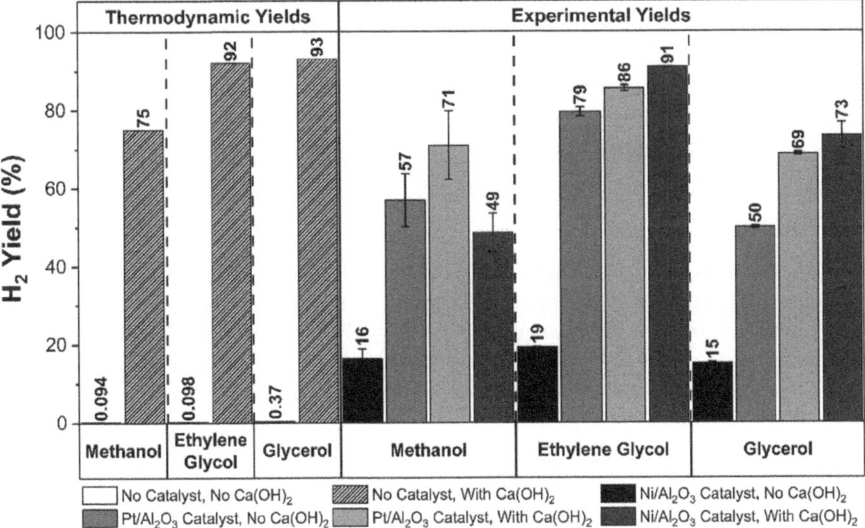

FIGURE 8.5 Thermodynamic and experimental H_2 yields resulting from hydrothermal conversion of 3 wt.% biomass oxygenates with and without inherent carbon removal using the $Ca(OH)_2$ sorbent. Data reported in Ochonma et al. (2021, 2023).

(Ochonma et al., 2023). Potentially limiting reaction kinetics and thermodynamics due to elevated pressures and hydrothermal conditions could also be resolved by harnessing catalysts to selectively cleave biomass oxygenate molecules. Catalysts such as Pt or Pd have been shown to have high activity for these reactions; however, the scarcity and high cost of these metals are key drivers motivating the development of other cost-effective and earth abundant nickel-based catalysts. H_2 yields as high as 64%, 79%, and 50% have been reported from 3 wt% methanol, ethylene glycol, and glycerol, respectively, using Pt/Al_2O_3 catalyst. However, lower than 19% H_2 yields are obtained using Ni/Al_2O_3 as shown in Figure 8.5. The relatively lower H_2 yields with Ni catalyst relative to Pt catalyst are due to lower H_2 selectivity due to methanation reactions (Ochonma et al., 2023; Vaidya & Lopez-Sanchez, 2017). This issue can be resolved by coupling oxygenate reforming with carbon mineralization which leads to relatively higher H_2 yields with Ni (73%–91%) compared to Pt (69%–85%).

8.5 CONCLUSIONS

The rising need for H_2 as a low-carbon fuel to facilitate industrial decarbonization motivates transformative scientific advances in the liquefaction of algae and the conversion of the liquefied products to H_2 and other high-value products. The feasibility of harnessing lipid-rich content and aqueous phases resulting from algal biomass liquefaction to produce H_2 facilitates the optimal use of algal resources. These approaches can be deployed in a distributed manner based on the regional availability of low-carbon and low-cost renewable electricity and solar thermal energy resources. Advances in novel catalytic materials enable energy-efficient electrochemical conversion of oxygenates to H_2 from aqueous oxygenates. Similarly, the coupling of endothermic thermochemical conversion of oxygenates to H_2 with exothermic carbon mineralization to capture and mineralize CO_2 emissions enhances the overall energy and material efficiency of these conversion routes. Scalable transformative advances require the development of distributed reactor systems that harness renewable energy in the form of electrical energy or solar thermal energy for on-demand and on-site algal liquefaction coupled to H_2 conversion.

REFERENCES

Abedini Najafabadi, H. et al. (2015) "Effect of various carbon sources on biomass and lipid production of *Chlorella vulgaris* during nutrient sufficient and nitrogen starvation conditions," *Bioresource Technology*, 180, pp. 311–317. doi:10.1016/j.biortech.2014.12.076.

Ahmed, F. et al. (2014) "Profiling of carotenoids and antioxidant capacity of microalgae from subtropical coastal and brackish waters," *Food Chemistry*, 165, pp. 300–306. doi:10.1016/j.foodchem.2014.05.107.

Alalayah, W.M. et al. (2017) "Kinetics of biological hydrogen production from green microalgae *Chlorella vulgaris* using glucose as initial substrate," *Energy Sources, Part A: Recovery, Utilization, and Environmental Effects*, 39(12), pp. 1210–1215. doi:10.1080/15567036.2017.1315755.

Alavijeh, R.S. et al. (2020) "Combined bead milling and enzymatic hydrolysis for efficient fractionation of lipids, proteins, and carbohydrates of Chlorella vulgaris microalgae," *Bioresource Technology*, 309, p. 123321. doi:10.1016/j.biortech.2020.123321.

Arumugam, M. (2022) "Editorial: advanced technologies and perspectives on sustainable microalgae production," *Frontiers in Bioengineering and Biotechnology*, 10, p. 841261. doi:10.3389/fbioe.2022.841261.

Alvear, M. et al. (2020) "Aqueous phase reforming of xylitol and xylose in the presence of formic acid," *Catalysis Science & Technology*, 10(15), pp. 5245–5255. doi:10.1039/D0CY00811G.

Barahoei, M., Hatamipour, M.S., & Afsharzadeh, S. (2021) "Direct brackish water desalination using Chlorella vulgaris microalgae," *Process Safety and Environmental Protection*, 148, pp. 237–248. doi:10.1016/j.psep.2020.10.006.

Barott, K.L. et al. (2012) "Natural history of coral–algae competition across a gradient of human activity in the Line Islands," *Marine Ecology Progress Series*, 460, pp. 1–12. doi:10.3354/meps09874.

Behrenfeld, M. (2011) "Uncertain future for ocean algae," *Nature Climate Change*, 1(1), pp. 33–34. doi:10.1038/nclimate1069.

Biddy, M.J. et al. (2013) *Whole algae hydrothermal liquefaction technology pathway*. Richland, WA: Pacific Northwest National Laboratory (PNNL). doi:10.2172/1073584.

Biller, P., & Ross, A.B. (2011) "Potential yields and properties of oil from the hydrothermal liquefaction of microalgae with different biochemical content," *Bioresource Technology*, 102(1), pp. 215–225. doi:10.1016/j.biortech.2010.06.028.

Boga, D. A., Liu, F., Bruijnincx, P. C., & Weckhuysen, B. M. (2016) "Aqueous-phase reforming of crude glycerol: effect of impurities on hydrogen production," *Catalysis Science & Technology*, 6(1), pp. 134–143. doi: 10.1039/C4CY01711K.

Borowitzka, M.A. (2013) "Energy from microalgae: A short history," in Borowitzka, M.A. & Moheimani, N.R. (eds.) *Algae for biofuels and energy*. Dordrecht: Springer Netherlands, pp. 1–15. doi:10.1007/978-94-007-5479-9_1.

Bothe, H., Distler, E., & Eisbrenner, G. (1978) "Hydrogen metabolism in blue-green algae," *Biochimie*, 60(3), pp. 277–289. doi:10.1016/s0300-9084(78)80824-4.

Castello, D., Pedersen, T., & Rosendahl, L. (2018) "Continuous hydrothermal liquefaction of biomass: A critical review," *Energies*, 11(11), p. 3165. doi:10.3390/en11113165.

Chandra, R., & Venkata Mohan, S. (2011) "Microalgal community and their growth conditions influence biohydrogen production during integration of dark-fermentation and photo-fermentation processes," *International Journal of Hydrogen Energy*, 36(19), pp. 12211–12219. doi:10.1016/j.ijhydene.2011.07.007.

Chen, G. Y., Li, W. Q., Chen, H., & Yan, B. B. (2015) "Progress in the aqueous-phase reforming of different biomass-derived alcohols for hydrogen production," *Journal of Zhejiang University-SCIENCE A*, 16(6), pp. 491–506. doi: 10.1631/jzus.A1500023.

Chen, Y. (2022) "Global potential of algae-based photobiological hydrogen production," *Energy & Environmental Science*, 15(7), pp. 2843–2857. doi:10.1039/D2EE00342B.

Cheng, J. et al. (2012) "Combination of dark- and photo-fermentation to improve hydrogen production from Arthrospira platensis wet biomass with ammonium removal by zeolite," *International Journal of Hydrogen Energy*, 37(18), pp. 13330–13337. doi:10.1016/j.ijhydene.2012.06.071.

Cherad, R. et al. (2016) "Hydrogen production from the catalytic supercritical water gasification of process water generated from hydrothermal liquefaction of microalgae," *Fuel*, 166, pp. 24–28. doi:10.1016/j.fuel.2015.10.088.

Chisti, Y. (2008) "Biodiesel from microalgae beats bioethanol," *Trends in Biotechnology*, 26(3), pp. 126–131. doi:10.1016/j.tibtech.2007.12.002.

Church, J. et al. (2017) "Effect of salt type and concentration on the growth and lipid content of Chlorella vulgaris in synthetic saline wastewater for biofuel production," *Bioresource Technology*, 243, pp. 147–153. doi:10.1016/j.biortech.2017.06.081.

Coronado, I. et al. (2017) "Aqueous-phase reforming of methanol over nickel-based catalysts for hydrogen production," *Biomass and bioenergy*, 106, pp. 29–37. doi:10.1016/j.biombioe.2017.08.018.

Cortright, R. D., & Bednarova, L. (2007) Hydrogen generation from biomass-derived carbohydrates via aqueous phase reforming (APR) Process. *US Department of Energy Hydrogen Program FY2007 Anual Progress Report*, pp. 56–59.

Cortright, R. D., Davda, R. R., & Dumesic, J. A. (2002) "Hydrogen from catalytic reforming of biomass-derived hydrocarbons in liquid water," *Nature*, 418(6901), pp. 964–967. doi:10.1038/nature01009.

Cui, Z. et al. (2020) "Co-hydrothermal liquefaction of wastewater-grown algae and crude glycerol: A novel strategy of bio-crude oil-aqueous separation and techno-economic analysis for bio-crude oil recovery and upgrading," *Algal research*, 51, p. 102077. doi:10.1016/j.algal.2020.102077.

Cui, J. et al. (2015) "Adaptability of free-floating green tide algae in the Yellow Sea to variable temperature and light intensity," *Marine Pollution Bulletin*, 101(2), pp. 660–666. doi:10.1016/j.marpolbul.2015.10.033.

Davda, R. R., & Dumesic, J. A. (2004) "Renewable hydrogen by aqueous-phase reforming of glucose," *Chemical Communications*, (1), pp. 36–37. doi: 10.1039/B310152E.

Davis, S. J. et al. (2018) "Net-zero emissions energy systems," *Science*, 360(6396), p. eaas9793. doi:10.1126/science.aas9793.

Dębowski, M., Zieliński, M., Grala, A., & Dudek, M. (2013) "Algae biomass as an alternative substrate in biogas production technologies," *Renewable and Sustainable Energy Reviews*, 27, pp. 596–604.

Déniel, M. et al. (2016) "Bio-oil production from food processing residues: Improving the bio-oil yield and quality by aqueous phase recycle in hydrothermal liquefaction of blackcurrant (*Ribes nigrum* L.) pomace," *Energy & Fuels : An American Chemical Society Journal*, 30(6), pp. 4895–4904. doi:10.1021/acs.energyfuels.6b00441.

Ding, L. et al. (2022) "The environmental adaptability and reproductive properties of invasive green alga Codium fragile from the Nan'ao Island, South China Sea," *Acta Oceanologica Sinica*, 41(3), pp. 70–75. doi:10.1007/s13131-021-1928-6.

Elliott, D. C. et al. (2015) "Hydrothermal liquefaction of biomass: Developments from batch to continuous process," *Bioresource technology*, 178, pp. 147–156. doi: 10.1016/j.biortech.2014.09.132.

Elliott, D.C. et al. (2013) "Process development for hydrothermal liquefaction of algae feedstocks in a continuous-flow reactor," *Algal Research*, 2(4), pp. 445–454. doi:10.1016/j.algal.2013.08.005.

Ellis, J. T., Hengge, N. N., Sims, R. C., & Miller, C. D. (2012) "Acetone, butanol, and ethanol production from wastewater algae," *Bioresource technology*, 111, p. 491–495. doi:10.1016/j.biortech.2012.02.002.

El Maghraby, D.M., & Fakhry, E.M. (2015) "Lipid content and fatty acid composition of Mediterranean macro-algae as dynamic factors for biodiesel production," *Oceanologia*, 57(1), pp. 86–92. doi:10.1016/j.oceano.2014.08.001.

Felix, C.B. et al. (2022) "A comprehensive review of thermogravimetric analysis in lignocellulosic and algal biomass gasification," *Chemical Engineering Journal*, 445, p. 136730. doi:10.1016/j.cej.2022.136730.

Figler, A. et al. (2019) "Salt tolerance and desalination abilities of nine common green microalgae isolates," *Water*, 11(12), p. 2527. doi:10.3390/w11122527.

Francoeur, S.N., Rier, S.T., & Whorley, S.B. (2013) "Methods for sampling and analyzing wetland algae," in Anderson, J.T. & Davis, C.A. (eds.) *Wetland techniques:* Volume 2: Organisms. Dordrecht: Springer Netherlands, pp. 1–58. doi:10.1007/978-94-007-6931-1_1.

Gadikota, G. (2021) "Carbon mineralization pathways for carbon capture, storage and utilization," *Communications Chemistry*, 4(1), p. 23. doi:10.1038/s42004-021-00461-x.

Gadikota, G., & Park, A.H.A., (2015) "Accelerated carbonation of Ca-and Mg-bearing minerals and industrial wastes using CO2," in Styring, P. Quadrelli, E.A. & Armstrong, K. (eds.) *Carbon dioxide utilization*. Amsterdam: Elsevier, pp. 115–137. doi:10.1016/B978-0-444-62746-9.00008-6.

Galadima, A., & Muraza, O. (2018) "Hydrothermal liquefaction of algae and bio-oil upgrading into liquid fuels: Role of heterogeneous catalysts," *Renewable and Sustainable Energy Reviews*, 81, pp. 1037–1048. doi:10.1016/j.rser.2017.07.034.

Gollakota, A.R.K., Kishore, N., & Gu, S. (2017) "A review on hydrothermal liquefaction of biomass," *Renewable and Sustainable Energy Reviews*, 81, pp. 1378–1392. doi:10.1016/j.rser.2017.05.178.

Gross, M., Jarboe, D., & Wen, Z. (2015) "Biofilm-based algal cultivation systems," *Applied Microbiology and Biotechnology*, 99(14), pp. 5781–5789. doi:10.1007/s00253-015-6736-5.

Gu, X. et al. (2019) "Comparative techno-economic analysis of algal biofuel production via hydrothermal liquefaction: One stage versus two stages," *Applied Energy*, p. 114115. doi:10.1016/j.apenergy.2019.114115.

Guiry, M.D. (2012) "How many species of algae are there?" *Journal of Phycology*, 48(5), pp. 1057–1063. doi:10.1111/j.1529-8817.2012.01222.x.

Guo, Y. et al. (2015) "A review of bio-oil production from hydrothermal liquefaction of algae," *Renewable and Sustainable Energy Reviews*, 48, pp. 776–790. doi:10.1016/j.rser.2015.04.049.

Hoang, T. M. C., Vikla, A. K. K., & Seshan, K. (2015) "Aqueous-phase Reforming of Sugar Derivatives: Challenges and Opportunities," in Murzin, D. & Simakova. O. (eds.) Biomass Sugars for Non-Fuel Application. Cambridge: The Royal Society of Chemistry, pp. 54–88. doi: 10.1039/9781782622079-00054.

Hoek, C. et al. (1995) *Algae: An introduction to phycology*. illustrated, reprint ed. Cambridge: Cambridge University Press.

Hognon, C. et al. (2014) "Comparison of steam gasification reactivity of algal and lignocellulosic biomass: influence of inorganic elements," *Bioresource Technology*, 164, pp. 347–353. doi:10.1016/j.biortech.2014.04.111.

Hoham, R. W. (1975) "Optimum Temperatures and Temperature Ranges for Growth of Snow Algae," *Arctic and Alpine Research*, 7(1), pp. 13–24. doi: 10.1080/00040851.1975.12003805.

Hussain, J. et al. (2020) "Using non-metric multi-dimensional scaling analysis and multi-objective optimization to evaluate green algae for production of proteins, carbohydrates, lipids, and simultaneously fix carbon dioxide," *Biomass and Bioenergy*, 141, p. 105711. doi:10.1016/j.biombioe.2020.105711.

Hwang, J.-H. et al. (2018) "Photosynthetic biohydrogen production in a wastewater environment and its potential as renewable energy," *Energy*, 149, pp. 222–229. doi:10.1016/j.energy.2018.02.051.

Ibrahim, A.F.M. et al. (2020) "Pyrolysis of hydrothermal liquefaction algal biochar for hydrogen production in a membrane reactor," *Fuel*, 265, p. 116935. doi:10.1016/j.fuel.2019.116935.

Jiang, L. et al. (2011) "Biomass and lipid production of marine microalgae using municipal wastewater and high concentration of CO2," *Applied Energy*, 88(10), pp. 3336–3341. doi:10.1016/j.apenergy.2011.03.043.

Johnson, M.B., & Wen, Z. (2010) "Development of an attached microalgal growth system for biofuel production," *Applied Microbiology and Biotechnology*, 85(3), pp. 525–534. doi:10.1007/s00253-009-2133-2.

Kang, D. et al. (2017) "Carbon capture and utilization using industrial wastewater under ambient conditions," *Chemical Engineering Journal*, 308, pp. 1073–1080. doi:10.1016/j.cej.2016.09.120.

Karthik, V. et al. (2021) "Hydrothermal production of algal biochar for environmental and fertilizer applications: A review," *Environmental Chemistry Letters*, 19(2), pp. 1025–1042. doi:10.1007/s10311-020-01139-x.

Kapdan, I. K., & Kargi, F. (2006) "Bio-hydrogen production from waste materials," *Enzyme and Microbial Technology*, 38(5), pp. 569–582. doi:10.1016/j.enzmictec.2005.09.015.

Kavitha, S. et al. (2023) "A review on current advances in the energy and cost effective pretreatments of algal biomass: Enhancement in liquefaction and biofuel recovery," *Bioresource Technology*, 369, p. 128383. doi:10.1016/j.biortech.2022.128383.

Kirilin, A., Wärnå, J., Tokarev, A., & Murzin, D. Y. (2014). "Kinetic modeling of sorbitol aqueous-phase reforming over Pt/Al2O3," *Industrial & Engineering Chemistry Research*, 53(12), 4580–4588. doi:10.1021/ie403813y.

Kröger, M., & Müller-Langer, F. (2012) "Review on possible algal-biofuel production processes," *Biofuels*, 3(3), pp. 333–349. doi:10.4155/bfs.12.14.

Lam, M.K., Lee, K.T., & Mohamed, A.R. (2012) "Current status and challenges on microalgae-based carbon capture," *International Journal of Greenhouse Gas Control*, 10, pp. 456–469. doi:10.1016/j.ijggc.2012.07.010.

Laurinavichene, T.V. et al. (2008) "Prolongation of H2 photoproduction by immobilized, sulfur-limited Chlamydomonas reinhardtii cultures," *Journal of Biotechnology*, 134(3–4), pp. 275–277. doi:10.1016/j.jbiotec.2008.01.006.

Lavrinovičs, A. et al. (2021) "Increasing phosphorus uptake efficiency by phosphorus-starved microalgae for municipal wastewater post-treatment," *Microorganisms*, 9(8). doi:10.3390/microorganisms9081598.

Lawton, R.J. et al. (2015) "The effect of salinity on the biomass productivity, protein and lipid composition of a freshwater macroalga," *Algal Research*, 12, pp. 213–220. doi:10.1016/j.algal.2015.09.001.

Leite, G.B., Abdelaziz, A.E.M., & Hallenbeck, P.C. (2013) "Algal biofuels: Challenges and opportunities," *Bioresource Technology*, 145, pp. 134–141. doi:10.1016/j.biortech.2013.02.007.

Lim, D.K.Y. et al. (2012) "Isolation and evaluation of oil-producing microalgae from subtropical coastal and brackish waters," *PLoS One*, 7(7), p. e40751. doi:10.1371/journal.pone.0040751.

Lim, S.-L., Chu, W.-L., & Phang, S.-M. (2010) "Use of Chlorella vulgaris for bioremediation of textile wastewater," *Bioresource Technology*, 101(19), pp. 7314–7322. doi:10.1016/j.biortech.2010.04.092.

Liu, F. et al. (2022) "Elevated-temperature bio-ethanol-assisted water electrolysis for efficient hydrogen production," *Chemical Engineering Journal*, 434, p. 134699. doi:10.1016/j.cej.2022.134699.

Liu, H. et al. (2020) "Evaluation of hydrogen yield potential from Chlorella by photo-fermentation under diverse substrate concentration and enzyme loading," *Bioresource Technology*, 303, p. 122956. doi:10.1016/j.biortech.2020.122956.

Liu, M., Hohenshil, A., & Gadikota, G., 2021 "Integrated CO2 capture and removal via carbon mineralization with inherent regeneration of aqueous solvents," *Energy & Fuels*, 35(9), pp. 8051–8068.doi: 10.1021/acs.energyfuels.0c04346.

Liu, M., & Gadikota, G. (2020) "Single-step, low temperature and integrated CO2 capture and conversion using sodium glycinate to produce calcium carbonate," *Fuel*, 275, p. 117887. doi:10.1016/j.fuel.2020.117887.

Lizotte, M.P. (2001) "The contributions of sea ice algae to Antarctic marine primary production," *American Zoologist*, 41(1), pp. 57–73. doi:10.1093/icb/41.1.57.

Maddi, B. et al. (2016) "Quantitative characterization of the aqueous fraction from hydrothermal liquefaction of algae," *Biomass and Bioenergy*, 93, pp. 122–130. doi:10.1016/j.biombioe.2016.07.010.

Mantzorou, A., & Ververidis, F. (2019) "Microalgal biofilms: A further step over current microalgal cultivation techniques," *The Science of the Total Environment*, 651(Pt 2), pp. 3187–3201. doi:10.1016/j.scitotenv.2018.09.355.

Marshall, A.T., & Haverkamp, R.G. (2008) "Production of hydrogen by the electrochemical reforming of glycerol-water solutions in a PEM electrolysis cell," *International Journal of Hydrogen Energy*, 33(17), pp. 4649–4654. doi:10.1016/j.ijhydene.2008.05.029.

Martínez-Rodríguez, G., Fuentes-Silva, A. L., Velázquez-Torres, D., & Picón-Núñez, M. (2022) "Comprehensive solar thermal integration for industrial processes," *Energy*, 239, 122332. doi: 10.1016/j.energy.2021.122332.

Melis, A., & Melnicki, M. (2006) "Integrated biological hydrogen production," *International Journal of Hydrogen Energy*, 31(11), pp. 1563–1573. doi:10.1016/j.ijhydene.2006.06.038.

Mincer, T.J., & Aicher, A.C. (2016) "Methanol production by a broad phylogenetic array of marine phytoplankton," *PloS one*, 11(3), p.e0150820. doi: 10.1371/journal.pone.0150820.

Moody, J.W., McGinty, C.M., & Quinn, J.C. (2014) "Global evaluation of biofuel potential from microalgae," *Proceedings of the National Academy of Sciences of the United States of America*, 111(23), pp. 8691–8696. doi:10.1073/pnas.1321652111.

Murzin, D. Y. et al. (2017) "Kinetics, modeling, and process design of hydrogen production by aqueous phase reforming of xylitol," *Industrial & Engineering Chemistry Research*, 56(45), 13240–13253. doi:10.1021/acs.iecr.7b01636.

Nagarajan, D., Lee, D.-J., Kondo, A., & Chang, J.-S. (2016) "Recent insights into biohydrogen production by microalgae - From biophotolysis to dark fermentation," *Bioresource Technology*, 227, pp. 373–387. doi:10.1016/j.biortech.2016.12.104.

Ochonma, P., Blaudeau, C., Krasnoff, R., & Gadikota, G., (2021) Exploring the thermodynamic limits of enhanced H2 recovery with inherent carbon removal from low value aqueous biomass oxygenate precursors. *Frontiers in Energy Research*, 9, p. 742323. doi:10.3389/fenrg.2021.742323.

Ochonma, P. et al. (2023) "Integrated low carbon H2 conversion with in situ carbon mineralization from aqueous biomass oxygenate precursors by tuning reactive multiphase chemical interactions," *Reaction Chemistry & Engineering*, 8(8), pp. 1943–1959. doi: 10.1039/D2RE00542E.

Onyeaka, H. et al. (2021) "Minimizing carbon footprint via microalgae as a biological capture," *Carbon Capture Science & Technology*, 1, p. 100007. doi:10.1016/j.ccst.2021.100007.

Peccia, J. et al. (2013) "Nitrogen supply is an important driver of sustainable microalgae biofuel production," *Trends in Biotechnology*, 31(3), pp. 134–138. doi:10.1016/j.tibtech.2013.01.010.

Rahman, M. et al. (2014) "Bimetallic Pt–Ni composites on ceria-doped alumina supports as catalysts in the aqueous-phase reforming of glycerol," *RSC advances*, 4(36), pp. 18951–18960. doi: 10.1039/C4RA00355A.

Ravichandran, S.R. et al. (2022) "A review on hydrothermal liquefaction of algal biomass on process parameters, purification and applications," *Fuel*, 313, p. 122679. doi:10.1016/j.fuel.2021.122679.

Renaud, S M., & Parry, D L. (1994) "Microalgae for use in tropical aquaculture II: Effect of salinity on growth, gross chemical composition and fatty acid composition of three species of marine microalgae," *Journal of Applied Phycology*, 6, pp. 347–356.

Rezvani, S., Moheimani, N.R., & Bahri, P.A. (2016) "Techno-economic assessment of CO2 bio-fixation using microalgae in connection with three different state-of-the-art power plants," *Computers & Chemical Engineering*, 84, pp. 290–301. doi:10.1016/j.compchemeng.2015.09.001.

Rochelle, G T. (2009) "Amine scrubbing for CO2 capture," *Science*, 325(5948), pp. 1652–1654. doi:10.1126/science.1176731.

Ruiz-López, E. et al. (2020) "Over-faradaic hydrogen production in methanol electrolysis cells," *Chemical Engineering Journal*, 396, p. 125217. doi:10.1016/j.cej.2020.125217.

Saifuddin, N., & Priatharsini, P. (2016) "Developments in bio-hydrogen production from algae: A review," *Research Journal of Applied Sciences, Engineering and Technology*, 12(9), pp. 968–982. doi:10.19026/rjaset.12.2815.

Sarangi, P.K., & Nanda, S. (2020) "Biohydrogen production through dark fermentation," *Chemical Engineering & Technology - CET*, 43(4), pp. 601–612. doi:10.1002/ceat.201900452.

Sekar, M. et al. (2021) "A review on the pyrolysis of algal biomass for biochar and bio-oil - Bottlenecks and scope," *Fuel*, 283, p. 119190. doi:10.1016/j.fuel.2020.119190.

Shabaker, J W. et al. (2003) "Aqueous-phase reforming of methanol and ethylene glycol over alumina-supported platinum catalysts," *Journal of Catalysis*, 215(2), pp. 344–352. doi:10.1016/S0021-9517(03)00032-0.

Shabaker, J W., Huber, G.W., & Dumesic, J.A. (2004) "Aqueous-phase reforming of oxygenated hydrocarbons over Sn-modified Ni catalysts," *Journal of Catalysis*, 222(1), pp. 180–191. doi:10.1016/j.jcat.2003.10.022.

Shakya, R. et al. (2015) "Effect of temperature and Na2CO3 catalyst on hydrothermal liquefaction of algae," *Algal Research*, 12, pp. 80–90. doi:10.1016/j.algal.2015.08.006.

Sharma, A.K., Sharma, C., Mullick, S.C., & Kandpal, T.C. (2017) "Solar industrial process heating: A review," *Renewable and Sustainable Energy Reviews*, 78, pp. 124–137. doi:10.1016/j.rser.2017.04.079.

Sheath, R.G., & Wehr, J.D. (2015) "Introduction to the freshwater algae," in *Freshwater algae of North America*. Academic Press, pp. 1–11. doi:10.1016/B978-0-12-385876-4.00001-3.

Show, K.-Y. et al. (2018) "Hydrogen production from algal biomass - Advances, challenges and prospects," *Bioresource Technology*, 257, pp. 290–300. doi:10.1016/j.biortech.2018.02.105.

Show, K.-Y., Yan, Y.-G., & Lee, D.-J. (2019) "Biohydrogen production from algae: Perspectives, challenges, and prospects," in *Biofuels from algae*. Elsevier, pp. 325–343. doi:10.1016/B978-0-444-64192-2.00013-5.

Singh, R., Bhaskar, T., & Balagurumurthy, B. (2015) "Effect of solvent on the hydrothermal liquefaction of macro algae *Ulva fasciata*," *Process Safety and Environmental Protection*, 93, pp. 154–160. doi:10.1016/j.psep.2014.03.002.

Singh, U.B., & Ahluwalia, A.S. (2013) "Microalgae: A promising tool for carbon sequestration," *Mitigation and Adaptation Strategies for Global Change*, 18(1), pp. 73–95. doi:10.1007/s11027-012-9393-3.

Smith, V.H. et al. (2010) "The ecology of algal biodiesel production," *Trends in Ecology & Evolution*, 25(5), pp. 301–309. doi:10.1016/j.tree.2009.11.007.

Starks, T.L., Shubert, L.E., & Trainor, F.R. (1981) "Ecology of soil algae: A review," *Phycologia*, 20(1), pp. 65–80. doi:10.2216/i0031-8884-20-1-65.1.

Stevens, S.E., Patterson, C.O.P., & Myers, J. (1973) "The production of hydrogen peroxide by glue-green algae: A survey," *Journal of Phycology*, 9(4), pp. 427–430. doi:10.1111/j.1529-8817.1973.tb04116.x

Sun, P. et al. (2019) "Criteria air pollutants and greenhouse gas emissions from hydrogen production in US steam methane reforming facilities," *Environmental Science & Technology*, 53(12), pp. 7103–7113. doi:10.1021/acs.est.8b06197.

Sunphorka, S. et al. (2015) "Biocrude oil production and nutrient recovery from algae by two-step hydrothermal liquefaction using a semi-continuous reactor," *Korean Journal of Chemical Engineering*, 32, pp. 79–87. doi: 10.1007/s11814-014-0165-5.

Swetha, A. et al. (2021) "Review on hydrothermal liquefaction aqueous phase as a valuable resource for biofuels, bio-hydrogen and valuable bio-chemicals recovery," *Chemosphere*, 283, p. 131248. doi:10.1016/j.chemosphere.2021.131248.

Takagi, M., Karseno, & Yoshida, T. (2006) "Effect of salt concentration on intracellular accumulation of lipids and triacylglyceride in marine microalgae Dunaliella cells," *Journal of Bioscience and Bioengineering*, 101(3), pp. 223–226. doi:10.1263/jbb.101.223.

Tamburic, B. et al. (2011) "Design of a novel flat-plate photobioreactor system for green algal hydrogen production," *International Journal of Hydrogen Energy*, 36(11), pp. 6578–6591. doi:10.1016/j.ijhydene.2011.02.091.

Thompson, G.A. (1996) "Lipids and membrane function in green algae," *Biochimica et Biophysica Acta*, 1302(1), pp. 17–45. doi:10.1016/0005-2760(96)00045-8.

Tian, C., Li, B., Liu, Z., Zhang, Y., & Lu, H. (2014) "Hydrothermal liquefaction for algal biorefinery: A critical review," *Renewable and Sustainable Energy Reviews*, 38, pp. 933–950. doi:10.1016/j.rser.2014.07.030.

Ugya, A.Y., Ajibade, F.O., & Hua, X. (2021) "The efficiency of microalgae biofilm in the phycoremediation of water from River Kaduna," *Journal of Environmental Management*, 295, p. 113109. doi:10.1016/j.jenvman.2021.113109.

Vaidya, P.D., & Lopez-Sanchez, J.A. (2017) "Review of hydrogen production by catalytic aqueous-phase reforming," *ChemistrySelect*, 2(22), pp. 6563–6576. doi:10.1002/slct.201700905.

Wang, L. et al. (2018) "Acclimation process of cultivating Chlorella vulgaris in toxic excess sludge extract and its response mechanism," *The Science of the Total Environment*, 628–629, pp. 858–869. doi:10.1016/j.scitotenv.2018.02.020.

Wang, X. et al. (2020) "Economically important red algae resources along the Chinese coast: History, status, and prospects for their utilization," *Algal Research*, 46, p. 101817. doi:10.1016/j.algal.2020.101817.

Wang, X., Sheng, L., & Yang, X. (2017) "Pyrolysis characteristics and pathways of protein, lipid and carbohydrate isolated from microalgae Nannochloropsis sp," *Bioresource Technology*, 229, pp. 119–125. doi:10.1016/j.biortech.2017.01.018.

Weaver, P.F., Lien, S., & Seibert, M. (1980) "Photobiological production of hydrogen," *Solar Energy*, 24(1), pp. 3–45. doi:10.1016/0038-092X(80)90018-3.

Winkler, M. (2002) "6Fe9-hydrogenases in green algae: photo-fermentation and hydrogen evolution under sulfur deprivation," *International Journal of Hydrogen Energy*, 27(11–12), pp. 1431–1439. doi:10.1016/S0360-3199(02)00095-2.

Xia, A. et al. (2015) "Fermentative hydrogen production using algal biomass as feedstock," *Renewable and Sustainable Energy Reviews*, 51, pp. 209–230. doi:10.1016/j.rser.2015.05.076.

Xiao, M. et al. (2015) "Onsite deposition of self-repairing biomimetic nanostructured ni catalysts with improved electrocatalysis toward glycerol oxidation for H2 production," *Electrochimica Acta*, 178, pp. 209–216. doi:10.1016/j.electacta.2015.07.185.

Xie, T. et al. (2017) "Nitrate concentration-shift cultivation to enhance protein content of heterotrophic microalga *Chlorella vulgaris*: Over-compensation strategy," *Bioresource Technology*, 233, pp. 247–255. doi:10.1016/j.biortech.2017.02.099.

Xu, X., Kentish, S.E., & Martin, G.J.O. (2021) "Direct air capture of CO2 by microalgae with buoyant beads encapsulating carbonic anhydrase," *ACS Sustainable Chemistry & Engineering*. doi:10.1021/acssuschemeng.1c01618.

Yang, Y.F., Feng, C.P., Inamori, Y., & Maekawa, T. (2004) "Analysis of energy conversion characteristics in liquefaction of algae," *Resources, Conservation and Recycling*, 43(1), pp. 21–33. doi:10.1016/j.resconrec.2004.03.003.

Yin, T., Yin, S., Srivastava, A., & Gadikota, G. (2022) "Regenerable solvents mediate accelerated low temperature CO2 capture and carbon mineralization of ash and nano-scale calcium carbonate," *Resources, Conservation and Recycling*, 180, p. 106209. doi:10.1016/j.resconrec.2022.106209.

Zheng, Q., Martin, G.J.O., & Kentish, S.E. (2016) "Energy efficient transfer of carbon dioxide from flue gases to microalgal systems," *Energy & Environmental Science*, 9(3), pp. 1074–1082. doi:10.1039/C5EE02005K.

Zhu, J., Lei, X., Quan, J., & Yue, X. (2019) "Algae growth distribution and key prevention and control positions for the middle route of the South-to-North water diversion project," *Water*, 11(9), p. 1851. doi:10.3390/w11091851.

Zou, G. et al. (2020) "Cultivation of Chlorella vulgaris in a Light-Receiving-Plate (LRP)-Enhanced raceway pond for ammonium and phosphorus removal from pretreated pig urine," *Energies*, 13(7), p. 1644. doi:10.3390/en13071644.

Section III

Biomass Biochemical Conversion Processes

9 Biophotolysis Process for Biomass Conversion to Hydrogen

Amizon Azizan and Rafidah Jalil

9.1 INTRODUCTION

Hydrogen fuel commercialization faces significant technological, economic, and environmental challenges. To ensure long-term sustainability, low-cost hydrogen (H_2) production technologies must be industrialized, which necessitates the optimization of crucial process parameters, genetic modifications, and metabolic engineering (Ahmed et al., 2021). Hydrogen is one of the most plentiful renewable energy sources. Additionally, the combustion of H_2 produces solely vapor. As a result, it is recognized as the greenest source of energy (Abdalla et al., 2018). Recently, several countries are becoming more interested in advancing the production of H_2 technology. This is due to harnessing H_2 as an energy carrier in conjunction with other potential local fuels as well as technologies that can improve long-term energy security, simultaneously reducing air pollution and greenhouse gas (GHG) emissions (Buitrón et al., 2017). Currently, the H_2 demand contributes of about 80% from the total energy until the year 2025 where H_2 is mostly used for fuels, electrical generation, and high-value-added chemical production (Melitos et al., 2021).

Hydrogen is a potentially valuable and useful fuel and energy carrier, nonpolluting with up to zero GHGs and environmentally friendly as well as green products (Benemann, n.d.; Melitos et al., 2021). Its breakdown produces no pollutants, in contrast to ethanol, where widespread usage is expected to emit enormous amounts of carcinogenic acetaldehyde while also producing huge quantities of pollution (Jose, 2021). Besides that, due to its high energy density and annual production of 55 million tons, H_2 is a crucial environmentally benign industrial feedstock. The H_2 market is expected to expand by up to 10% annually to more than \$191.80 billion by 2024 (Ahmed et al., 2021). The benefits of H_2 include zero GHG emissions if it is made from green resources as well as a high energy content ranging from 120 to 142 MJ/kg. The practicality and application of H_2 necessitate an investigation of factors such as storage capacity, energy content, transportation, versatility, and environmental impact (Abdalla et al., 2018).

Although H_2 is used in a variety of industrial applications, most of it is created using non-free processes such as steam reforming or electrolysis and water splitting by water electrolysis and photoelectrolysis, where more than 85% of H_2 available in the market is produced by steam reforming of methane (Hallenbeck, 2013; Buitrón et al., 2017).

DOI: 10.1201/9781003382270-12

9.2 BIOPHOTOLYSIS

Biophotolysis, or the synthesis of H_2 gas from water and solar energy using microalgae, has been a subject of applied research since the early 1970s (Benemann, 2000). Biophotolysis is a process similar to plant-type photosynthesis that produces H_2 through water breakdown under the influence of photosynthesis microorganisms such as microalgae and cyanobacteria. These microorganisms use solar energy as a source of light energy to produce either carbohydrates through photosynthesis or H_2 by splitting water through biophotolysis for hydrogen generation under anaerobic circumstances (Wukovits and Schnitzhofer, 2009; Hallenbeck, 2013; Azwar et al., 2014). Moreover, H_2 production via biophotolysis methods by microalgae and cyanobacteria has emerged as an alternate way of gaseous energy recovery with the potential to be used in renewable energy production (Azwar et al., 2014). Generally, biophotolysis is divided into direct and indirect processes. Figure 9.1 shows one of the optional routes generally using water (H_2O) in a direct and indirect biophotolysis process with different route processes.

Direct biophotolysis is a method that uses solar energy to turn H_2O into H_2 and O_2 with the presence of microorganisms such as microalgae which are rich in carbohydrate and protein as feedstock. It is similarly named as a photosynthetic reaction mechanism directly using solar energy (strictly light dependent) to produce H_2 (Ahmed et al., 2021; Ghiasian, 2019). The method is particularly sensitive to O_2 (the enzyme) even though the sunlight conversion efficiency is over 80% (Ahmed et al., 2021). The cost of producing biohydrogen is yet sustainable due to the high cost ranging between \$2.13 and \$7.54 for every kilogram of H_2O for direct biophotolysis and \$1.42 for every kilogram of H_2O if using indirect biophotolysis (Ahmed et al., 2021).

Because of the sensitivity of O_2, H_2 and O_2 are separated during the indirect biophotolysis by separating the stages beginning with photosynthesis, followed by dark

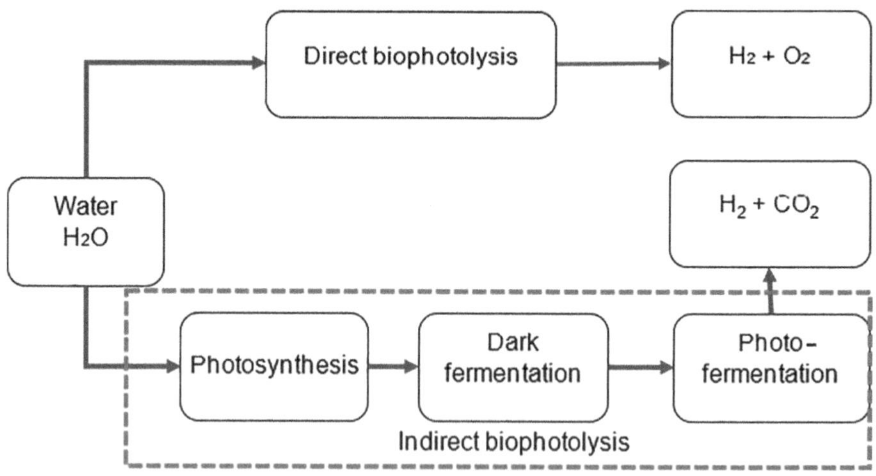

FIGURE 9.1   General route using water (H_2O) in a direct and indirect biophotolysis process.

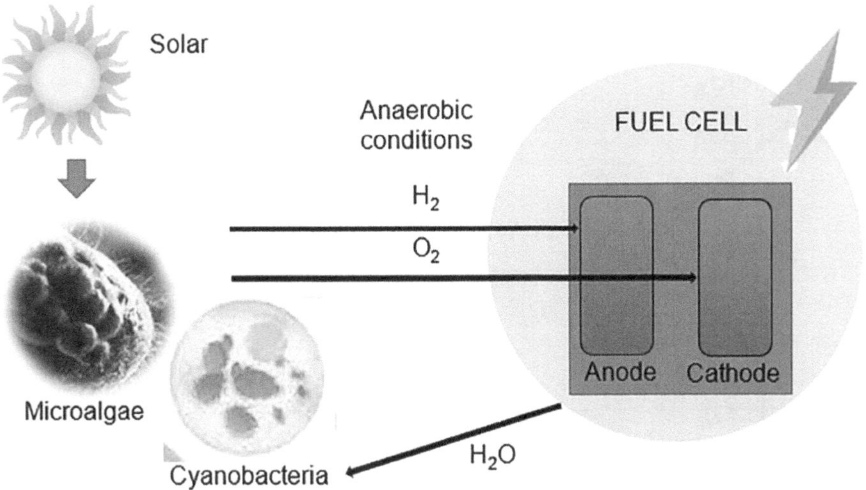

FIGURE 9.2 Utilization of hydrogen for fuel cells to generate electricity.

fermentation and photo-fermentation (Wukovits and Schnitzhofer, 2009). In other words, indirect biophotolysis is defined as an anaerobic fermentation degrading any starch (could come from microalgae) and glycogen (could come from cyanobacteria) accumulated after the carbon dioxide (CO_2) fixation and the photosynthesis mechanism (with photosystem I (PSI) or II (PSII) cells under the presence of light), the light-driven mechanisms, in which the anaerobic fermentation can be conducted by the dark-driven mechanisms (without light) (Ghiasian, 2019). Photo-fermentation is considered a better process compared to that of dark fermentation, with the former producing a maximum of 141.42 mL/g total substrate, while the latter (dark fermentation) produces a minimum H_2 yield of 36.08 mL/g total substrate (Ahmed et al., 2021).

According to Wukovits and Schnitzhofer (2009), under anaerobic conditions, microalgae can divide H_2O into H_2 and O_2 with the use of light energy. Microalgae photosynthesis is separated into two reactions: light-driven reaction and dark-driven reaction. With the availability of numerous resources, such as H_2O as a substrate and solar energy as a source of energy, H_2 production by microalgae and cyanobacteria is one of the methods that produce green energy that does not release GHG effect. As demonstrated in Figure 9.2, the H_2 gas produced might be used in a fuel cell to generate energy (Azwar et al., 2014).

9.2.1 Direct Biophotolysis

According to Wukovits and Schnitzhofer (2009) and Azwar et al. (2014), direct biophotolysis is a biological process that can produce H_2 directly from dissociation of H_2O or splitting of H_2O in PSII (Ahmed et al., 2021; Vargas et al., 2014) using photosynthetic microorganisms to convert solar energy into chemical energy in the form of H_2, producing the low potential reductant or ferredoxin (Fd) (Razu et al., 2019).

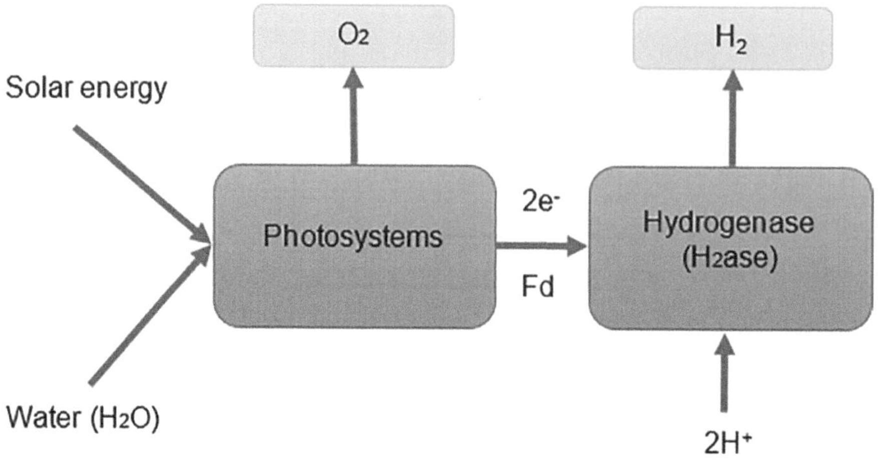

FIGURE 9.3   Direct biophotolysis of hydrogen production in PSI and PSII mechanisms.

The chemical energy is then transferred through the electron transport chain, moving through the membrane-bound electron transport chain consisting of PSI and PSII (Razu et al., 2019), reducing ferredoxin (Fd), providing electrons to the hydrogenase (H_2ase) enzyme (Azwar et al., 2014). The reaction of the dissociation of H_2O to O_2 with the aid of solar energy is described as equation (9.1) (Ahmed et al., 2021; Azwar et al., 2014), which is also described in Figure 9.3.

$$2H_2O + Fd_{Oxidized} + \text{solar energy} - \rightarrow 4H^+ + Fd_{reduced}\left(4e^-\right) + O_2 \qquad (9.1)$$

Many types of photosynthetic microorganisms isolated from soil or natural H_2O such as microalgae and cyanobacteria can be used in the biophotolysis process to produce H_2 (Mona et al., 2020; Azwar et al., 2014). Generally, direct biophotolysis occurs in the photosynthetic system, PSI and PSII, where both processes operate in series. PSI and PSII assimilate light energy and electrons which are being generated from the endogenous substrates via catabolism (Ahmed et al., 2021) and are transferred to the plastoquinone (PQ) pool in PSI and PSII themselves.

In photosynthetic-type microorganisms, chlorophyll a type that is contained is the predominant light-absorbing pigment in PSI (Ghiasian, 2019). The PSI phase generates reductants for CO_2 reduction, while the PSII phase is the process of separating H_2O molecules using solar energy and releasing O_2 into the atmosphere (Azwar et al., 2014; Abdalla et al., 2018; Ahmed et al., 2021). PSII has phycobilins as the predominant energy collectors in PSII to the photosynthetic reaction center with a small number of chlorophyll types (Ghiasian, 2019).

The solar energy received by PSI and PSII oxidizes O_2 molecules by transferring energized electrons (e^-) that are boosted in energy to Fd via the electron transport chain. During this process, NADP+ is reduced to nicotinamide NADPH by plastoquinol produced through an electron transfer (Ahmed et al., 2021). Then, the reduced

FD acts as an electron donor, transporting electrons to the H_2ase enzyme, which catalyzes the reversible conversion of a proton to H_2 where this proton is needed to generate ATP through ATP synthase. The PSII process creates no harmful GHGs that affect the environment (Hallenbeck, 2013; Benemann, 2000; Mona et al., 2020; Ahmed et al., 2021; Benemann, n.d.). However, once the O_2 level exceeds 0.1%, H_2ase's ability to generate H_2 is limited. As a result, the O_2 content should be maintained (Abdalla et al., 2018). Green algae and cyanobacteria are oxygenic (producing O_2) photosynthetic microorganisms which can split H_2O and convert sunlight to chemical energy with the former as the only microorganism which with the absence of O_2 allows the biophotolysis to occur (Ahmed et al., 2021).

In elaboration, the H_2O to plastoquinone oxidoreductase (H_2O-PQOR) which exists in the chloroplast as situated in PSII energizes the electrons under the presence of solar energy, transferring the electrons to PSI, causing the Fd to be reduced (Mona et al., 2020) by Fd-oxidoreductase, that then donates an electron to FeFe-H_2ase. With this, the H_2 protons H^+ to H_2 molecules (H_2) and produces O_2 as shown in equation (9.1) at the PSI (Ahmed et al., 2021). This Fd reduction in PSI is explained by equation (9.2). The protons of H+ released via the photochemical oxidation of H_2O are the terminal acceptors of the electrons in the chloroplast, enabling the biotransformation of H_2O to O_2 and H_2 at the same time (El-Dalatony et al., 2020). Combining equations (9.1–9.3) is the summarized equation representing the direct biophotolysis.

$$4H^+ + Fd_{reduced}\left(4e^-\right) 2H_2 + Fd_{oxidized} \tag{9.2}$$

$$2H_2O + Solar\ energy \ -\rightarrow 2H_2 + O_2 \tag{9.3}$$

The generation of O_2 via the splitting of H_2O becomes a limiting factor (El-Dalatony et al., 2020) to produce H_2 due to the O_2 sensitivity of the H_2ase, and to increase the yield of H_2, one of the ways is to investigate the suppression of O_2 during the biophotolysis, allowing H_2 production.

9.2.2 BIOPHOTOLYSIS OF WATER BY GREEN MICROALGAE

Hydrogen is among the most potential areas of bioenergy that may be produced from microalgae via biological pathways that occur at ambient temperature and atmospheric pressure, making it a low-energy-intensive process and low production cost. Microalgae biomass is also widely used to produce a variety of bioenergy (Ahmed et al., 2021; Bechara et al., 2021). The mechanism of H_2 generation by biophotolysis is H_2 generated from H_2O utilizing sunlight and CO_2 as the main energy supply via the H_2ase enzyme produced by microalgae (Azwar et al., 2014). Microalgae have 10–15 times higher capability for CO_2 fixation with 183 tons of CO_2 being fixated for every 100 tons of microalgae cell biomass being generated, in comparison with that of terrestrial plants' capability (Bhatia et al., 2021).

Hydrogen production from microalgae is an attractive and eco-friendly process with no accumulation of CO_2 with solar energy efficiency of less than 80% (Mona et al., 2020; Jose, 2021). The most popular microalgae for producing H_2 is

TABLE 9.1

Comparison of Hydrogen Production from Various Species of Microalgae

Bacterial Strains	H$_2$ Production Rate (mL/L·h)	Reference
Chlamydomonas MGA 161	4.48	
Chlorella reinhardtii	2.80 – 2.90	Mona et al. (2020)
Chlamydomonas reinhardtii D239–40	7.10	Buitrón et al. (2017)
Chlamydomonas reinhardtii V219M	6.28	Buitrón et al. (2017)
C. reinhardtii D239–40	2.92	Melitos et al. (2021)
C. reinhardtii D239–40	2.31	Melitos et al. (2021)
Chlamydomonas reinhardtii CC-124	7.50	
Chlamydomonas reinhardtii CC-1036	9.20	

Chlamydomonas reinhardtii followed by Scenedesmus obliquus and *Chlorella fusca* (Buitrón et al., 2017; Mona et al., 2020). Melitos et al. (2021) reported that H$_2$ is produced by using microalgae ranging from 25°C to 30°C at pH range of 7–8 with light intensity ranging from 100 to 160 μE/m^2/s with H$_2$ production rate varies from 0.80 to 9.20 mL/L·h. Ahmed et al. (2021) reported that H$_2$ is produced from microalgae via biophotolysis process ranging from 0.015 to 1.084 mmol/L·h. Table 9.1 shows the production of H$_2$ from various species of microalgae.

Like higher plants, microalgae use light energy and carry out oxygenic photosynthesis (under anoxic condition), with the assistance of PSI and PSII. The H$_2$ase uses the decreased Fd's e to reduce H+ and produce H$_2$ when there is a lack of O$_2$ (Mona et al., 2020) via photochemical oxidation of H$_2$O (El-Dalatony et al., 2020). According to Abdalla et al. (2018), a larger proportion of H$_2$ production can be reached by using mutant-derived microalgae as it can tolerate O$_2$.

Benemann (2000) reported that under highly controlled laboratory parameter conditions, the light conversion efficiency of such a sustained direct biophotolysis process was proven to be as high as 22% using the microalgae *Chlamydomonas reinhardtii*, practically the theoretical maximum. *Chlamydomonas* mutants produce H$_2$ at O$_2$ partial pressures considerably greater than those tolerated by the parent strains that have been identified. Nevertheless, it indicates to be respiratory mutants that drastically lower O$_2$ concentrations in the culture broth, which is harmful to H$_2$ generation. Buitrón et al. (2017) summarized that the decreased antenna size, high photosynthetic capacity, high respiration-to-photosynthesis ratio, and fast induction of the xanthophyll cycle, which shields PSII from deactivation, are the major characteristics of high H$_2$-producing *C. reinhardtii* strains.

Under photosynthesis growth conditions, sulfur-deprived *Chlamydomonas reinhardtii* was able to generate photobiological H$_2$ synthesis, and H$_2$ production was artificially continued by limiting photosynthesis process. Depletion of sulfate slows down the photosynthesis and less O$_2$ is generated than is required for respiration, causing the culture to become anaerobic and shift from carbon fixation to a mixture of H$_2$ generation and starch breakdown where it helps to use the small quantity of O$_2$ that remains. So, the culture begins to accumulate H$_2$ gas (Jose, 2021).

9.2.3 Biophotolysis of Water by Cyanobacteria

Photosynthesis-based fermentation using cyanobacteria cell biomass which contains chlorophyll a, utilizing solar energy, is aided by PSI and PSII. In PSII, the photosystem absorbs solar energy at a wavelength of less than 680 nm (P680) producing strong oxidant which converts H_2O to O_2 gas, protons, and electrons The most popular cyanobacteria for producing H_2 is *Anabaena variabilis, Cyanothece* sp., and *Synechocystis* PCC (Ahmed et al., 2021). Melitos et al. (2021) reported that the H_2 is produced by using cyanobacteria ranging from 26°C to 30°C with light intensity ranging from 20 to 456 $\mu E/m^2/s$ with H_2 production rate varies from 0.90 to 14.90 mL/L·h. Table 9.2 shows the production of H_2 from various strains of cyanobacteria.

According to Melitos et al. (2021), cyanobacteria show considerably higher production efficiency than microalgae and many cyanobacterial strains demand low light intensity (compared to microalgae) and thus significantly lower energy demands.

Many cyanobacteria are capable of fixing nitrogen (N_2) from the atmosphere (Ghiasian, 2019). Theoretically, biophotolysis with cyanobacteria works when the nitrogenase (N_2ase) which exists in blue-green algae (cyanobacteria) is the prokaryotic bacteria carrying N_2ase. It drives N_2 fixation to catabolize substrates (i.e., cyanobacteria in H_2O) and produces H_2, which is sensitive to O_2 or free O_2 (being inactivated). Endogenous substrates catabolize and produce electrons which are being used in biophotolysis by bacteria, for instance, a wild isolated cyanobacteria from nature. Cyanobacteria can also possess multiple H_2-producing enzymes (for instance, H_2ase or N_2ase) and can produce H_2 under the presence of light (light-driven) or without light (dark-driven) conditions (Ghiasian, 2019).

The performance of producing H_2 of the cyanobacteria may differ even with the same genetic origin even from the wild. Engineeringly, this is because there are some cyanobacteria with N_2 fixation capability which are sensitive to the presence of either N_2, O_2, or CO_2. Any wild isolated cyanobacteria, for instance, do still depend on physiological and chemical factors like light, nutrient salt, biomass density, temperature, pH, O_2, N_2, and CO_2 content during the production of H_2 (Ghiasian, 2019; Kamshybayeva et al., 2023), and it is found that N_2ase enzymes in cyanobacteria are more sensitive to O_2 than those in H_2ase.

TABLE 9.2

Examples for Hydrogen Production from Various Strains of Cyanobacteria: Mostly *Anabaena* Species

Bacterial Strains	H_2 Production Rate (mL/L.hr)	Reference
Anabaena variabilis ATCC 29413	13.00	
Anabaena PCC7120 and AMC 414	14.90	
Arthrospira sp. PCC 8005.	5.91	
Anabaena PCC7 120	14.90	Mona et al. (2020)
Anabaena variabilis	20.00	Mona et al. (2020)
Anabaena variabilis ATCC 29413	14.90	Azwar et al. (2014)
Anabaena AMC 414	14.90	
ChroococcidiopsiSs thermalis CALU758	4.03	

The isolated wild cyanobacteria strains, for instance, there exist the so-called heterocyst cyanobacteria which can choose to be either in the state of vegetative state or the heterocyst state depending on the state of substrates' presence, namely as the heterocyst cyanobacteria strain scheme of H_2. Heterocyst filamentous diazotrophic *Anabaena* species as earlier shown in Table 9.2 is shown to have the N_2-fixing (N_2-starving) cyanobacteria strain (isolated from the wild) which is effective to synthesize H_2 micro-aerobically, happens to be producing the highest H_2 production capacity in a short time, and the yield of the H_2 produced from N_2-fixing cyanobacteria rises over time (Kamshybayeva et al., 2023).

For instance, *Anabaena* sp., the phototrophic bacteria that were isolated from rice paddies in the Republic of Kazakhstan, grown in BG-11 culture media, are ecologically prone to the cellular protein mechanism of H_2ase and N_2ase routes. When the N_2ase state is activated in the heterocyst state (as secluded in a thick protective layer of O_2), the glycogen reserves in heterocyst cells are depleted, and at the same time, at the vegetative state where the H_2ase enzyme exists, they are hampered or deactivated by the high concentration of O_2 which is produced via PSII. At this stage of the depletion of the cellular energy reserves, in the heterocyst state where the N_2ase is activated or in other words as in N_2 fixation, it needs 16 adenosine triphosphate (ATP) molecules for hydrolysis.

Interestingly, in the presence of carbon contents like the CO_2 concentration, N_2 fixation can increase the yield of H_2 production as the rate of the photosynthesis and its N_2 fixation in cyanobacteria cell increase to a certain maximum peak before it reduces due to the reduction of nutritional contents like carbon and nitrogen contents.

9.3 ADVANTAGES AND DISADVANTAGES OF DIRECT AND INDIRECT BIOPHOTOLYSIS

The advantage of biophotolysis is that there is no need to supply substrate as nutrients. Water is the principal electron donor required for H_2 gas generation. Solar energy and CO_2 are the primary inputs required to grow microalgae or cyanobacteria via the H_2ase enzyme during the biophotolysis process. Also, it can remove N_2 from the environment (Azwar et al., 2014). Direct photolysis has the benefits of high solar conversion of microalgae, clean H_2 generation, and low CO_2 consumption when compared to other H_2 production techniques. Still, this technology requires more research to overcome obstacles and discover ideal operating conditions before it can be used at commercial scale (Bechara et al., 2021).

According to Mona et al. (2020), when compared to plants, the direct photolysis process may produce H_2 directly from H_2O and solar, with solar conversion energy increasing by tenfold. Unfortunately, it has significant limitations, including a high sensitivity of H_2ase to O_2 and a low light conversion efficiency. To deal with this problem, O_2 can be utilized as an absorbance and H_2ase resistant through light input optimization in a photobioreactor (PBR).

One of the disadvantages of biophotolysis is the inhibition of H_2 synthesis driven by the accumulation of O_2 (Table 9.3). Other disadvantages are the risk of utilizing O_2, the need for strong light, and having poor photochemical efficiency. Some additional obstacles include the separation and accumulation of cell biomass, photosynthetic capacity ratio, and respiration (Ahmed et al., 2021). Hence, for the sustainability of

TABLE 9.3

Advantages and Disadvantages of Hydrogen Production via Direct and Indirect Biophotolysis

Process	Advantages	Disadvantages
Direct biophotolysis	Very abundant, inexpensive substrate, solar energy capture, high solar flux	Oxygen evolved by photosynthesis incompatible with hydrogen-evolving catalyst
	Hydrogen production directly from water and solar with maximum efficiency single-stage process (Ghiasian, 2019)	High intensity of light
	Separate O_2 generation from H_2 production requirement	O_2 can be dangerous to the system and lower photochemical efficiencies and H_2ase enzyme generates carbon dioxide and provides a low yield of hydrogen
	Solar conversion energy increased by tenfold as compared to trees and crops	When needing PBRs, a large hydrogen impermeable one is required (Ghiasian, 2019)
	Metabolite by-product conversion to H_2	Hydrogenate uptake is reduced and uptake of hydrogenates can be removed
Indirect biophotolysis	Fixing nitrogen from the atmosphere	No waste utilization
	Possibly can reduce the need of PBR when the separations of the incompatible evolved oxygen and hydrogen reaction is possible	Energy loss possibly occurs between stages and during production
	Oxygen sensitivity problem is overcome	Expensive separation system in the hydrogen isolation is needed

Source: Antonopoulou et al. (2011), Hallenbeck (2013), Ahmed et al. (2021), Ghiasian (2019), Nikolaidis and Poullikkas (2017), and Hitam and Jalil (2020).

H_2 production, it is required to retain the O_2 content at a low level which is below 0.1% (Antonopoulou et al., 2011).

Another challenge for the commercial scale of H_2 production is the modeling and simulation of photolytic systems to aid the system design and optimization process (Antonopoulou et al., 2011).

9.4 INDIRECT BIOPHOTOLYSIS OF WATER

Indirect biophotolysis of H_2O uses green algae or cyanobacterium via photosynthesis via solar energy to split H_2O molecules, decreasing ferredoxin, H_2ase, or N_2ase which are also sensitive to O_2 (Azwar et al., 2014). These are the mechanisms for the production of H_2 for the indirect biophotolysis which are shown in equations (9.4) and (9.5).

$$6H_2O + 6C_2O \; - \rightarrow \left(C_6H_{12}O_6\right)_n + 6O_2 \; \left(\text{stage 1}\right) \qquad (9.4)$$

$$\left(C_6H_{12}O_6\right)_n + 12H_2O - \rightarrow 12H_2 + 6CO_2 \; \left(\text{stage 2}\right) \qquad (9.5)$$

For the indirect biophotolysis, there are two stages: stage 1 and stage 2. In stage 1 (Acar and Dincer, 2018; Ghiasian, 2019; Hitam and Jalil, 2020; Kothari et al., 2010; Nikolaidis and Poullikkas, 2017), CO_2 fixation occurs, producing reduced substrates (carbohydrate accumulations) either starch from microalgae or glycogen from cyanobacteria releasing the electrons and ferredoxin (Fd) as well as producing O_2 (Ahmed et al., 2021). The functioning of the electron transport chain stops when the O_2 concentration is reduced and faces shortages, causing anaerobic condition to occur. Then, these reduced substrates are used in the second stage in a dark-driven fermentation anaerobically using the carbon reserve from stage 1, activating the O_2-sensitive H_2ase, assisting indirect biophotolysis of H_2O in producing H_2, and CO_2 being re-evolved (Ahmed et al., 2021; Jiménez-Llanos et al., 2020). Here, for instance, for a green microalga, the H_2ase activation in chloroplast depends on the influence by the presence of either the external or internal photosynthetic O_2, inhibiting and blocking the H_2 production genes in the metabolism, thus helping under strict anoxic conditions.

Figure 9.4 shows the indirect biophotolysis mechanisms for stage 1 and stage 2 which can be conducted by microalgae or cyanobacteria using H_2O. In stage 1,

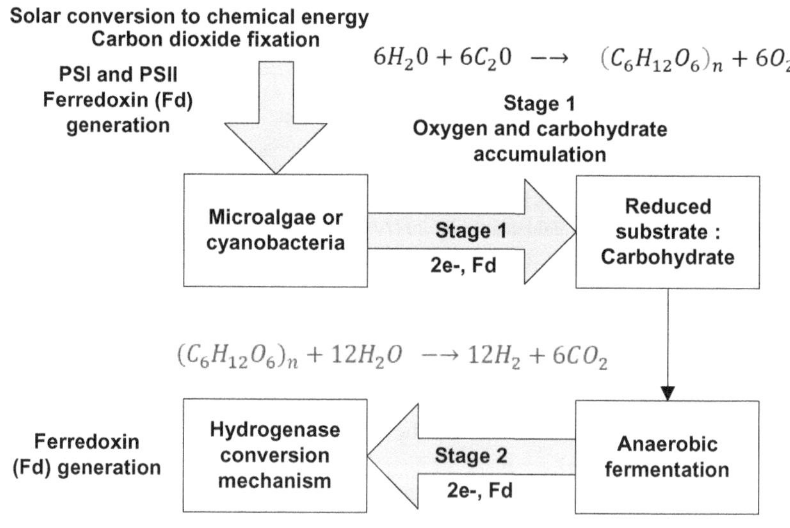

FIGURE 9.4 Indirect biophotolysis stage 1 and stage 2 mechanisms for a possible reversible H_2ase-based mechanism.

electrons are extracted from H_2O (via oxidation), producing O_2 with the aid of sunlight (solar energy) allowing photosynthesis to occur in the photosynthetic reaction site namely PSII (in the chloroplast).

At this stage 1, the microorganisms (i.e., phototrophic, heterotropic, mixotropic, or photoheterotrophic (Jiménez-Llanos et al., 2020)) are cultivated under nutrient-rich media to form lipids, reduced substrate carbohydrate types, and other organic molecules, and with that, the biomass can be used to be transformed via solar energy, allowing CO_2 fixation (Hitam and Jalil, 2020), allowing oxygenic photosynthesis, and increasing the biomass concentrations.

H_2ase mechanism occurs in PSI in stage 2 as the O_2 depletes (inhibiting PSII; Jiménez-Llanos et al., 2020), starting the anaerobic fermentation stage 2, converting cell stores to H_2 from accumulated carbohydrates and cells (from stage 1) to H_2 and CO_2 productions in a dark-driven or light-driven based fermentation. In PS1, catabolism of endogenous substrates in the citric acid cycle and glycolytic pathway produces electrons in the H_2ase mechanism to produce H_2 (Hitam and Jalil, 2020), depending on which species has either nitrogen fixation type or non-nitrogen fixation type. There are species which have both the H_2ase and N_2ase activities (Mona et al., 2020). The strategy to deplete the O_2 may occur via many ways, for instance, by the usage of O_2 scavengers, PSII inhibitors, or inert gas purging strategy (Jiménez-Llanos et al., 2020), apart from physically sealing the bioreactor to allow cell biomass to consume all O_2 concentrations (lowering the O_2 partial pressure) to finally create the anaerobic condition in the bioreactor.

In stage 2, the culture media are prepared to initiate sulfur limitation conditions as the influencing limiting factor, inducing anaerobic state and activating H_2ase reaction-active site to allow the degradation of the endogenous substrates produced from stage 1 to produce H_2. Since sulfur (essentially cysteine and methionine—amino acid types) is partly the important essential of the protein structures in PSII, naturally, when a sulfur limitation state is induced, the PSII activity decreases, inhibiting H_2O biophotolysis and thus causing a reduced generation of O_2, or alternatively, the microbes consume dissolved O_2 under light and finally expressing H_2ase. Gradually, the culture media are anaerobically induced and the produced H_2ase initiates gradually the production of more H_2 photosynthetically (Vargas et al., 2018) via endogenous catabolism of the biomass cells, by the release of electrons to convert to H_2 (Grechanik et al., 2020).

Indirect biophotolysis using cyanobacteria produces H_2 of approximately 10%–15% yield with a 16.30% efficiency in light conversion (Ahmed et al., 2021), and the biomass of cyanobacteria can be improved 7.30 times using a two-stage indirect biophotolysis in comparison with that of only a single-stage system.

9.5 DIRECT AND INDIRECT BIOPHOTOLYSIS WITH PBR ENGINEERING DEVELOPMENT

Biophotolysis becomes a sustainable and potentially H_2-producing process with the aid of large-scale compacting PBRs cultivating energy plant, possibly green algae, blue algae (cyanobacteria), or heterocystous cyanobacteria, for instance, to conduct photosynthesis, fixing CO_2 into carbohydrates, in which the next step is to be used to produce H_2 or to fix N_2 with N_2ase or fixation using reversible H_2ase-based indirect biophotolysis (Vargas et al., 2014). Here, the biophotolysis by photosynthesis is not inhibited by O_2, focusing on producing increasing biomass cells firstly with aerobic conditions with air supply without the deprivation of sulfur, attaining an exponential phase growth (Bhatia et al., 2021; Vargas et al., 2018) and then secondly in an anaerobic condition (without air supply) for stabilizing H_2 supply, under sulfur-induced limited conditions (Bhatia et al., 2021; Vargas et al., 2018) in an anaerobic PBR, for instance. In the second stage of anaerobically conditioning, the H_2ase which is sensitive to O_2 can be overcome by physically creating the anaerobic condition or alternatively by creating an anaerobic due to the sulfur limitation state, inducing the decrease in PSII activity during the photosynthesis (Vargas et al., 2018) in the PBR.

Alternatively, an effective design of the H_2-producing bioreactors is also needed (Mona et al., 2020). It was shown that the two stages of cultivation of aerobic and anaerobic biophotolysis were effective in producing H_2, for instance, for *Chlamydomonas* species, enabling the generation of approximately 5.95 mol/mg of H_2 (Vargas et al., 2018).

There are various examples of PBR designs being developed for both direct and indirect biophotolysis producing biohydrogen, for instance, tubular- and panel-based PBRs (Oncel and Kose, 2014) (*Chlamydomonas reinhardtii*), modified hanging bang with sparger (Cui et al., 2022) (*Chlamydomonas reinhardtii*), hollow-fiber PBRs (*Chlamydomonas reinhardtii*), and glass coaxial cylinders with internal mixers (Grechanik et al., 2020, 2021) (*Chlamydomonas reinhardtii*). Various culture mediums were also used for the aerobic and anaerobic cultivations of *Chlamydomonas reinhardtii* strains, for instance, HS medium, Tris-acetate-phosphate (TAP) medium (Cui et al., 2022; Grechanik et al., 2021, 2020; Oncel and Kose, 2014; Vargas et al., 2018) and high-salt medium (HSM) (Grechanik et al., 2021). In conclusion, with the use of green gold which are algae and cyanobacteria, many challenges for environmental sustainability and feasibility need to be further investigated, for instance, fulfilling the international standards for the environment and the life cycle assessment (LCA) for a carbohydrate-rich microalgal biomass-based biohydrogen production. The pilot-scale plant simulation should be foreseen actively particularly on the issues of photolight sources for an effective biohydrogen production strategy. Few pilot plants are still undergoing investigations for renewable biohydrogen production and yet published. These should be seen coming with better technological readiness level (TRL) status.

9.6 CONCLUSION

The biohydrogen potential from biophotolysis has greater potential in conditions considering sustainable supply of water and sunlight, with an aiding nutrients' supply of feasible microorganisms, targeting toward sustainable efficiency during production. For improved prospects in the future, continuous fundamental understanding toward highly efficient metabolic pathways in biophotolysis toward engineering optimizations is highly recommended, presenting effective circular economic feasibility.

REFERENCES

Abdalla, A.M., Hossain, S., Nisfindy, O.B., Azad, A.T., Dawood, M., Azad, A.K., 2018. Hydrogen production, storage, transportation and key challenges with applications: A review. *Energy Conversion and Management* 165, 602–627. https://doi.org/10.1016/j. enconman.2018.03.088

Acar, C., Dincer, I., 2018. *Hydrogen production. Comprehensive Energy Systems.* https://doi. org/10.1016/B978-0-12-809597-3.00304-7

Ahmed, S.F., Rafa, N., Mofijur, M., Badruddin, I.A., Inayat, A., Ali, M.S., Farrok, O., Yunus Khan, T.M., 2021. Biohydrogen production from biomass sources: Metabolic pathways and economic analysis. *Frontiers in Energy Research* 9, 1–13. https://doi.org/10.3389/ fenrg.2021.753878

Antonopoulou, G., Ntaikou, I., Stamatelatou, K., Lyberatos, G., 2011. Biological and fermentative production of hydrogen. *Handbook of Biofuels Production: Processes and Technologies.* Woodhead Publishing Limited. https://doi.org/10.1533/9780857090492.2.305

Azwar, M.Y., Hussain, M.A., Abdul-Wahab, A.K., 2014. Development of biohydrogen production by photobiological, fermentation and electrochemical processes: A review. *Renewable and Sustainable Energy Reviews* 31, 158–173. https://doi.org/10.1016/j. rser.2013.11.022

Bechara, R., Azizi, F., Boyadjian, C., 2021. Process simulation and optimization for enhanced biophotolytic hydrogen production from green algae using the sulfur deprivation method. *International Journal of Hydrogen Energy* 46, 14096–14108. https://doi.org/10.1016/j. ijhydene.2021.01.115

Benemann, J.R., 2000. Hydrogen production by microalgae. *Journal of Applied Phycology* 12, 291–300. https://doi.org/10.1023/a:1008175112704

Benemann, J.R., n.d. *Process analysis and economics of biophotolysis of water. IEA technical report from the IEA Agreement on the Production and Utilization of Hydrogen.* United States: N. p., 1998. Web. doi:10.2172/776255..

Bhatia, S.K., Mehariya, S., Bhatia, R.K., Kumar, M., Pugazhendhi, A., Awasthi, M.K., Atabani, A.E., Kumar, G., Kim, W., Seo, S.O., Yang, Y.H., 2021. Wastewater based microalgal biorefinery for bioenergy production: Progress and challenges. *Science of the Total Environment* 751, 141599. https://doi.org/10.1016/j.scitotenv.2020.141599

Buitrón, G., Carrillo-Reyes, J., Morales, M., Faraloni, C., Torzillo, G., 2017. Biohydrogen production from microalgae. *Microalgae-Based Biofuels and Bioproducts: From Feedstock Cultivation to End-Products,* 209–234. https://doi.org/10.1016/B978-0-08-101023-5.00009-1

Cui, J., Purton, S., Baganz, F., 2022. Characterisation of a simple 'hanging bag' photobioreactor for low-cost cultivation of microalgae. *Journal of Chemical Technology and Biotechnology* 97, 608–619. https://doi.org/10.1002/jctb.6985

El-Dalatony, M.M., Zheng, Y., Ji, M.K., Li, X., Salama, E.S., 2020. Metabolic pathways for microalgal biohydrogen production: Current progress and future prospectives. *Bioresource Technology* 318, 124253. https://doi.org/10.1016/j.biortech.2020.124253

Ghiasian, M., 2019. Biophotolysis-based hydrogen production by cyanobacteria. *Prospects of Renewable Bioprocessing in Future Energy Systems, Biofuel and Biorefinery Technologies* 10. Springer International Publishing, 161–184. https://doi. org/10.1007/978-3-030-14463-0_5

Grechanik, V., Naidov, I., Bolshakov, M., Tsygankov, A., 2021. Photoautotrophic hydrogen production by nitrogen-deprived Chlamydomonas reinhardtii cultures. *International Journal of Hydrogen Energy* 46, 3565–3575. https://doi.org/10.1016/j.ijhydene.2020.10.215

Grechanik, V., Romanova, A., Naydov, I., Tsygankov, A., 2020. Photoautotrophic cultures of Chlamydomonas reinhardtii: Sulfur deficiency, anoxia, and hydrogen production. *Photosynthesis Research* 143, 275–286. https://doi.org/10.1007/s11120-019-00701-1

Hallenbeck, P.C., 2013. Fundamentals of biohydrogen. *Biohydrogen*, 1st ed. Elsevier BV. https://doi.org/10.1016/B978-0-444-59555-3.00002-7

Hitam, C.N.C., Jalil, A.A., 2020. A review on biohydrogen production through photo-fermentation of lignocellulosic biomass. *Biomass Conversion and Biorefinery*. https://doi.org/10.1007/s13399-020-01140-y

Jiménez-Llanos, J., Ramírez-Carmona, M., Rendón-Castrillón, L., Ocampo-López, C., 2020. Sustainable biohydrogen production by *Chlorella sp.* microalgae: A review. *International Journal of Hydrogen Energy* 45, 8310–8328. https://doi.org/10.1016/j.ijhydene.2020.01.059

Jose, V., 2021. Photobiological hydrogen as a renewable fuel. *The Holistic Approach to Environment* 11, 67–71. https://doi.org/10.33765/thate.11.2.4

Kamshybayeva, G.K., Kossalbayev, B.D., Sadvakasova, A.K., Bauenova, M.O., Zayadan, B.K., Bozieva, A.M., Alharby, H.F., Tomo, T., Allakhverdiev, S.I., 2023. Screening and optimisation of hydrogen production by newly isolated nitrogen-fixing cyanobacterial strains. *International Journal of Hydrogen Energy*, 1–14. https://doi.org/10.1016/j.ijhydene.2023.01.163

Kothari, R., Tyagi, V.V., Pathak, A., 2010. Waste-to-energy: A way from renewable energy sources to sustainable development. *Renewable and Sustainable Energy Reviews* 14, 3164–3170. https://doi.org/10.1016/j.rser.2010.05.005

Melitos, G., Voulkopoulos, X., Zabaniotou, A., 2021. Waste to sustainable biohydrogen production via photo-fermentation and biophotolysis – A systematic review. *Renewable Energy and Environmental Sustainability* 6, 45. https://doi.org/10.1051/rees/2021047

Mona, S., Kumar, S.S., Kumar, V., Parveen, K., Saini, N., Deepak, B., Pugazhendhi, A., 2020. Green technology for sustainable biohydrogen production (waste to energy): A review. *Science of the Total Environment* 728, 138481. https://doi.org/10.1016/j.scitotenv.2020.138481

Nikolaidis, P., Poullikkas, A., 2017. A comparative overview of hydrogen production processes. *Renewable and Sustainable Energy Reviews* 67, 597–611.

Oncel, S., Kose, A., 2014. Comparison of tubular and panel type photobioreactors for biohydrogen production utilizing Chlamydomonas reinhardtii considering mixing time and light intensity. *Bioresource Technology* 151, 265–270. https://doi.org/10.1016/j.biortech.2013.10.076

Razu, M.H., Hossain, F., Khan, M., 2019. Advancement of bio-hydrogen production from microalgae. *Microalgae Biotechnology for Development of Biofuel and Wastewater Treatment*. https://doi.org/10.1007/978-981-13-2264-8_17

Vargas, J.V.C., Mariano, A.B., Corrêa, D.O., Ordonez, J.C., 2014. The microalgae derived hydrogen process in compact photobioreactors. *International Journal of Hydrogen Energy* 39, 9588–9598. https://doi.org/10.1016/j.ijhydene.2014.04.093

Vargas, S.R., dos Santos, P.V., Giraldi, L.A., Zaiat, M., Calijuri, M. do C., 2018. Anaerobic phototrophic processes of hydrogen production by different strains of microalgae *Chlamydomonas sp. FEMS Microbiology Letters* 365. https://doi.org/10.1093/femsle/fny073

Wukovits, W., Schnitzhofer, W., 2009. Fuels - hydrogen production | Biomass: Fermentation. *Encyclopedia of Electrochemical Power Sources*, 268–275. https://doi.org/10.1016/B978-044452745-5.00312-9

10 Biological Water–Gas Shift Reaction Process for Hydrogen Production

Uday Kumar Nutakki, Mohamed Al Zarooni,
Suresh Kuppireddy, Sara Faiz, and
Mohamed Abdalla Omar

10.1 INTRODUCTION

Energy is a fundamental necessity for the development of humans in social civilization and economic development. Over the past decades, energy production has primarily relied on fossil fuels, including coal, petroleum, and natural gas. Fossil fuels form approximately 81% of the world's total primary energy production, with oil contributing 31.7%, coal contributing 28.1%, and natural gas contributing 21.6% (Agency IE, 2017). Fossil fuels cause climate change due to environmental pollutants emitted during their combustion. Pollutants such as ash, COx, SOx, NOx, CxHy, soot, and organic substances are emitted into the atmosphere (Akkarman et al., 2002; Lee et al., 2002, 2003). Considering the decline of fossil fuels as nonrenewable sources of energy in comparison with the increase in energy requirements and carbon dioxide emissions, there is a need to look for renewable sources of energy that emit less or no carbon. Recently, sustainable and regenerative energy sources, that is, biomass, have been regarded as fossil fuel substitutions to minimize environmental deterioration. Further, extensive research and development have been conducted on the use of biomass in the manufacture of hydrogen (Kondo et al., 2002). Hydrogen presents itself as a promising, sustainable, and regenerative source of energy that can be used for energy provision. Due to hydrogen's properties, that is, large capacity for storing energy, significant power conversion, ecologically friendly, renewable energy production, tremendous quantities of specific energy, zero emissions, a variety of sources, dependability, and simple rejuvenation, future of the energy is predicted to depend on hydrogen (Vanzin et al., 2002; Winkler et al., 2002). Fossil and naturally occurring energy sources can both be used to make hydrogen from either renewable or nonrenewable sources of energy. To support the zero carbon emissions and greenness of the hydrogen produced, renewable energy sources should be used in the production processes (Zhang et al., 2021; Martino et al., 2021).

The greenness or cleanliness of hydrogen depends on the manufacturing pathway; hence, the extent of cleanliness can be presented by color code. The color code level of cleanliness depends on the amount of greenhouse gases emitted during

DOI: 10.1201/9781003382270-13

TABLE 10.1

The Color Code of Hydrogen and Their Cleanliness

Source	Color Code	Possible Process	CO_2 Emissions	Environmental Impact	Level of Cleanliness of Hydrogen
Fossil fuels	Turquoise hydrogen	Natural gas—methane pyrolysis with production of solid carbon	Low	Medium	Medium
	Gray hydrogen	Natural gas—steam reforming	High	High	Low
	Blue hydrogen	Steam reforming with carbon capture and CO_2 storage	Medium	Medium	Medium
	Brown hydrogen	Lignite coal—coal gasification	High	High	Low
	Black hydrogen	Bituminous coal—coal gasification	High	High	Low
Renewable sources	Green hydrogen	Renewable sources	Low	Low	High
	Yellow hydrogen	Water electrolysis by power from grid electric mix	Low	Medium	High
	Pink hydrogen	Nuclear power as energy source, that is, water splitting	Low	Medium	High
	White hydrogen	Natural occurring or deposit	Low	Medium	High

manufacturing. The color codes from the hydrogen production operation are statements for sustainability and indications of the kind of hydrogen supplied to the consumers (Germscheidt et al., 2021). Table 10.1 shows various color codes of hydrogen and their cleanliness levels (Paniˊc et al., 2022).

There are multiple biological sources for producing green hydrogen. Water serves as an example of the best biohydrogen source. Water covers more than 70% of the earth, and about 11.11% of it is hydrogen, thus making it a very abundant source of biohydrogen. On the other hand, the current interest in biohydrogen production also emphasizes the biochemical process that generates hydrogen using carbon monoxide. It is a very promising reaction pathway since it serves the purpose of producing clean hydrogen while reducing carbon monoxide. Thus, the existing harmful fossil fuel-powered energy system may eventually be replaced by a hydrogen-based energy and financial system owing to the developments in biohydrogen. Also, hydrogen has an abundance of potential promise in biochemical processes as an electron donor in numerous reductive biochemical and biotechnological processes (Sipma et al., 2004). The present work aims to provide insights into hydrogen production through biochemical mechanisms for converting renewable resources to biological hydrogen. Further, this chapter will highlight the formation of hydrogen through the biological water–gas shift (BWGS) reaction.

10.2 BIOLOGICAL HYDROGEN PRODUCTION PATHWAYS

Hydrogen may be generated from different biological materials through different biological pathways. The use of biological renewable and waste sources to produce hydrogen reduces the hydrogen production cost, contributes to the reduction of land-filled waste, and enables the recovery of resources from the waste (Ni et al., 2005). Additionally, there exist many biological sources for hydrogen production, such as the following (Eker and Sarp, 2017):

I. **Energy crops:** marine corps, wood, agricultural, industrial, and herbaceous
II. Waste from farming: waste from crops and animals.
III. **Waste from vegetation:** shrub remains, mill wood waste, logging remains, and trees.
IV. Both commercial and municipal garbage.

Biological hydrogen production techniques have emerged as practical processes to produce hydrogen with high efficiency and less environmental impact, hence being considered alternative pathways for sustainable hydrogen supply (Wu and Chang, 2007). Among many biohydrogen reduction pathways, bio-photolysis, fermentation, and the BWGS reaction are prominent (Ljunggren et al., 2011; Saxena et al., 2009; Sinharoy et al., 2019; Su et al., 2018; Van Niel et al., 2003).

10.2.1 FERMENTATION

Fermentation is a chemical alteration in biological substrates brought on by microbial activity. The fermentation process can be categorized into light-independent (dark fermentation) and light-dependent (photo-fermentation) processes.

10.2.1.1 Dark Fermentation

The dark fermentation process produces hydrogen through a series of biochemical processes wherein anaerobic microorganisms function as accelerators for biomass conversion to produce hydrogen in the absence of light. Dark fermentation is linked to greater productivity and low output (Zhang et al., 2021). Biomass conversion to hydrogen may occur in a single stage or multiple stages. The single-stage anaerobic method combines all hydrolysis, acidogenesis, and acetogenesis in one reactor unit. The balanced handling of microorganisms throughout the entire process makes one-stage systems more complicated than multistage systems (Talapko et al., 2023). The multistage systems use separate reactors for the hydrolysis/acidogenesis and acetogenesis phases (Ergal et al., 2020). The multistage processes are designed to improve each stage of digestion and reach operational stability, greater organic loading capability, and greater resilience to contaminants and inhibitors (Aziz et al., 2021; Nanqi et al., 2011). Clostridium bacteria are considered the microorganisms that generate hydrogen most efficiently by the dark fermentative technique (Martino et al., 2021). The yield of the dark fermentation depends on pH levels, pressure, temperature, the length of the hydraulic retention period, the kind of feed material, the type of biological catalyst used, and the existence of other elements within the feed material. Additionally, some pretreatment methods may also improve the production process of hydrogen (Kumar et al., 2020).

The dark fermentation pathways show the following limitations (Kayfeci et al., 2019):

I. Thermodynamically, the reaction tends to lose its stability as the output increases.
II. There is low conversion efficiency of the substrate used, hence low hydrogen production.

10.2.1.2 Photo-Fermentation

The photo-fermentation pathway produces hydrogen by a series of biochemical processes, whereas photosynthetic bacteria function as accelerators of biomass conversion in biochemical processes to generate hydrogen in the presence of light. In contrast to dark fermentation, photo-fermentation is linked to low productivity and greater output (Zhang et al., 2021; Aziz et al., 2021; Tsygankov et al., 2002). Photo-fermentation also involves three stages like anaerobic conversion but through the action of nitrogenase. In the presence of sunlight, the nitrogenase is reduced, and subsequently, the protons are reduced to biohydrogen by hydrolyzing adenosine triphosphate (ATP). The simple mechanic is presented in Figure 10.1 (Kumar et al., 2020).

Rhodobium, Rhodospirillum, and Rhodopseudomonas are examples of photosynthetic bacteria that generate hydrogen via the photo-fermentation pathway (Martino et al., 2021). The effectiveness of the generation of hydrogen is highly impacted by lighting and mixing. The distribution of light to fulfill the needs of specific substrates limits hydrogen production. The mixing enhances the mass transfer; nonetheless, the necessity for light intensity increases with increasing mixing speed

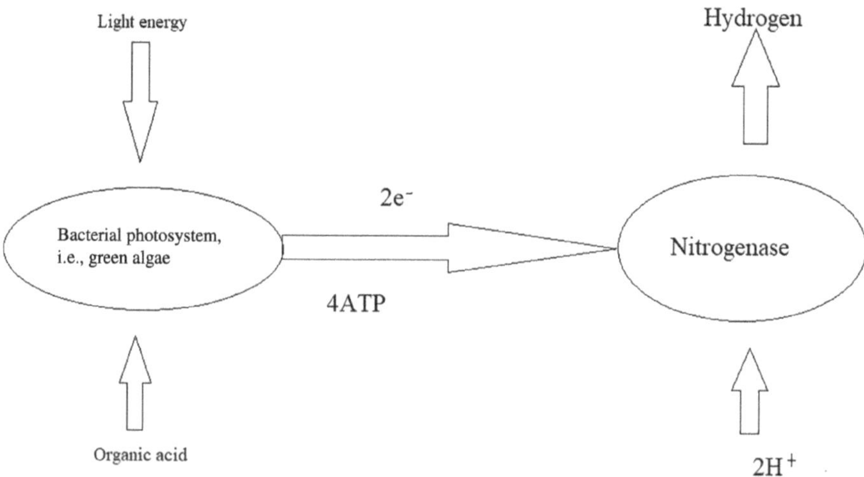

FIGURE 10.1 Mechanism of the photo-fermentation.

(Techtmann et al., 2012). Stirring at an irregular interval has been demonstrated to promote the generation of hydrogen compared to continuous stirring (Martino et al., 2021). Further, the hybridization of biosorption and ultrasonication pretreatment steps could also improve hydrogen generation by the photo-fermentative pathway (Martino et al., 2021). pH has a significant impact on hydrogen generation. Decreasing the pH value (operating in acidic conditions) becomes a hindering element on producing hydrogen by using the photo-fermentation pathway. The photo-fermentation process is also highly limited, with the photochemical efficiency ranging between 3% and 10%. This is because the distribution of light in the reactor cannot be homogeneous, which contributes to the efficiency drop of the overall light conversion process (Kayfeci et al., 2019).

10.2.1.3   Carbon Monoxide Gas Fermentation

Carbon monoxide gas fermentation can be defined as the biological hydrogen production process in which hydrogen is generated when carbon monoxide and water react at anaerobic conditions in the presence of photosynthetic bacteria. Homoacetogenesis consumption of hydrogen limits the process, and the investigations on the effect of pH and carbon monoxide loading reported an increase in hydrogen production with the rise in pH from 5 to 8 (Liu et al., 2020). Chao Liu et al. (2020) investigated the pH effect on hydrogen production with glucose as a substrate in different phases. In the first phase (no carbon monoxide presence) and the fourth phase (with a considerable concentration of carbon monoxide dosed but no harmful effect), the studies were run under constant operating conditions. The first and fourth phases' hydrogen consumption rates at a steady pH of 5 were comparable (less than 2%), but at pH 6, both phases' ratios rose to over 15%. The hydrogen consumption rates were also raised to 43%–48% in the first phase and 56%–58% in the fourth phase, respectively, as the process pH level altered to 7 and 8. Further evidence indicates that the injection of carbon monoxide had an impact on glucose's ability to produce hydrogen (Liu et al., 2020).

10.2.2   Bio-Photolysis

Bio-photolysis is known as a mechanism for generating hydrogen powered by photonics that converts organic compounds to hydrogen by photosynthetic bacteria (cyanobacteria and blue-green algae) in the presence of solar energy. Normally, by harnessing sunlight for photosynthesis, hydrogen from water or other organic substances is extracted. In plants, carbon dioxide reduction occurs since no hydrogenase is present (Kayfeci et al., 2019). In contrast, algae can dissociate water during photosynthesis and give hydrogen due to the presence of hydrogenases and cyanobacteria (Kumar et al., 2020; Kayfeci et al., 2019). The bio-photolysis process to produce hydrogen can be classified into two groups: indirect bio-photolysis and direct bio-photolysis. Sunlight and carbon dioxide are the two main inputs used to grow organisms on the photobiological pathway facilitated via the hydrogenase enzyme. Photosynthetic organisms have chlorophyll pigments in their thylakoid membrane that enable them to absorb sunlight and produce atomic oxygen throughout photosynthesis. Through the aid of ATP and nicotinamide adenine dinucleotide phosphate

(NADPH), enzymatic mechanisms convert carbon dioxide into triose phosphate (Brentner et al., 2010). Cyanobacteria and microalgae have been reported to be widely used hydrogen production microorganisms (Aziz et al., 2021).

10.2.2.1 Direct Bio-Photolysis

The direct bio-photolysis conversion generates hydrogen directly from water in the presence of microorganisms such as algae and in the presence of light. Normally, direct bio-photolysis occurs in one step (Aziz et al., 2021). The process begins with green algae breaking down water molecules and results in hydrogen ions and oxygen under the influence of light (via photosynthesis). The hydrogenase enzyme subsequently transforms the resulting hydrogen ions into hydrogen gas (Nikolaidis and Poullikkas, 2017). Figure 10.2 shows the mechanism of the direct bio-photolysis pathway. It is reported that the enzymes have excessive sensitivity to oxygen; hence, it is essential to keep the lower levels of oxygen within the process, that is, below 0.1% (Ni et al., 2005). The direct bio-photolysis reactions can be sustained for a noticeably brief time, and the rate of hydrogen produced is incredibly low (less than one-tenth of other photosynthetic reactions) (Kayfeci et al., 2019).

10.2.2.2 Indirect Bio-Photolysis

Indirect bio-photolysis is the hydrogen conversion pathway wherein carbon dioxide is absorbed by photosynthetic microorganisms in the presence of solar energy to generate carbohydrate biomass, which is then converted to hydrogen. For indirect bio-photolysis conversion, the production of hydrogen proceeds in two stages. The first stage involves photosynthesis and the accumulation of carbohydrates. In the second stage, accumulated carbohydrates are fermented in the presence of hydrogenase and nitrogenase enzymes, resulting in the generation of hydrogen (Martino et al., 2021). Indirect bio-photolysis has a low maximum light conversion efficiency due to the multiple steps involved in converting solar energy to hydrogen and the ATP required by the nitrogenase enzyme. Figure 10.3 shows the mechanism of the indirect bio-photolysis process.

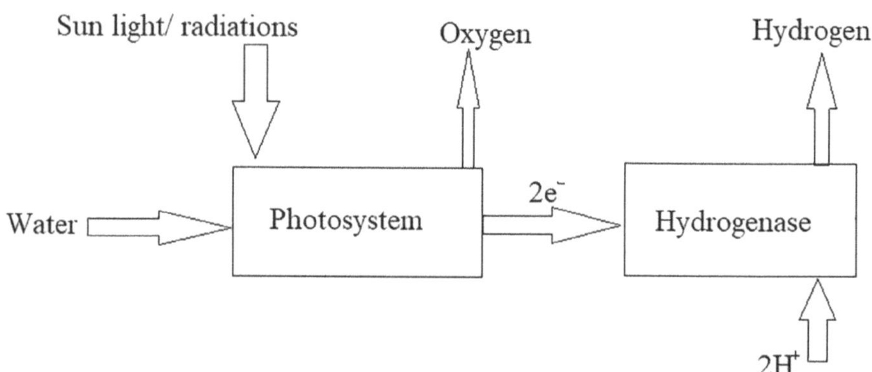

FIGURE 10.2 Mechanism of the direct bio-photolysis.

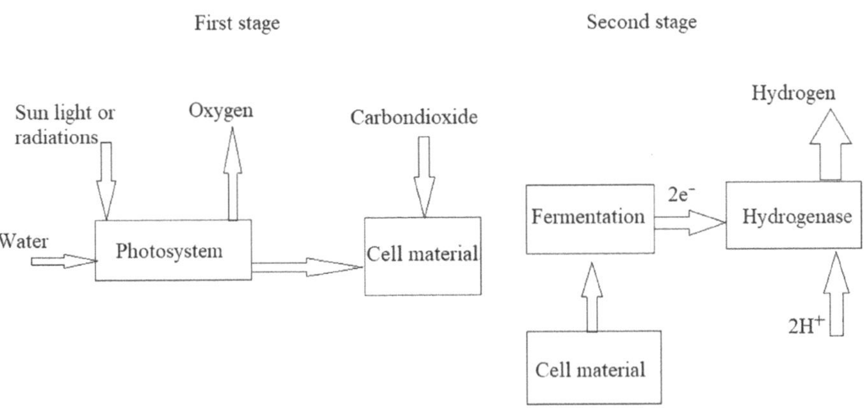

FIGURE 10.3 Mechanism of the indirect bio-photolysis.

10.2.3 WATER–GAS SHIFT (WGS) REACTION

The WGS reaction is an exothermic reaction that generates hydrogen and carbon dioxide from the reversible reaction of water and carbon monoxide, as shown in the chemical equation (10.1) below (Alfano and Cavazza, 2018). The WGS reaction can either be conventional or biological.

$$CO + H_2O \rightarrow CO_2 + H_2 \tag{10.1}$$

10.2.3.1 Conventional Water–Gas Shift Reaction

A conventional WGS reaction is the hydrogen production mechanism that utilizes a catalyst of metallic or chemical origin in a heterogeneous gas-phase reaction to generate hydrogen and carbon dioxide from carbon monoxide and steam. The WGS reaction is the basic technique for the majority of the industries in the world that use methane (CH_4) through steam methane reforming to produce hydrogen. As shown in Figure 10.4, the conventional WGS reaction process is further classified based on the use of catalysts in the process (Henstra and Stams, 2011). The catalytic reactions at high and low temperatures are those WGS reactions driven in the presence of catalyst, while noncatalytic WGS reactions are observed in a handful of circumstances, for example, plasma systems and supercritical WGS reactions. Since the WGS reaction is an exothermic reaction, the system should work at low temperatures for better operation and efficient production (Ragsdale et al., 2012). However, initial research on the WGS reaction reported faster reaction kinetics at elevated temperatures, and hence, the conventional WGS reaction was conducted in a high-temperature shift (HTS) reactor system in the temperature range of 350°C–370°C (Amos, 2004). In recent years, a lot of attempts have been made on development of the catalyst to carry out the WGS reaction at a lower temperature with the target of improving the purity of hydrogen by improving the carbon monoxide conversion. Recently, nonthermal dielectric barrier discharge (DBD) has been reported to promote the activities

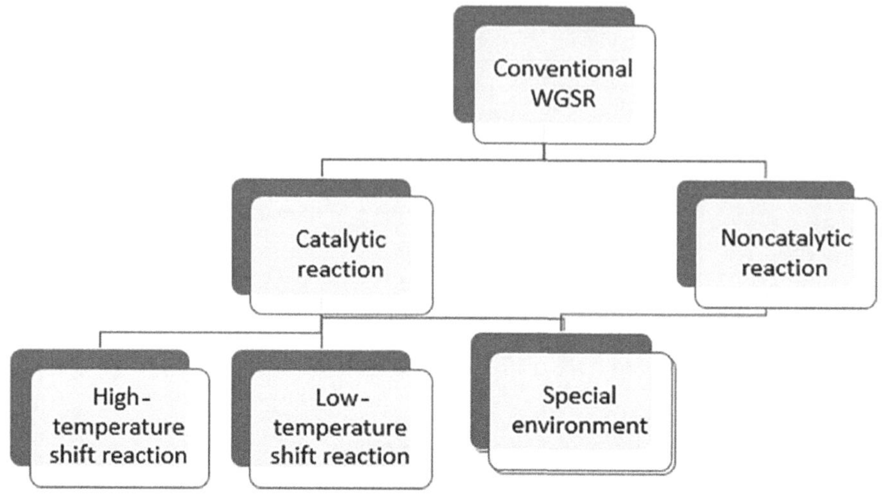

FIGURE 10.4 The classification of conventional WGSR.

of the low-temperature WGS (LT-WGS) reaction (Wangkawong et al., 2020). DBD was proven to get high conversions in many reactions, including nitric oxide decomposition (Tang et al., 2014), toluene oxidation (Magureanu et al., 2013), and methane oxidation (Gibson et al., 2017). The catalytic reactions can be conducted in special environments as well, such as photocatalyst, microwave, and HiGee WGS reactions (Henstra and Stams, 2011).

A few limitations of the conventional WGS reaction are as follows:

I. The conventional WGS reaction systems require several steps to restore the heat, whereas the same quantity and quality of heat energy could be recovered in one step in the BWGS reaction (Amos, 2004).
II. The technique requires operation at an elevated temperature, which is thermodynamically unfavorable for the removal of carbon monoxide (Henstra and Stams, 2011).
III. DBD has been reported to have low energy efficiency (Wangkawong et al., 2020).

10.2.3.2 Biological Water–Gas Shift Reaction

The BWGS reaction is the hydrogen production reaction process that uses a biological catalyst, enzymes, or microorganisms with water vapor (steam) and carbon monoxide, to give carbon dioxide and hydrogen. The catalysts or enzymes used in the BWGS reaction are known as hydrogenogens. Hydrogenogens are known to be advantageous since they can run the WGS reaction at lower temperatures, which favors the thermodynamic removal of carbon monoxide (Henstra and Stams, 2011), for example, **Rubrivivax gelatinosus** CBS, which is a purple, nonsulfur, photosynthetic bacteria that can undergo a similar chemical catalytic WGS reaction at an ambient temperature of 25°C and atmospheric pressure in anaerobic conditions.

Anaerobic fermentation, aerobic heterotrophic metabolic activities, and photosynthesis provide energy to **Rubrivivax gelatinosus** CBS (Amos, 2004). Normally, microorganisms use WGSR to obtain energy that can be used in metabolic and growth processes. The biological WGS reaction is a representation of an anaerobic, dark-phase pathway that has lower energy production for metabolic activities in comparison with the aerobic or photosynthesis pathway, which results in a slow cellular growth rate and hence extends the duration of time required for achieving the state of equilibrium, although it minimizes the mass generation of waste cells (Amos, 2004). A few advantages of using biocatalysts are highlighted below.

I. Some biological catalysts, such as C. hydrogenoformans, used in the BWGS reactions maintain low level of carbon monoxide concentration in outlet gases, which stands as the limiting step for some technologies due to catalyst being poisoned by carbon monoxide (carbon monoxide catalyst poisoning) (Alfano and Cavazza, 2018).
II. The bioresources for the reaction pathways are easy to obtain and cheap.
III. They provide highly efficient catalytic activities (Demirel and Yenigu̇n, 2002).
IV. Aerobic bacteria live on carbon monoxide as their sole energy and carbon source (Demirel and Yenigu̇n, 2002).
V. Hydrogenases are highly electroactive enzymes (Demirel and Yenigu̇n, 2002).
VI. They accelerate and control oxidation–reduction reactions (Demirel and Yenigu̇n, 2002).

BWGS reaction is catalyzed by carboxydotrophic microorganisms using carbon monoxide as an electron donor (Wolfrum and Watt, 2002). The catalyst has an intricate enzymatic system and establishes a carbon monoxide oxidation reaction to carbon dioxide by a carbon monoxide dehydrogenase (CODH) (Svetlitchnyi et al., 2001; Wolfrum and Watt, 2002; Zhao et al, 2011). Further, the reaction ends with the reduction of two protons to form hydrogen by the hydrogenase enzymes (Alfano and Cavazza, 2018; Zhao et al, 2011). The following chemical reaction equations represent the mechanism of the BWGS reactions, as shown in Figure 10.5.

First Step Reaction: by CODH

$$CO + H_2O \rightarrow CO_2 + 2e^- + 2H^+ \tag{10.2}$$

Second Step Reaction: by hydrogenase

$$2e^- + 2H^+ \rightarrow H_2 \tag{10.3}$$

Overall Reaction: CODH with hydrogenase

$$CO + H_2O \rightarrow CO_2 + H_2 \tag{10.4}$$

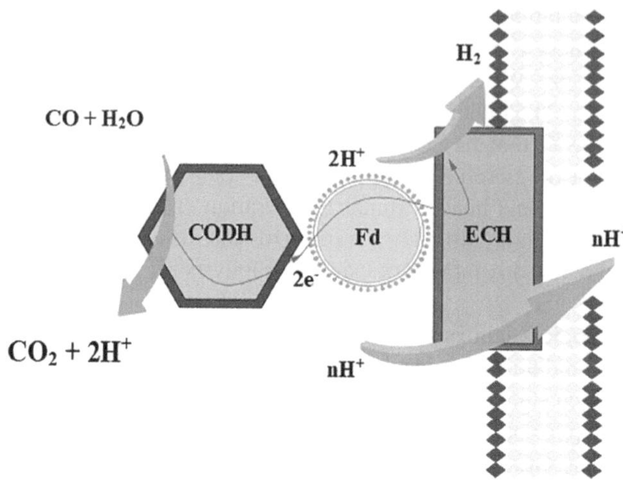

FIGURE 10.5 Systematic mechanism of the BWGS reaction.

Ferredoxin (Fd) in Figure 10.5 is a small protein that has iron and sulfur atoms organized as iron–sulfur clusters.

In general, there are a total of five CODH-encoding genes identified in Carboxydothermus hydrogenoformans that perform different suggested functions in the context of their genomics (Alfano and Cavazza, 2018; L. Domnik et al., 2017). Those genes are as follows:

I. CODH-I: participate in the WGS reaction.
II. CODH-II: associated with the production of NADH.
III. CODH-III: connected to acetyl-Co-enzyme A (acetyl-CoA) synthase (ACS) in the acetyl CoA pathway.
IV. CODH-IV: participate in the reaction (response) to oxidative stress.
V. CODH-V: No function has been found.

Rhodospirillum rubrum bacteria possess CODH-I, which is involved in the WGS reaction (Alfano and Cavazza, 2018).

Further, hydrogenases are categorized into three classes with no phylogenetic connection according to the metal composition of their active site, namely, [Fe], [FeFe], and [NiFe] hydrogenases. Molecular hydrogen is reversibly oxidized to electrons and protons by [FeFe] and [NiFe] hydrogenases (Fontecilla-Camps et al., 2009). Researchers also reported that the BWGS reaction mechanism can be controlled by mainly two types of biological metal cation-co-factored enzymes known as metalloenzymes, which are ***energy-conserving [NiFe] hydrogenase*** and ***monofunctional [NiFe] CODH*** (Alfano and Cavazza, 2018; Kung and Drennan, 2011).

Monofunctional [NiFe] CODH: Ni-dependent CODHs perform a crucial part in the global carbon cycle in an anaerobic microorganism since they catalyze the oxidation of carbon monoxide reversibly to carbon dioxide. CODH can either form a monofunctional enzyme or a bifunctional complex with acetyl-CoA. The CODHs that are commonly referred to are the CODH/ACS complex from *Moorella thermo-acetica* (MtCODH/ACS) and the monofunctional CODHs that are in *Rhodospirillum rubrum* and *Carboxydothermus hydrogenoformans* (Alfano and Cavazza, 2018; Doukov et al., 2002). Recently, the monofunctional CODH with an unidentified physiology function was traced and expressed in Desulfovibrio vulgaris (Alfano and Cavazza, 2018; Hadj-Saïd et al., 2015).

Energy-conserving [NiFe] hydrogenases: Bacteria and Archaea [NiFe] hydrogenases are grouped into four main groups, which are as follows (Alfano and Cavazza, 2018):

I. Group 1: They are membrane-associated hydrogen uptake hydrogenases.
II. Group 2: Composed of soluble ingestion hydrogenase and sensitivity hydrogenase.
III. Group 3: They possess cytoplasmic hydrogenases with heteromultimeric structures and a reducible cofactor (NADPH, NAD(P), or F420). F420 is a dependent enzyme that is an essential coenzyme of methanogenesis. F420 is found in many microorganisms and can catalyze a redox reaction even on substrates that are otherwise recalcitrant to enzyme-mediated reductions.
IV. Group 4: These are hydrogenases that transform energy.

The **carbon-monoxide-induced hydrogenases** used in the WGS reaction process are part of the ECH hydrogenase family from *Methanosarcina barkeri* (Alfano and Cavazza, 2018). Most hydrogenase family members have been discovered to be purified in their entirety, but not the ECH enzyme from *Methanosarcina barkeri,* which was homogeneously purified with a high yield (Jörn et al., 2001).

10.3 BIOCATALYSTS OR ORGANISMS FOR BWGS REACTION

10.3.1 BIOCATALYSTS FOR WGSR

The biocatalysts are microorganisms that have the ability to alter and perform WGSR in a comparable manner to nonbiological catalysts. Different catalysts may be adjusted to different conditions to conduct the WGS reaction for hydrogen production. There are hundreds of biocatalysts that are classified into many groups. Some are classified as photosynthetic bacterium (e.g., *Rhodobacter sp., Rubrivivax gelatinosus, and Rhodospirillum rubrum*) and anaerobic bacterium (e.g., Methanosarcina barkeri) (Alfano and Cavazza, 2018; Gibson et al., 2017; Younesi et al., 2008; Henstra and Stams, 2011). Alfano and Marillo et al. (2018) presented the classification of the biocatalyst into three separate groups. Those groups are as follows (Alfano and Cavazza, 2018):

I. **Mesophilic bacteria,** for example, Rubrivivax gelatinosus, Rhodospirillum rubrum, and Rhodopseudomonas palustris.
II. **Thermophilic bacteria,** for example, Carboxydothermus hydrogenoformans, Desulfotomaculum carboxydivorans, Caldanaerobacter subterraneus pacificus, and Carboxydocella pertinax.

III. **Archaea** such as Thermococcus onnurineus and Thermococcus strain AM4.

Mesophilic and thermophilic species are found among the hydrogenogenic. Thermophilic species were discovered to exist in high numbers compared to mesophilic species, suggesting that increased temperature may aid hydrogenogenic carbon monoxide metabolism by accelerating gas diffusion rates (Alfano and Cavazza, 2018). Some of the microorganisms that can perform the BWGS reaction are presented below with their abilities.

I. **Rhodospirillum rubrum (R. rubrum)** is a purple, nonsulfur photosynthetic bacteria that belong to the mesophilic bacteria family group (Alfano and Cavazza, 2018). Based on the number of factors in growing conditions, R. rubrum can engage in either anaerobic or aerobic respiration, photosynthesis, and acid-mixed fermentation (Alfano and Cavazza, 2018). R. rubrum is said to be capable of directly generating biohydrogen from syngas by catalyzing the WGS reaction (Younesi et al., 2008).
II. **Carboxydothermus hydrogenoformans (C. hydrogenoformans)** is an incredibly intriguing candidate for study as a potential biohydrogen production source because it develops quickly (doubling in 2 hours) in the dark while using carbon monoxide as both carbon and energy source for catalyzing the WGS reaction. C. hydrogenoformans can produce hydrogen with a low carbon monoxide concentration, even below 2 ppm (Alfano and Cavazza, 2018;; Zhao et al., 2011).
III. **Rubrivivax gelatinosus CBS** is known as a nonsulfur, purple photosynthetic bacteria that can perform the same BWGS reaction as metallic catalysts but under an ambient temperature of 25°C in anaerobic conditions and atmospheric pressure (Amos, 2004).
IV. **Parageobacillus thermoglucosidasius** has a high oxygen tolerance when catalyzing the WGS reaction. They can perform the WGS reactions well even under oxygen depletion conditions (Table 10.2; Alfano and Cavazza, 2018; Díaz et al., 2022; Mohr et al., 2018).

10.3.2 Biocatalyst Growth Kinetics

The catalyst growth rate is a key factor for bioreactor startups to avoid and recover from process upsets. Before the startup of the reactor, the reactor is fed with a sterile seed culture. In most cases, the newly grown or produced cells are the ones converting the carbon monoxide to hydrogen after inoculation. Thus, the generation rate increases and reaches the full generation capacity as the growth rate increases, and vice versa. On the other side, cell growth should be controlled to reduce excess cell waste production within the system. The reactor operating system will be erratic if the rate of growth is slower than the rate at which cells die naturally within the system (Amos, 2004).

Sudden changes in operating conditions (disturbance), such as operating temperature or pH, may result in damaging the cells and thus losing their biological activities. Therefore, biocatalyst growth needs to be controlled, and to increase the production

TABLE 10.2

List of Different Biocatalyst Species, Their Origin, Growth Conditions, and the Best Substrate for H_2 Production

Species	Origin	Growth Conditions			Substrate	Ref.
		Temperature (°C)	Light	pH.		
R. rubrum	Brackish ditches	30	Yes	6.5	Acetate	Alfano and Cavazza (2018); Ismail et al. (2008); Kim et al. (2020)
Rubrivivax gelatinosus	Lake	34	No	6.7	CO	Alfano and Cavazza (2018); Kim et al. (2020); Wolfrum and Watt (2002)
Rubrivivax gelatinosus	Lake	30	No	6.7	Malate	Najafpour et al. (2006); Koku et al. (2002).
Rhodopseudomonas palustris	Anaerobic sludge digester	30	Yes	7.0 ± 0.1	CO	Alfano and Cavazza (2018); Kim et al. (2020); Oh et al. (2005)
Carboxydothermus. hydrogenoformans	Hypothermal Springs	70	No	6.9–7.8	CO	Alfano and Cavazza (2018); Kim et al. (2020), Haddad et al. (2014); Zhao et al. (2011)
Thermococcus onnurineus	Deep-sea hypothermal vent	80	Not reported	6.1–6.2	CO	Alfano and Cavazza (2018); Schut et al. (2016); Kim et al. (2020); M. et al. (2015)
Citrobacter sp. Y19	Human and animal feces	30 to 37	No	7	CO	Jung et al. (2002); Kim et al. (2020); Oh et al. (2005); Oliveira et al. (2016); Sokolova et al. (2004)

and growth rate, nutrients (i.e., an acetate or sugar source) and oxygen need to be provided to boost the production energy while taking into consideration the conditions required for cell growth without contamination or the development of unwanted organisms (Amos, 2004).

Kinetic growth depends on the type of microorganisms used in the system and the conditions of operation. For example, Amos (2004) reported that ***Rubrivivax gelatinosus*** can produce about 1.4 g of cell mass for each mole of carbon monoxide at aerobic conditions, but if grown aerobically on oxygen and acetate, 2.0 g of cell mass will be produced. Biohydrogen production requires a medium formulation that could satisfy the element requirements for cell growth and catabolite production (Das and Veziroglu, 2001; Kalil et al., 2009; Kumar and Das, 2000; Zhao et al., 2011). A simple batch growth kinetic model is reported based on the experimental growth rate (Najafpour et al., 2006). The following equations can be used to figure out cell growth rates.

$$\frac{dx}{dt} = \mu x \tag{10.5}$$

where μ is the specific growth rate (in hrs⁻¹), t is time (in hrs.), and x is the cell dry weight concentration (in g/L). The specific growth of the population can be estimated by the following equation (Najafpour et al., 2006):

$$\mu = \mu_m \left(1 - \frac{x}{x_m}\right) \tag{10.6}$$

in which μ_m is the maximum specific growth rate (in hrs⁻¹) and x_m is the maximum cell dry weight concentration (in g/L). Combining the above two equations, upon integration, results in the following equation:

$$x = \frac{x_0 e^{\mu_m t}}{1 - \left(\dfrac{x_0}{x_m}\right)\left(1 - e^{\mu_m t}\right)} \tag{10.7}$$

where x_0 is the initial cell dry weight concentration (in g/L). If the effect of the substrate in the process is considered, the specific growth rate can be evaluated by the following equation (Najafpour et al., 2006):

$$\mu = \mu_m \left(1 - e^{-\frac{s^n}{k_1}}\right) \tag{10.8}$$

where n is the number of units, s is substrate concentration (in g/L), and k_1 is Tessier substrate rate constant (in g/L). The Monod rate constant (in g/L) presented by (k_2) can be added to the equation to make it more fit for experimental data, which can also be expressed in terms of purpose of the cell dry mass concentration and biomass concentration's role in preventing growth. The equation will then be modified as follows (Koku et al., 2002):

$$\mu = \mu_m \left(\frac{1}{1 + k_2 s^{-n}} \right) = \mu_m \left(\frac{s^n}{Bx + s^n} \right) \tag{10.9}$$

where B is the Contois equation's constant. According to the Monod growth model, the rate of growth is correlated with the concentration of a single growth-controlling carbon supply source (substrate). If n is equal to unity in equation (10.9), then the equation is identified as the Contois equation. The Monod constant that is possessed in the Contois equation is directly proportional to biomass concentration (Table 10.3; Najafpour et al., 2006).

10.4 REACTORS AND KINETICS OF BWGS REACTION

10.4.1 REACTORS USED IN BWGS REACTION

The BWGS reaction to be completed needs a reactor where the reaction process can occur. There are several types of reactors that operate in different modes (either continuous or noncontinuous reactors). Some of the reactors that can be used on the BWGS reactions are as follows: batch reactors (Koku et al., 2002), continued stirred tank reactor (CSTR) (Alfano and Cavazza, 2018; Nikolaidis and Poullikkas, 2017), trickle bed reactors (TBRs) both counter-current and co-current (Najafpour et al., 2006; Haddad et al., 2014; Koku et al., 2002), bubble column reactor (BCR) (Haddad et al., 2014), hollow fiber membrane bioreactor (HFMBR) (Najafpour et al., 2006; Zhao et al., 2011), gas lift reactor (GLR) (Najafpour et al., 2006; Haddad et al., 2014), packed (fixed)-bed bioreactor, and moving-bed bioreactor (Inharoy and Pakshirajan, 2020; Akkerman et al., 2002).

10.4.2 MASS TRANSFER LIMITATIONS AND KINETICS

The chemical reactor's overall reaction rate is limited by two factors: **the mass transfer rate** and **the intrinsic reaction rate**. The peak performance rate of a chemical reaction is referred to as the intrinsic reaction rate, and it depends on the reactant's concentration and temperature.

If the rate at which mass is transferred is not a limiting factor, the overall rate of reaction will be roughly equal to the intrinsic reaction rate. And if the mass transfer rate is a limited factor, then the reaction will proceed slower than the reactant's possibilities of being transported to the site of the reaction. Despite this circumstance, the reaction may still be fast if the reactor cannot be supplied quickly enough, leading to a slower overall reaction. For the BWGS reaction, the mass transfer of the carbon monoxide from the gas phase to the liquid phase acts as the factor that limits most of the reactor arrangements, which means that the reactions supposed to happen after the carbon monoxide reaches the organisms happen when the carbon monoxide transfers to the solution, while the mass transfer rate is expected to be slow (Amos, 2004; Younesi et al., 2004).

In general, the reactor arrangement is either batch or continuous culture. The two options have different mass transfer approaches.

TABLE 10.3

BWGS Reaction-Based Hydrogen Production Rates, Reactors Used, and Operation Time

Species	Cell Density (gcell/L)	Hydrogen Production Rate (mmol/g/h)	Reactor Type	Operation Time (hrs.)	CO Conversion (%)	Ref.
R. rubrum	<0.9	10.0–16.1	CSTR	<100	38–58	Alfano and Cavazza (2018); Ismail et al. (2008); Kim et al. (2020)
Rubrivivax gelatinosus	1.35–1.65	1.3–33	TBR	120	<75	Alfano and Cavazza (2018); Kim et al. (2020); Wolfrum and Watt (2002)
Rubrivivax gelatinosus	1.35–1.65	15	TBR counter-current	120	87	Najafpour et al. (2006); Koku et al. (2002)
Rhodopseudomonas palustris	10	41	CSTR	450	61–100	Alfano and Cavazza (2018); Kim et al. (2020); Oh et al. (2005)
Carboxydothermus. hydrogenoformans	6.1–8.12	16.3–83.3	HFMBR and GLR	Not reported	57.1	Alfano and Cavazza (2018); Kim et al. (2020), Haddad et al. (2014); Zhao et al. (2011)
Thermococcus onnurineus	Not reported	108.7–151.3	CSTR	Not reported	64.9	Alfano and Cavazza (2018); Schut et al. (2016); Kim et al. (2020); Kim. et al. (2013)
Citrobacter sp. Y19	1.5	20	CSTR	<70	<30	Jung et al. (2002); Kim et al. (2020); Oh et al. (2005); Oliveira et al. (2016); Sokolova et al. (2004)

10.4.2.1 Batch Culture Approach

The following equation governs the transfer of mass from the gas phase to the liquid phase (Amos, 2004; Younesi et al., 2004):

$$N = K_L a \left(C^* - C_{\text{liquid}} \right) = \frac{K_L a}{H} \left(P_{\text{CO,liquid}} - P_{\text{CO-gas}} \right) \tag{10.10}$$

where N is mass transfer rate (in number of moles per time), H is Henry's law constant, $K_L a$ is the overall coefficient of mass transfer rate, C^* is the concentration of carbon monoxide in equilibrium at the interface of the gas and liquid, C_{liquid} is the concentration of carbon monoxide in the liquid phase, $P_{\text{CO,liquid}}$ is a partial pressure related to the solubility of carbon monoxide in the liquid phase, and $P_{\text{CO-gas}}$ is a partial pressure of carbon monoxide in the gas phase.

Cell density times cell-specific growth rate is reported to equal the cell balance for carbon monoxide and biomass in the gaseous and liquid phases. Also, the fermentation broth's rate of carbon monoxide uptake varies with the carbon monoxide pressure gradient of the gaseous and liquid phases; thus, equation (10.10) is expressed as follows (Younesi et al., 2004):

$$-\frac{1}{V} \frac{dn_{\text{CO}}}{dt} = K_L a \left(C^* - C_{\text{liquid}} \right) = \frac{K_L a}{H} \left(P_{\text{CO,liquid}} - P_{\text{CO-gas}} \right) \tag{10.11}$$

As stated before, the mass transfer process regulates the reaction's pace; the carbon monoxide must be dissolved in the liquid phase first, and then, it has to get inside the cell to be moved to the reaction site so that microorganisms have access to utilize it. Though this fact might not hold true for newly introduced inoculums in liquid media, as the growth of the microorganisms continues, the concentration of carbon monoxide in the gas phase will fall since active organisms will dominate the culture. Thus, the carbon monoxide concentration in the liquid phase of a medium dominated by microorganisms at massive growth phase is decreasing approximately to zero. This justifies the assumption that $P_{\text{CO, liquid}}$ or $C_{\text{liquid}} = 0$ (Younesi et al., 2004). Thus, the equation (10.11) will become

$$-\frac{1}{V} \frac{dn_{\text{CO}}}{dt} = K_L a \, C^* = -\frac{K_L a}{H} P_{\text{CO,liquid}} \tag{10.12}$$

10.4.2.2 Continuous Culture Approach

Unlike batch systems, continuous systems always have inlets and outlets, with no accumulation of products within the system. The following equation governs the transfer of mass from the gas phase to the liquid phase (Najafpour et al., 2006):

$$\frac{dC_{\text{CO}}}{dt} = K_L a \left(C^* - C_{\text{liquid}} \right) - qX \tag{10.13}$$

where q is a specific uptake rate of carbon monoxide and X is the cell concentration.

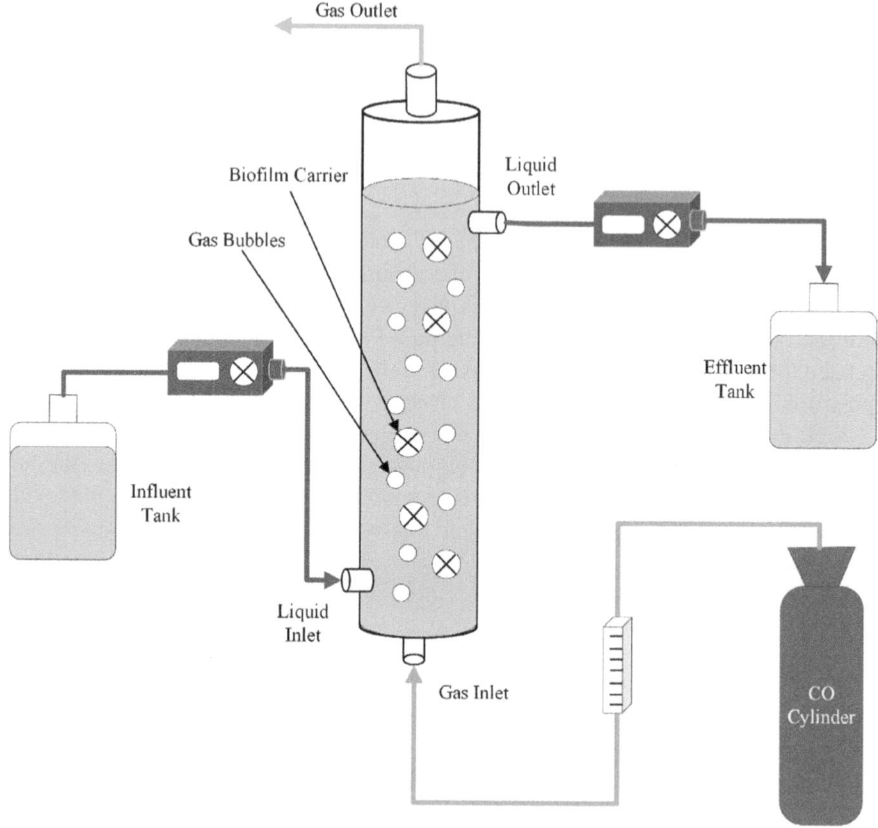

FIGURE 10.6 Setup of a moving-bed biofilm reactor used for continuous biohydrogen production.

Researchers reported that a direct correlation exists between the hydrogen production rate per unit cell and the dilution rate in their study (Najafpour et al., 2006). Thus, considering effect of the dilution rate, the equation (10.13) becomes (Najafpour et al., 2006)

$$\frac{dC_{CO}}{dt} = K_L a \left(C^* - C_{\text{liquid}} \right) - DC_{\text{liquid}} - qX \tag{10.14}$$

where D is the dilution rate (in hrs^{-1}).

10.5 FACTORS AFFECTING THE PRODUCTION OF HYDROGEN BY BWGS REACTION

Several factors can affect the operation of biohydrogen generation through the BWGS reaction. The factor might be physical- or mechanical-related. Some of the factors that can affect hydrogen production are listed below and discussed.

10.5.1 EFFECT OF SUBSTRATE IN CONSUMPTION OF THE CARBON MONOXIDE AND BIOHYDROGEN GENERATION

Substrate is the carbon source on which the catalyst or enzyme acts to produce hydrogen. There are diverse sources of substrates, which are classified as carbon-hydrated and organic acid substrates (Koku et al., 2002).

    I. **Carbon-hydrated substrates** such as glucose, fructose, sucrose, etc.
   II. **Organic acid substrates** such as formate, acetate, malate, etc.

These substrates can be used for the cell growth of any biological catalyst. To have a better hydrogen yield from the BWGS reaction, the choice of a suitable substrate plays a role. Different substrates display different abilities in different microorganisms when consuming carbon monoxide to producing hydrogen. The substrate consumption with respect to time can be expressed in first-order differential equation (Koku et al., 2002, 2003).

$$-\frac{ds}{dt} = k_s s \rightarrow S = S_0 e^{-k_s t} \tag{10.15}$$

where s is the substrate concentration (in g/L), t is the time, k_s is the substrate consumption rate constant, and S_0 is the initial substrate concentration (in g/L).

The substrate conversion efficiency can be expressed as follows (Koku et al., 2003):

$$\eta = \frac{100P}{6VS_0} \tag{10.16}$$

where P is the number of moles of hydrogen produced until the measuring time and V is the culture volume in liters.

R. Gelatinosus was reported to show about 80% carbon monoxide conversion on malate (Oh et al., 2005). Experiments conducted on R. ***rubrum*** using different substrates showed that ***R. rubrum*** showed high carbon monoxide conversion when acetate was used as a substrate (carbon source), and when acetate was implemented as the energy source, the rate of uptake was growth-related. The maximum hydrogen yield was also observed when R. *rubrum* was cultivated on acetate, followed by formate (Koku et al., 2002).

10.5.2 EFFECT OF TEMPERATURE ON GROWTH AND HYDROGEN PRODUCTION

The process is an exothermic one that needs to remove heat from the process or operate at a lower temperature. For the purity of the hydrogen that is produced, the process should operate at a lower temperature. Fortunately, with a biological catalyst, BWGS reaction can operate at a lower temperature which thermodynamically favors the removal of carbon monoxide (Henstra and Stams, 2011). The process is preferred to operate at temperatures between 25°C and 80°C for better growth of microorganisms and the best hydrogen yield (Alfano and Cavazza, 2018).

However, the temperature range might vary from one microorganism to another. For example, ***Thermincola ferriacetica*** grows at a temperature of 60°C, while ***Thermincola carboxydiphila*** grows at a temperature of 73°C (Alfano and Cavazza, 2018). It has been reported that most biological reactions multiply their biological activity rates with a raise of 10°C in the reaction temperature, though depending on the environment to which a microorganism has adapted, it has a different ideal temperature (Amos, 2004; Vargas et al., 2018).

10.5.3 EFFECT OF pH ON GROWTH AND HYDROGEN PRODUCTION

The pH value affects the BWGS reaction, depending on the composition of the medium. For instance, the composition of $PO^{3-}_4 = 1.0\,mM$, $HCO^{-1}_3 = 5.0\,mM$, $Ca^{+2} = 0.7\,mM$, and $Mg^{+2} = 2.5\,mM$ when added in neutral medium (pH = 7.0) may shift the pH to approximately 7.8 due to the alkalinity of the contents (Zhao et al., 2011). The pH affects the cell mineral contents, and the biological activities and solubility of the mineral contents varied with the pH, independent of the ions in the culture medium (Zhao et al., 2011). It is reported that ***Carboxydothermus hydrogenoformans*** was able to produce hydrogen at a reasonable yield at pH values ranging between 7.5 and 8.9, and the best hydrogen yield was obtained in the neutral pH range (6.8–7.3) (Zhao et al., 2011). The best operating pH also depends on the microorganisms used under the specified conditions. For example, ***Carboxydocella sporoproducens*** grows at a pH of 6.8, while ***Carboxydocella pertinax*** grows at a pH of 6.0–6.5 (Alfano and Cavazza, 2018).

10.5.4 EFFECT OF CARBON MONOXIDE INLET FLOW RATE ON HYDROGEN PRODUCTION

As the carbon monoxide flow rate increases in a continuous reactor system, the duration of retention is lowered, which consequently minimizes the likelihood of contact between water and carbon monoxide and thereby causes a low reaction rate (Jeong et al., 2015). An experiment aimed at evaluating the performance of a packed-bed reactor operated with various carbon monoxide inlet flow rates discovered that carbon monoxide conversion reduces with the increase of the gas flow rate (Mohammed and Ali, 2013).

10.5.5 EFFECT OF AGITATION RATE ON HYDROGEN PRODUCTION

Increasing the agitation rate of the reactor causes an increase in the mass transfer rate coefficient and thus improves the transfer rate of carbon monoxide and results in more conversion. Thus, increasing the transfer rate improves the provision of gas to organisms and hence improves production. However, higher agitation speeds can cause turbulence within the system and cause an increase in the rate of aeration. Mohammed and Ali (2013) conducted an experiment to examine the performance of a packed-bed reactor operated at different circulation rates of water. When the speed was increased from 200 to 800 mL/min, the conversion was increased, but there was no conversion beyond 800 mL/min of circulation rate. These findings are also in agreement with those found by McIlveen-Wright et al. (2006), where the effect of circulation rate was studied in a fluidized bed.

10.5.6 EFFECT OF PRESSURE ON PRESSURIZED SYSTEMS

Applying pressure (pressurize) to a reactor raises the equilibrium concentration of carbon monoxide at the gas–liquid interface and hence increases the mass transfer of carbon monoxide to the liquid phase. Adding pressure within the system also raises the gas concentration within the system and thereby increases the carbon monoxide concentration in the gas phase. This will result in an increase in the equilibrium concentration and the mass transfer rate. For the reactor, where the mass transfer limits their activities, running at elevated pressures will boost the mass transfer. However, high pressures might result in mass transfer rates above the intrinsic biological reaction rate. When the reactor works beyond its intrinsic biological reaction rate, the liquid-phase CO content builds up and hampers the biological reaction, and therefore, hydrogen generation falls (Amos, 2004). Since the reaction rate in the bioreactors highly relies on the mass of the functioning cells, increasing the cell mass could enable operation at elevated pressures and faster mass transfer rates.

10.6 CONCLUSION

It has been established that hydrogen is a trustworthy source of energy. The thermochemical WGS reaction performed using heterogeneous metal catalysts at high temperatures is a well-known process that converts carbon monoxide and water into carbon dioxide and hydrogen. As compared to that, the BWGS reaction is a biologically mediated WGS reaction that employs microorganisms or their enzymes to catalyze the conversion under mild conditions. Several microorganisms, including ***Rubrivivax gelatinosus, Rhodospirillum rubrum,*** and ***Carboxydothermus hydrogenoformans***, among others, can perform the BWGS reaction. This study highlights the current state of knowledge on the BWGS reaction, including the various organisms and enzymes involved, the underlying mechanisms, kinetics, mass transfer limitations, and factors affecting the BWGS reaction process. Despite the promising potential of the BWGS reaction for hydrogen production over the traditional thermochemical routes, several challenges need to be addressed to enhance its feasibility and scalability.

- **Enzyme and Substrate Stability:** CODH enzyme activity might be decreased when hydrogen is present in the reaction media, limiting the efficiency of the BWGS reaction. Further, high concentrations of carbon monoxide can be toxic to microorganisms and limit the choice of suitable biocatalysts for the BWGS reaction, and hence, developing carbon-monoxide-tolerant strains or identifying naturally occurring carbon-monoxide-tolerant microorganisms is necessary for efficient hydrogen production.

- **Optimized Reactor Design:** The design and development of bioreactors with improved mass transfer capabilities and high surface area contact between the gas and liquid phases are crucial for enhancing the efficiency of the BWGS reaction. The optimized design of bioreactors tailored for the BWGS reaction and novel bioreactor designs, such as membrane-based systems or immobilized cell reactors, should be considered for improved performance.
- **Scalability and Economic Feasibility:** Scaling up BWGS processes to industrial levels and ensuring their economic viability remains a significant challenge. Further research on bioreactor design, process optimization, and cost reduction is necessary.

In conclusion, the BWGS reaction presents a promising alternative for sustainable hydrogen production with potential environmental benefits compared to traditional thermochemical methods. However, challenges such as low reaction rates, enzyme stability, CO toxicity, and gas–liquid mass transfer need to be addressed to improve the feasibility and scalability of this approach (Lu et al., 2020). Future research directions should focus on enzyme and metabolic engineering, novel bioreactor designs, and integrated processes to enhance the overall efficiency of the BWGS reaction.

REFERENCES

IEA (2017), Key World Energy Statistics 2017, IEA, Paris, https://doi.org/10.1787/key_energ_stat-2017-en.Akkerman, I., Janssen, M., Rocha, J., and Wijffels, R. 2002. Photobiological hydrogen production: Photochemical efficiency and bioreactor design. *Int. J. Hydrogen Energy* 27:1195–1208.

Alfano, M., and Cavazza, C. 2018. The biologically mediated water-gas shift reaction: Structure, function and biosynthesis of monofunctional [NiFe]-carbon monoxide dehydrogenases. *Sustain. Energy Fuels* 2:1653–1670.

Amos, W. A. 2004. *Biological Water Gas Shift Conversion of Carbon Monoxide to Hydrogen.* Milestone report. Available from National Renewable Energy Laboratory, Golden, CO.

Aziz, M., Darmawan, A., and Juangsa, F. B. 2021. Hydrogen production from biomasses and wastes: A technological review. *Int. J. Hydrogen Energy* 46(68):33756–33781. https://doi.org/10.1016/j.ijhydene.2021.07.189

Brentner, L. B., Jordan, P. A., and Zimmerman, J. B. 2010. Challenges in developing biohydrogen as a sustainable energy source: Implications for a research agenda. *Environ. Sci. Technol.* 44:2243e54. https://doi.org/10.1021/es9030613

Das, D., and Veziroglu, T. N. 2001. Hydrogen production by biological processes: A survey of literature. *Int. J. Hydrogen Energy* 26:13e28.

Demirel, B., and Yenigu¨n, O. 2002. Two-phase anaerobic digestion processes: A review. *J Chem Technol Biotechnol* 77:743e55. https://doi.org/10.1002/jctb.630.

Díaz, D. B., Neumann, A., and Aliyu, H. 2022. Thermophilic water gas shift reaction at high carbon monoxide and hydrogen partial pressures in Parageobacillus thermoglucosidasius KP1013. *Fermentation* 8:596. https://doi.org/10.3390/fermentation8110596

Domnik, L., Merrouch, M., Goetzl, S., Jeoung, J. H., L´eger, C., Dementin, S., Fourmond, V., and Dobbek, H. 2017. *Angew. Chem. Int. Ed.* 56:15466–15469.

Doukov, T. I., Iverson, T. M., Seravalli, J., Ragsdale, S. W., and Drennan, C. L. 2002. A Ni-Fe-Cu center in a bifunctional carbon monoxide dehydrogenase/acetyl-CoA synthase. *Science* 298:567–572.

Eker, S., and Sarp, M. 2017. Hydrogen gas production from waste paper by dark fermentation: Effects of initial substrate and biomass concentrations. *Int. J. Hydrogen Energy* 42(4):2562e8.

Ergal, I., Gräf, O., Hasibar, B., Steiner, M., Vukoti´c, S., Bochmann, G., Fuchs, W., Rittmann, and S. K. M. R. 2020. Biohydrogen production beyond the Thauer limit by precision design of artificial microbial consortia. *Commun. Biol.* 3:443.

Fontecilla-Camps, J. C., Amara, P., Cavazza, C., Nicolet, Y., and Volbeda, A. 2009. Structure-function relationships of anaerobic gas-processing metalloenzymes. *Nature* 460:814.

Germscheidt, R. L., Moreira, D. E., Yoshimura, R. G., Gasbarro, N. P., Datti, E., dos Santos, P. L., and Bonacin, J. A. 2021. Hydrogen environmental benefits depend on the way of production: An overview of the main processes production and challenges by 2050. *Adv. Energy Sustain. Res.* 2(10):2100093. https://doi.org/10.1002/aesr.202100093

Gibson, E. K., Stere, C. E., Curran-McAteer, B., Jones, W., Cibin, G., Gianolio, D., Goguet, A., Wells, P. P., Catlow, C. R. A., Collier, P., Hinde, P., and Hardacre, C. 2017. Probing the role of a non-thermal plasma (NTP) in the hybrid NTP catalytic oxidation of methane. *Angew. Chem. Int. Ed.* 56:9351–9355. https://doi.org/10.1002/anie.201703550

Haddad, M., Cimpoia, R., and Guiot, S. R. 2014. Performance of Carboxydothermus hydrogenoformans in a gas-lift reactor for syngas upgrading into hydrogen. *Int. J. Hydrogen Energy* 39:2543–2548.

Hadj-Saïd, J., Pandelia, M. E., Léger, C., Fourmond, V., and Dementin, S. 2015. The Carbon Monoxide Dehydrogenase from Desulfovibrio vulgaris. *Biochim. Biophys. Acta, Bioenerg.* 1847:1574–1583.

Henstra, A. M., and Stams, A. J. M. 2011. Deep conversion of carbon monoxide to hydrogen and formation of acetate by the anaerobic thermophile Carboxydothermus hydrogenoformans, Hindawi Publishing Corporation. *Int. J. Microbiol.* https://doi.org/10.1155/2011/641582

Inharoy, A., and Pakshirajan, K. 2020. Methane free biohydrogen production from carbon monoxide using a continuously operated moving bed biofilm reactor. *Int. J. Hydrogen Energy.* https://doi.org/10.1016/j.ijhydene.2020.09.250

Ismail, K. S. K., Najafpour, G., Younesi, H., Mohamed, A. R., and Kamaruddin, A. H. 2008. Biological hydrogen production from CO: Bioreactor performance. *Biochem. Eng. J.* 39:468–477.

Jeong, J., Bertsch, J., Hess, V., Choi, S., Choi, I. G., Chang, I. S., and Müller, V. 2015. Energy conservation model based on genomic and experimental analyses of a carbon monoxide-utilizing, butyrate-forming acetogen, Eubacterium limosum KIST612. *Appl. Environ. Microbiol.* 81(14):4782–4790. https://doi.org/10.1128/AEM.00675-15

Jörn, M., Stefan, B., Jürgen, K., Andreas, K., and Reiner, H. 2001. *Eur. J. Biochem.* 265:325–335.

Jung, G. Y., Kim, J. R., Park, J.-Y., and Park, S. 2002. Hydrogen production by a new chemoheterotrophic bacterium *Citrobacter sp.* Y19. *Int. J. Hydrogen Energy* 27:601–610.

Kalil, M. S., Alshiyab, H. S. S., and Yusoff, W. M. W. 2009. Media improvement for hydrogen production using C. acetobutylicum NCIMB13357. *Am. J. Appl. Sci.* 6:1158e68.

Kayfeci, M., Keçebaş, A., and Bayat, M. 2019. Chapter 3 - Hydrogen production. *Environmental Science\Physical science and Mathematics*, 45–83. https://doi.org/10.1016/B978-0-12-814853-2.00003-5

Kim, M.-S., Bae, S. S., Kim, Y. J., Kim, T. W., Lim, J. K., Lee, S. H., Choi, A. R., Jeon, J. H., Lee, J.-H., Lee, H. S., and Kang, S. G. 2013. CO-Dependent H_2 Production by Genetically Engineered Thermococcus onnurineus NA1. *c.*

Kim, T. W., Bae, S. S., Lee, S. M., Lee, H. S., Lee, J.-H., Na, J.-G., and Kang, S. G. 2020. Long-term operation of continuous culture of the hyperthermophilic archaeon, *Thermococcus onnurineus* for carbon monoxide-dependent hydrogen production. *Biotechnol. Bioproc. E* 25:485–492. https://doi.org/10.1007/s12257-020-0005-x

Koku, H., Eroglu, I., Gunduz, U., Yucel, M., and Turker, L. 2002. Aspects of the metabolism of hydrogen production by *Rhodobacter sphaeroides. In. J. Hydrogen Energy* 27:1315–1329.

Koku, H., Eroglu, I., Gunduz, U., Yucel, M., and Turker, L. 2003. Kinetics of biological hydrogen production by the photosynthetic bacterium *Rhodobacter sphaeroides* O.U. 001. *Int. J. Hydrogen Energy* 28:381–388.

Kondo, T., Arakawa, M., Hirai, T., Wakayama, T., Hara, M., and Miyake, J. 2002. Enhancement of hydrogen production by a photosynthetic bacterium mutant with reduced pigment. *J. Biosci. Bioeng.* 39(2):145–150.

Kumar, N., and Das, D. 2000. Enhancement of hydrogen production by Enterobacter cloacae IIT-BT 08. *Process Biochem.* 35:589e93.

Kumar, R., Kumar, A., and Pal, A. 2020. An overview of conventional and non-conventional hydrogen production methods. *Materials Today: Proceedings.* https://doi.org/10.1016/j.matpr.2020.08.793

Kung, Y., and Drennan, C. L. 2011. A role for nickel-iron cofactors in biological carbon monoxide and carbon dioxide utilization. *Curr. Opin. Chem. Biol.* 15:276–283.

Lee, C.-M., Chen, P.-C., Wang, C.-C., and Tung, Y.-C. 2002. Photohydrogen production using purple nonsulfur bacteria with hydrogen fermentation reactor effluent. *Int. J. Hydrogen Energy* 27:1309–1313.

Lee, J., Phung, N. T., Chang, I. S., Kim, B. H., and Sung, H. C. 2003. Use of acetate for enrichment of electrochemically active microorganisms and their 16rdna analyses. *FEMS Microbiol. Lett.* 223:185–191.

Liu, C., Shi, Y., Liu, H., Ma, M., Liu, G., Zhang, R., and Wang, W 1. 2020. Insight of co-fermentation of carbon monoxide with carbohydraterich wastewater for enhanced hydrogen production: Homoacetogenic inhibition and the role of pH. *J. Clean. Prod.* 267:122027. https://doi.org/10.1016/j.jclepro.2020.122027

Ljunggren, M., Willquist, K., Zacchi, G., and van Niel, E. W. 2011. A kinetic model for quantitative evaluation of the effect of hydrogen and osmolarity on hydrogen production by *Caldicellulosiruptor saccharolyticus. Biotechnol. Biofuels* 4:31.

Lu, C., Tahir, N., Li, W., Zhang, Z., Jiang, D., Guo, S., Wang, J., Wang, K., & Zhang, Q.. 2020. Enhanced buffer capacity of fermentation broth and biohydrogen production from corn stalk with Na_2HPO_4/NaH_2PO_4. *Biores. Technol.* 313:123783.

Magureanu, M., Dobrin, D., Mandache, N. B., Cojocaru, B., and Parvulescu, V. I. 2013. Toluene oxidation by non-thermal plasma combined with palladium catalysts. *Front. Chem.* 1:1–6. https://doi.org/10.3389/fchem. 2013.00007

Martino, M., Ruocco, C., Meloni, E., Pullumbi, P., and Palma, V. 2021. Main hydrogen production processes: An overview. *Catalysts* 11:547. https://doi.org/10.3390/catal11050547

McIlveen-Wright, D. R., Pinto, F., Armesto, L., and Caballero, M. A. 2006. A comparison of circulating fluidised bed combustion and gasification power plant technologies for processing mixtures of coal, biomass and plastic waste. *Fuel Proc. Technol.* 87:793–801.

Mohammed, A. K., and Ali, S. A. 2013. Bio production of hydrogen by water - gas shift reaction using packed bed reactor. *Int. J. Biol. Pharma. Res.* 4(8):564–567.

Mohr, T., Aliyu, H., Küchlin, R., Zwick, M., Cowan, D., Neumann, A., and de Maayer, P. 2018. Comparative genomic analysis of Parageobacillus thermoglucosidasius strains distinct hydrogenogenic capacities. *BMC Genom.* 19:880.

Najafpour, G., Younesi, H., and Mohamed, A. M. 2006 A survey on various carbon sources for biological hydrogen production via the water-gas reaction using a photosynthetic bacterium (*Rhodospirillum rubrum*). *Energy Sources, Part A: Recovery, Utilization, and Environmental Effects* 28(11):1013–1026, https://doi.org/10.1080/009083190910541

Nanqi, R., Wanqian, G., Bingfeng, L., Guangli, C., and Jie, D. 2011. Biological hydrogen production by dark fermentation: Challenges and prospects towards scaled-up production. *Curr. Opin. Biotechnol.* 22(3):365–370. https://doi.org/10.1016/j.copbio.2011.04.022

Ni, M., Leung, D. Y. C., Leung, M. K. H., & Sumathy, K. 2005. An overview of hydrogen production from biomass. *Fuel Process. Technol.* 87:461e72. https://doi.org/10.1016/j.fuproc.2005.11.003

Nikolaidis, P., and Poullikkas, A. 2017. A comparative overview of hydrogen production processes. *Renew. Sustain. Energy Rev.* 67:597–611.

Oh, Y. K., Kim, Y. J., Park, J. Y., Lee, T. H., Kim, M. S., and Park, S. 2005. Biohydrogen production from carbon monoxide and water by Rhodopseudomonas palustris P4. *Biotechnol. Bioprocess Eng.* 10:270–274.

Oliveira, H., Pinto, G., Oliveira, A., Oliveira, C., Faustino, M. A., Briers, Y., Domingues, L., and Azeredo, J. 2016. Characterization and genome sequencing of a *Citrobacter freundii* phage CfP1 harboring a lysin active against multidrug-resistant isolates. *Appl. Microbiol. Biotechnol.* 100(24):10543–10553. https://doi.org/10.1007/s00253-016-7858-0

Panić, I., Cuculić, A., and Ćelić, J. 2022. Color-coded hydrogen: Production and storage in maritime sector. *J. Mar. Sci. Eng.* 10:1995. https://doi.org/10.3390/jmse10121995

Ragsdale, S. W., Yi, L., Bender, G., Gupta, N., Kung, Y., Yan, L., Stich, T. A., Doukov, T., Leichert, L., Jenkins, P. M., Bianchetti, C. M., George, S. J., Cramer, S. P., Britt, R. D., Jakob, U., Martens, J. R., Phillips, G. N., and Drennan, C. L. 2012. Redox, haem and CO in enzymatic catalysis and regulation. *Biochem. Soc. Trans.* 40:501–507.

Saxena, R. C., Adhikari, D. K., and Goyal, H. B. 2009. Biomass-based energy fuel through biochemical routes: A review. *Renew. Sustain. Energy Rev.* 13(1):167–178. https://doi.org/10.1016/j.rser.2007.07.011

Schut, G. J., Lipscomb, G. L., Nguyen, D. M. N., Kelly, R. M., and Adams, M. W. W. 2016. Heterologous production of an energy-conserving carbon monoxide dehydrogenase complex in the hyperthermophile pyrococcus furiosus. *Front. Microbiol.* 7:1–9.

Sinharoy, A., Baskaran, D., and Pakshirajan, K. 2019. Sustainable biohydrogen production by dark fermentation using carbon monoxide as the sole carbon and energy source. *Int. J. Hydrogen Energy* 44:13114–13125.

Sipma, J., Meulepas, R. J. W., et al. 2004. Effect of carbon monoxide, hydrogen and sulfate on thermophilic (55°C) hydrogenogenic carbon monoxide conversion in two anaerobic bioreactor sludges, Springer-Verlag, *Appl. Microbiol. Biotechnol.* 64:421–428. https://doi.org/10.1007/s00253-003-1430-4

Sokolova, T. G., Jeanthon, C., Kostrikina, N. A., Chernyh, N. A., Lebedinsky, A. V., Stackebrandt, E., and Bonch-Osmolovskaya, E. A. 2004. The first evidence of anaerobic CO oxidation coupled with H2 production by a hyperthermophilic archaeon isolated from a deep-sea hydrothermal vent. *Extremophiles* 8:317–323.

Su, X., Zhao, W., and Xia, D. 2018. The diversity of hydrogen-producing bacteria and methanogens within an in-situ coal seam. *Biotechnol. Biofuels* 11:245.

Svetlitchnyi, V., Peschel, C., Acker, G., and Meyer, O. 2001. Two membraneassociated NiFeS-carbon monoxide dehydrogenases from the anaerobic carbon-monoxide-utilizing eubacterium Carboxydothermus hydrogenoformans. *J. Bacteriol.* 183:5134e44.

Talapko, D., Talapko, J., Erić, I., Škrlec, I. 2023. Biological hydrogen production from biowaste using dark fermentation, storage and transportation. *Energies* 16:3321. https://doi.org/10.3390/en16083321

Tang, X., Gao, F., Wang, J., Yi, H., and Zhao, S. 2014. RSC advances nitric oxide decomposition using atmospheric pressure dielectric barrier discharge reactor with different adsorbents. *RSC Adv.* 4:58417–58425. https://doi.org/10.1039/C4RA0 8447K

Techtmann, S. M., Lebedinsky, A. V., Colman, A. S., Sokolova, T. G., Woyke, T., Goodwin, L., and Robb, F. T. 2012. Evidence for horizontal gene transfer of anaerobic carbon monoxide dehydrogenases. *Front. Microbiol.* 3:132. https://doi.org/10.3389/fmicb.2012.00132

Tsygankov, A. A., Fedorov, A. S., Kosourov, S. N., and Rao, K. K. 2002. Hydrogen production by cyanobacteria in an automated outdoor photobioreactor under aerobic conditions. *Biotechnol. Bioeng.* 80(7):777–783.

Van Niel, E. W. J., Claassen, P. A. M., and Stams, A. J. M. 2003. Substrate and product inhibition of hydrogen production by the extreme thermophile, *Caldicellulosiruptor saccharolyticus. Biotechnol. Bioeng.* 81:255–262.

Vanzin, G. F., Huang, J., Smolinski, S., Kronoveter, K., and Maness, P.-C. 2002. Biological hydrogen from fuel gases. *Proceedings of the U.S. DOE Hydrogen Program Review, Denver, USA.*

Vargas, S. R., dos Santos, P. V., Zaiat, M., and Calijuri, M. do C. 2018. Optimization of biomass and hydrogen production by *Anabaena sp.* (UTEX 1448) in nitrogen-deprived cultures. *Biomass Bioenergy* 111:70e6. https://doi.org/10.1016/j.biombioe.2018.01.022

Wangkawong, K., Phanichphant, S., Inceesungvorn, B., Stere, C. E., Chansai, S., Hardacre, C., and Goguet, A. 2020. Kinetics of water gas shift reaction on Au/CeZrO$_4$: A comparison between conventional heating and dielectric barrier discharge (DBD) plasma activation. *Top Catal.* 63:363–369. https://doi.org/10.1007/s11244-020-01245-8

Winkler, M., Hemschemeier, A., Gotor, C., Melis, A., and Happe, T. 2002. [Fe]-hydrogenases in green algae: Photo-fermentation and hydrogen evolution under sulfur deprivation. *Int. J. Hydrogen Energy* 27:1431–1439.

Wolfrum, E. J., and Watt, A. S. 2002. Bioreactor design studies for a hydrogen-producing bacterium. *Appl. Biochem. Biotech.* 98/100:611e25. https://doi.org/10.1385/ABAB:98-100:1-9:611

Wu, K.-J., and Chang, J.-S. 2007. Batch and continuous fermentative production of hydrogen with anaerobic sludge entrapped in a composite polymeric matrix. *Process Biochem.* 42:279–284.

Younesi, H., Najafpour, G., Ku Ismail, K. S., Mohamed, A. R., and Kamaruddin, A. H. 2008. Biohydrogen production in a continuous stirred tank bioreactor from synthesis gas by anaerobic photosynthetic bacterium: Rhodopirillum rubrum. *Bioresour. Technol.* 99:2612–2619.

Younesi, H., Najafpour, G., and Mohamed, A. R. 2004. Effect of organic substrate on hydrogen production from synthesis gas using *Rhodospirillum rubrum*, in batch culture, Universiti Sains Malaysia. *Biochem. Eng. J.* 21:123–130.

Zhang, B., Zhang, S.-X., Yao, R., Wu, Y.-H., and Qiu, J.-S. 2021. Progress and prospects of hydrogen production: Opportunities and challenges. *J. Electron. Sci. Technol.* https://doi.org/10.1016/j.jnlest.2021.100080

Zhao, Y., Cimpoia, R., Liu, Z., and Guiot, S. R. 2011. Orthogonal optimization of Carboxydothermus hydrogenoformans culture medium for hydrogen production from carbon monoxide by biological water-gas shift reaction, *Int. J. Hydrogen Energy* 36(17):10655–10665. https://doi.org/10.1016/j.ijhydene.2011.05.134

11 The Fermentation Process for Biomass Conversion to Hydrogen

Sadia Fida, Zeshan Sheikh, and Zia Ullah

11.1 INTRODUCTION

Fossil energy consumption has significantly harmed the environment and polluted it. In addition, severe issues about fossil fuel (FF) availability and security are appearing. Hence, there is a need for the energy shift from a high-carbon energy system to a more sustainable one.

The coronavirus disease of 2019 (COVID-19) pandemic outbreak caused a pause in global energy trade and changes in energy resource use. Global conditions, meantime, are complex and can have an impact on the spot markets for FFs such as crude oil, gasoline, heating oil, and natural gas (Wang et al., 2022). One of the main objectives countries worldwide have set for 2050 to lessen the effects of climate change is decarbonizing the planet. Global issues like oil scarcity and climate change influence new developments in fuel markets (Vickers, 2017). Interest in generating renewable energy to reduce reliance on FF and achieve sustainable energy generation is rising (Wei et al., 2016).

Renewable energy, such as alternative fuels and biofuels, would lower overall CO_2 emissions. Hydrogen (H_2) could replace FFs due to its clean nature and sustainability. H_2 can be produced from nonrenewable as well as renewable sources. H_2 production from nonrenewable sources will still burden the available sources (Kayfeci et al., 2019). Therefore, H_2 production from renewable sources helps reduce dependence on FFs and carbon emissions. Among renewable ways of H_2 production, methods using biological processes, such as fermentation, have merit over others as they are cost-effective and natural. Thus, biohydrogen (bioH_2) is considered one of the future biofuels (Dawood et al., 2020). This chapter presents insights into the fermentation process for biomass conversion into H_2.

11.1.1 Hydrogen and Sustainability

H_2 can be utilized as a "clean" fuel because water vapors are the only by-product of its burning. According to market projections, H_2 output will rise sharply over the coming years by 5%–10% per year from 50 to 82 Mt by 2050. Currently, fossil resources account for more than 98% of the H_2 production, either through natural gas

DOI: 10.1201/9781003382270-14

steam reforming or other means (Qureshi et al., 2022). About 2% of H_2 production comes from renewable sources, mainly water electrolysis, a relatively new technique. H_2 produces 122 kJ/g of energy, 2.75 times more than FF (Arun et al., 2022).

11.1.2 Green Hydrogen Sources

The globe is facing a significant challenge in reducing and mitigating the consequences of carbon dioxide and other gas emissions brought on by global warming. To change the established rules of energy production by depending on green energy sources, green H_2 production is quickly emerging as a significant technology (Mikhaylov et al., 2020). H_2 is a "versatile energy carrier," according to the International Energy Agency (IEA), with a variety of uses in industries and sectors like transportation, industry, and heating that still primarily rely on FFs (Elzinga et al., 2014; IEA, 2014; Bhagwat and Olczak, 2020). It is understood that a quick increase in green H_2 will be necessary to reach climate-neutrality goals by the middle of the century (Ajanovic et al., 2022; Noussan et al., 2020).

11.2 HYDROGEN PRODUCTION METHODS

The two primary sources of H_2 production are renewable and nonrenewable (Dawood et al., 2020; Milani et al., 2020). Till 2020, about 95% of H_2 was generated from nonrenewable sources, that is, FFs through steam reforming of natural gas, and the remainder was generated from renewable resources through water electrolysis (Mosca et al., 2020).

H_2 production through renewable resources includes processes using biomass or water as renewable resources. The processes in which biomass is used as a fuel can be divided into two main subclasses: thermochemical and biological. Biological methods include direct and indirect biophotolysis, dark fermentation (DF), photofermentation (PF), and sequential DF and PF. At the same time, thermochemical technology mainly involves pyrolysis, gasification, combustion, and liquefaction (Yadav et al., 2018). Figure 11.1 illustrates the several H_2 generation processes.

11.2.1 Hydrogen Generation from Nonrenewable Sources

The viability of any H_2 production pathway (HPP) can be determined by how clean, affordable, effective, and efficient it is.

The presence of three elements is required for the H_2 production pathway (HPP) to function: H_2-containing substance (hydrocarbons or non-hydrocarbons), energy source, and catalyst material (El-Shafie et al., 2019). Any material with H_2 in its structure can be used for H_2 production, whether from renewable or nonrenewable sources. Hydrocarbon reforming and pyrolysis are conventional methods for generating H_2 from FFs.

H_2 production from a hydrocarbon fuel through reforming reactions is known as hydrocarbon reforming. Other substances, such as steam or CO_2, are employed as reactants in the hydrocarbon-improving processes (Lamb et al., 2020).

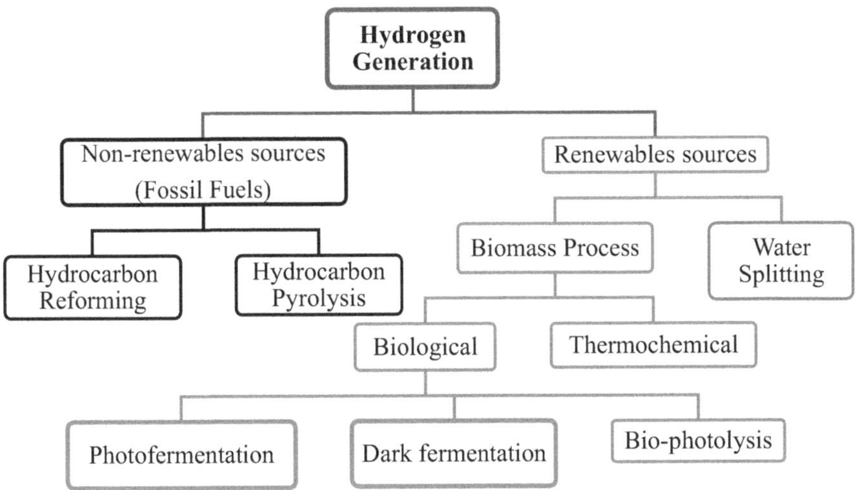

FIGURE 11.1 H_2 generation methods.

H_2 can be produced by pyrolysis, a well-known method of making it, that is, by breaking down hydrocarbons. Hydrocarbons undergo heat degradation during pyrolysis because they are H_2-containing molecules. In addition to producing H_2, this technique yields solid carbon, an essential by-product. Compared to the traditional methods of hydrocarbon reforming, this procedure is far less developed for the hydrocarbons found in FFs. Pyrolysis-based technologies are often not considered sustainable unless the energy source is low carbon due to the pyrolysis reaction's highly endothermic character (Bakhtyari et al., 2019).

11.2.2 Hydrogen Production from Renewable Sources

Renewable energy sources only account for a small portion of H_2 production, aiming to create environmentally friendly and pollution-free H_2. The advancement of renewable energy sources, such as biomass and water splitting, has supported the development of more efficient H_2 production systems (Acar and Dincer, 2019).

11.2.2.1 Electrolysis

Electrolysis is a known and effective method for producing H_2 by splitting water. Consequently, electricity is used to provide the required amount of input energy (Wang et al., 2021). Two cathode and anode electrodes submerged in an electrolyte made into a typical electrolyzer device. In general, water dissociates into H_2 and oxygen when an electric current is given to the system. While oxygen is created at the anode, H_2 evolution occurs at the cathodic electrode (equation (11.1)):

$$H_2O + Electricity\,(337.2\ kJ/mol) + Heat\ (48.6\ kJ/mol) \rightarrow 2H_2 + 1/2O_2 \quad (11.1)$$

The mentioned process (equation (11.1)) requires 1.23 V theoretical thermodynamic cell voltage to split water into H_2 and oxygen at room temperature. The needed cell voltage for effective water splitting has been estimated experimentally to be 1.48 V (Lim et al., 2021). The global portion of H_2 produced through electrolysis is around 5% (Yodwong et al., 2020).

11.2.2.2 Thermochemical Decomposition of Water

Thermochemical decomposition of water, known as thermolysis, is a high-temperature water decomposition method. A zero Gibbs energy $(G) = 0$ condition, which is one of the requirements for water decomposition. The essential high temperature of this process cannot be met with renewable energy sources. Several thermochemical water decomposition cycles have thus far been developed to lower temperature and boost overall efficiency (Acar et al., 2016).

11.2.2.3 Photovoltaic Electrolysis

This method is very analogous to electrolysis. Photovoltaic (PV) electrolysis generates electrical current via PV panels by converting solar energy to electricity. The PV electrolysis unit comprises PV panels, a direct current (DC) bud bar, an alternating current (AC) grid, a battery set, an electrolyzer, and a H_2 storage device. PV electrolysis has a current efficiency of roughly 16%. As PV efficiency improves, production costs will fall (Burton et al., 2021).

11.2.2.4 PEC Method

PEC generates H_2 from solar energy via electrochemical reactions. This method involves either one or two photo-electrodes made of semiconductor materials employed to generate electrical energy utilizing solar energy. The PEC technique has extensive use, including the production of electrical current, the generation of H_2, and the treatment of toxic materials. Several semiconductor materials have been investigated and tested in PEC systems (Eskandari, 2019).

11.2.2.5 Biomass Sources

Among renewable energy sources, biomass was the first option for humans to use for energy, and it has been used even now (Correa and Kruse, 2018). Biomass is derived from various sources, including agricultural products, wood, plant, and animal wastes and is thus regarded as a viable replacement for FFs (Sivabalan et al., 2021). This chapter focuses on H_2 production using biomass as feedstock.

11.3 HYDROGEN PRODUCTION FROM BIOMASS SOURCES

The research has focused on biomass lignocellulosic materials such as wood, agricultural leftovers, energy crops. A biological process in which biogas is formed by employing microorganisms (MOs) on organic material, also known as anaerobic digestion (AD), is a method of producing H_2 from biomass (Sivabalan et al., 2021). H_2 production from biomass is bioH_2 depending on its composition, source, and pretreatments, as elaborated below.

11.3.1 Biomass Types and Properties

The main factors influencing material selection in biohydrogen generation are accessibility, affordability, content of carbohydrate (CH), and biodegradation. Simple sugars that are readily biodegradable and favored substrates for the generation of H_2 include glucose, sucrose, and lactose. On the other hand, pure CH sources are pricey starting points for the synthesis of H_2. The principal wastes that may be utilized to create H_2 gas are discussed in the following sections.

Various categories of biomasses, such as lignocellulosic biomass, wastewater (WW) sludge, microalgae, and food waste, are promising substrates for bioH$_2$ generation as they contain content of high organic matter (OM), less nutrient dependence, and higher potential of energy (Dinesh et al., 2018, Show et al., 2019).

11.3.1.1 Food Industry and Agricultural Residues

Several agricultural, agro-industrial, and food residues have starch, cellulose, and high CH content. These include sugarcane bagasse, cornstalk, corn stover, corn cobs, corn bran, wheat straw, sorghum rusk, sorghum leaves, sorghum stover, rice straw, rice bran, rice husk, oat straw, forestry waste, wood, and grass. The complex nature of these waste materials may harm biodegradability. Solid wastes containing starch are less complicated to break down for CH and H_2 gas production. Agricultural residues containing cellulose must be treated before the fermentation process. Agricultural residues must be grounded and dignified mechanically or chemically before fermentation. Such wastes' cellulose and hemicellulose content may be hydrolyzed to CHs, which can be processed to yield organic acid and H_2 gas. It has been determined that lignin concentration has an adverse relationship with the efficiency of agricultural waste enzymatic hydrolysis (Varejão, 2022).

11.3.1.2 Carbohydrate-Rich Industrial WWs

WW derived from the dairy industries and breweries are a few examples of biodegradable, nontoxic industrial effluents that include CHs and can be used as raw materials to produce bioH$_2$. To remove unwanted elements and balance nutrition, specific WW may require pretreatment. CH-rich food sector effluents are further treated using the necessary bioprocessing methods to transform the CH content into organic acids and H_2 gas (Unni et al., 2022).

11.3.1.3 Sludge from Effluent Treatment Plants (ETPs)

Waste sludge in ETPs includes significant quantities of protein and CHs converted to fuels like CH_4 or H_2 gas. AD of excess sludge can be accomplished in two processes. In the acidogenic phase, organic materials will be transformed into organic acids, and photoheterotrophic bacteria will utilize these acids to produce H_2 gas in the later stage (Kim et al., 2022).

11.3.1.4 Kitchen Waste

The production of aqueous and gaseous biofuels like bioethanol, bio-butanol, and bioH$_2$ from kitchen waste is very feasible due to its high concentration of organic chemicals. At the consumer level, the food waste produced has around 52% of grain residues, 21% of vegetables and fruits, roots 6% of meat and fish, 7%, 12% of dairy, and 2% of oil waste (Xu et al., 2018; Vavouraki et al., 2013). Anaerobic fermentation can be utilized to produce bioH$_2$ from kitchen waste. Yet certain additives, such as municipal waste, can significantly boost the amount of bioH$_2$ generated, with a concentration of 13%–19% (Srivastava et al., 2021).

11.3.2 Biomass Pretreatments to Enhance Hydrogen Production

Challenges with bioH$_2$ from biomass include limited production and substrate deterioration (El Bari et al., 2022). Pretreatment can boost the productivity of H$_2$ generation while also speeding up the breakdown of biomass.

A pretreatment method helps degrade the crystal structure of macromolecules. Lignocellulose is one of the waste biomass's most resistant components, and that of its constituents is cellulose. Pretreatment lowers the polymerization degree, enabling the conversion of lignocellulosic material into fermentable compounds that are accessible to the majority of MOs. The types of pretreatments that are used are physical, chemical, biological, or a combination of these, and they are often dictated by the nature of the substrates that will be fermented.

11.3.2.1 Acid and Base Pretreatment

Chemical pretreatments have received the most significant research attention because of their simple use and low energy need. To speed up the chemical breakdown of lignocellulosic biomass and boost the production of bioH$_2$, chemical reagents can be utilized, including acids, bases, organosols, ionic liquids, metal chlorides, and others. Both acidic and basic approaches to cellulosic biomass degradation are particularly effective because they can convert most cellulose biomass into soluble sugars that MOs may utilize to make bioH$_2$ (Roy et al., 2020). Moreover, the feed's initial pH should be considered for the abovementioned pretreatments since it can impact the bioH$_2$ yield. Sulfuric acid, hydrochloric acid, boric acid, and nitric acid are often used for pretreatment (Yakaboylu et al., 2021).

Concentrated acids (30%–70%) at moderate temperatures (100°C) or dilute acids (0.1%–10%) at high temperatures (100°C–250°C) can be used for acid pretreatment. Although concentrated acid pretreatment may significantly boost sugar conversion rates (by more than 90%), most concentrated acids are extraordinarily harmful and corrosive, demanding high operating and maintenance costs. They also cause undesired cellulose breakdown, creating inhibitory compounds (Solarte-Toro et al., 2019).

Martinez et al. used olive tree biomass as feedstock for biofuel production. Olive tree biomass was pretreated with dilute sulfuric acid to achieve significant sugar recoveries from cellulosic and hemicellulosic fractions. Temperature (160°C–200°C), acid concentration (0–8 g acid/100 g extracted raw material), and solid loading (15%–35% w/v) were chosen as operating variables. The best acid pretreatment conditions were 160°C, 4.9 g sulfuric acid/100 g biomass, and 35% solid loading (w/v). These optimized conditions resulted in a sugar concentration of 79.8 g/L, equating to a total sugar yield of 39.8 g/100 g extracted from olive tree biomass (Martínez-Patiño et al., 2017). In another study, feedstocks, that is, stem, leaf, flower, cob, and husk, were pretreated with 2% H_2SO_4 at a temperature of 121°C for 60 minutes with 10% biomass loadings. The cob had the highest sugar content, generating 94.2% glucose (Li et al., 2016).

However, calcium hydroxide $(Ca\,(OH)_2)$, sodium hydroxide (NaOH), and potassium hydroxide (KOH) are widely used bases for the pretreatment of biomass. According to a study on the alkaline pretreatment of rice straw for biomethane production, 1% NaOH at room temperature for 3 hours reduces hemicellulose and lignin concentration while keeping cellulose content constant. Compared to untreated rice straw, this resulted in a 34% increase in methane output (Shetty et al., 2017). A low-temperature alkali/urea solution pretreatment approach has been proposed for better cellulose $bioH_2$ generation. Various alkaline solutions, including and without containing urea, were studied. The NaOH/urea pretreatment operated remarkably well at temperatures ranging from −8°C to −20°C. At low temperatures, NaOH/urea-pretreated rice straw produced the most significant quantity of H_2, 22.08 mmol/L, and an energy conversion efficiency of 9.76% (Dong et al., 2018).

It is commonly said that acid pretreatment is better than essential pretreatment for enhancing organic H_2 generation (Yakaboylu et al., 2021).

11.3.2.2 Biological Pretreatment

Aeration, enzymatic hydrolysis, and fungal pretreatment are biological pretreatment techniques for $bioH_2$. The quick and precise breakdown of lignocellulosic substrates is their significant advantage. Techniques for biological pretreatment offer certain advantages over those for chemical and physical pretreatment since they do not require as much force, pressure, or potent chemicals (Ravindran and Jaiswal, 2016).

The biological method pretreats the lignin with microorganisms, consortiums, or microbial enzymes as catalysts and uses fewer chemicals; therefore, it is a more environmentally friendly (Deivayanai et al., 2022).

Srivastava et al. found that hydrolysis of rice straw with cellulose enzyme produces 54.18 g/L sugar and 2.58 LH_2/L H_2 (Srivastava et al., 2017). In another study, wood biomass and *Streptomyces griseus*, isolated and subcultured from litter leaves, degraded around 24% of the lignin content (Deivayanai et al., 2022).

This pretreatment method is unsuitable for industrial applications because of the long residence time and monitoring of biological processes. For direct hydrolysis, industrial enzymes, including xylanase, arabinase, hemicellulase, and amylase, can be employed (Banu et al., 2020).

11.3.2.3 Physical Pretreatment

Physical pretreatment methods include mechanical and thermal methods, which can alter the lignocellulosic material's structure. The most widely used mechanical procedures for biomass are grinding, ball milling, screw pressing, microwaves, and sonication. The lignocellulosic structure, as well as the cell dissolution of the particle, is broken down by ultrasonic waves at a frequency range from 10 to 20 kHz. For example, the generation of $bioH_2$ was considerably boosted by 424%, and the optimal ultrasonic processing of paper and pulp mill waste was attained at an amplitude of 60% for 45 minutes (Wang and Yin, 2018). Water absorption at 400% (w/w) during ball milling at 80°C for 30 minutes on corn stover biomass enhances the yield of glucose up to 66.69% compared to 100°C along with ball milling process, boosting sustainability and making the process efficient (Shukla et al., 2023).

Much emphasis has been given to the hydrolysis of substrates via heat pretreatment, which breaks down complex macromolecule components in waste (Ma et al., 2018). Thermal pretreatments are of two types based on operating temperature: low temperature (100°C) and high temperature (>100°C) (Kondusamy and Kalamdhad, 2014). At high temperatures, most of the carbohydrates in biomass become soluble, hastening anaerobic fermentation. Thermally pretreating wastes eradicate harmful bacteria that obstruct AD's capacity to create biogas at a higher temperature. The application of heat pretreatment in a DF process boosts the generation of H_2. It decreases the dosages of additional inoculum by blocking H_2 absorption and sloughing off non-H_2-generating bacteria. Heat pretreatment raises the spore-forming bacteria and lowers rapidly accumulating organic acids (Karthikeyan et al., 2018). The most utilized temperature for H_2 production is 121°C for 20–30 minutes; however, for waste-activated sludge, the temperature ranges from 100°C to 175°C for 15–60 minutes. The temperature range of cellulose-based compounds is 50°C–220°C, and the period is 20 minutes to 24 hours. Thermal pretreatment is always combined with other pretreatment methods (Chozhavendhan et al., 2020).

Pagliaccia et al. reported that the impact of temperature, which was 20 minutes of thermal pretreatment at 134°C, boosted H_2 generation by up to 30% (Pagliaccia et al., 2016).

11.3.2.4 Combined Pretreatment

Pretreatments have increased the generation of bioH_2, but researchers have also looked into and evaluated the effects of combining two pretreatments (Mishra et al., 2020; Yin et al., 2018).

Yin and Wang examined H_2 production using a combination pretreatment approach that included both microwave and acid pretreatment. Polysaccharide consumption for fermentation was much lower with a single microwave pretreatment compared to a combined microwave and 2% sulfuric acid pretreatment, which was around 3,500 mg/L (Yin and Wang, 2018).

In a study, chemical and biological pretreatments (with NaOH, HCl, $CO(NH_2)_2$, and cellulase) were used to pretreat rice straw at ambient temperature (about 20°C) to improve its biodegradability and increase anaerobic biogas production. Cumulative biogas production yields increased by 16%–103% and 25%–122% for NaOH- and cellulase-pretreated rice straw substrates, respectively, compared to untreated rice straw substrates (Dai et al., 2018).

Every pretreatment method has merits and demerits of its own. Consequently, the performance of the present pretreatment procedures can be improved, leading to higher rates of bioH_2 production (Banu et al., 2020).

11.3.2.5 Limitations of Pretreatments

Pretreatment has various drawbacks related to the procedures utilized, even though it helps to increase the synthesis of bioH_2.

- The availability of resources, location, and environment all affect the biomass's composition, including minerals, CHs, protein, and fat. It is difficult to recommend a single pretreatment technique for all biomass sources. As a result, care must be taken while choosing an appropriate pretreatment method.
- Pretreatment can decrease biomass size to get the best product recovery. Further research will be necessary because there is currently no optimal particle size for increased $bioH_2$ synthesis.
- Pretreatment method optimization procedures are usually conducted in batch mode, but the results are not suited for pilot-sized systems due to the absence of comparable results.
- The pretreatment procedure must eradicate undesirable bacteria to produce $bioH_2$ at the anticipated rates. Higher reactor stability will result in understanding and maintaining the balance between H_2-generating bacteria and their favored MOs (Banu et al., 2020).

11.3.3 Methods for BioH$_2$ Production

A few biological methods are used for H_2 generation (Figure 11.2). These methods require less energy. However, systems must be monitored for pH, temperature, and nutrients.

$BioH_2$ production through biological methods can further be classified based on path (Figure 11.3). In light-dependent paths/routes, biochemical reactions occur in the presence of sunlight. However, no sunlight is required in light-independent methods for biochemical reactions and bio-electrochemical methods where microbes degrade organics through microbe–electrode interaction (Gautam et al., 2023).

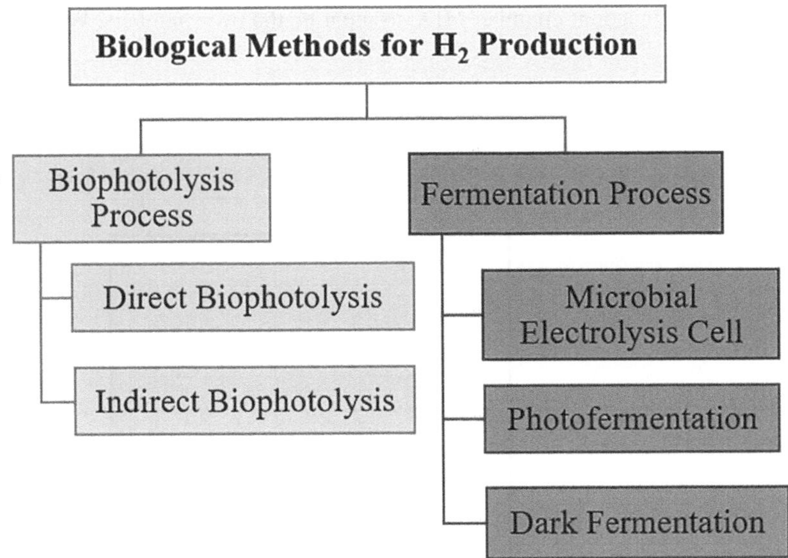

FIGURE 11.2 Biological methods for H_2 production.

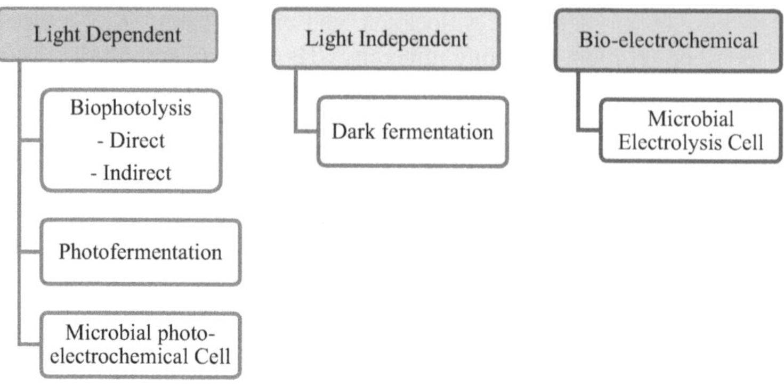

FIGURE 11.3   Bio H_2 production pathways.

11.3.3.1   Bio-Electrochemical Pathway for BioH$_2$ Production

11.3.3.1.1   Microbial Electrolysis Cell (MEC)

MEC is a method that integrates bacterial digestion to electrochemistry. MECs are retrofitted microbial fuel cells (MFCs) that generate bioH$_2$ gas from different organic substrates using an external voltage. MEC has evolved as a new anaerobic fermentation reactor recently, and it has drawn much interest as an intriguing method for increasing H$_2$ production from organic materials (Huang et al., 2020).

MEC consists of two electrodes: anode and cathode, which may be installed in one chamber (single-chamber MEC) or two different chambers (dual-chamber MEC), as shown in Figures 11.4 and 11.5, respectively. A proton exchange membrane (PEM) is commonly used in a dual-chamber MEC to separate the two chambers. WW is filled

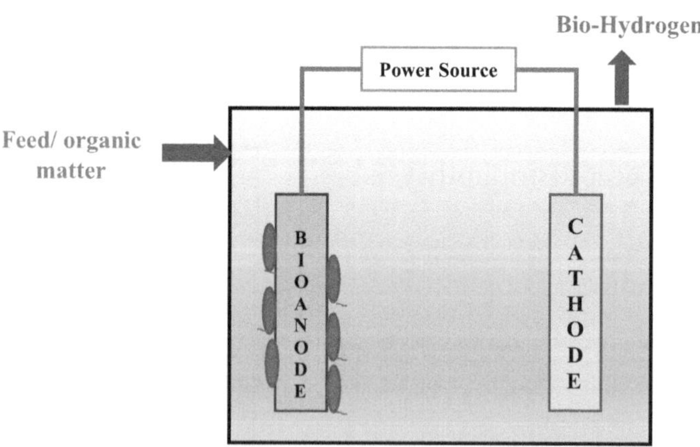

FIGURE 11.4   Single-chamber MEC.

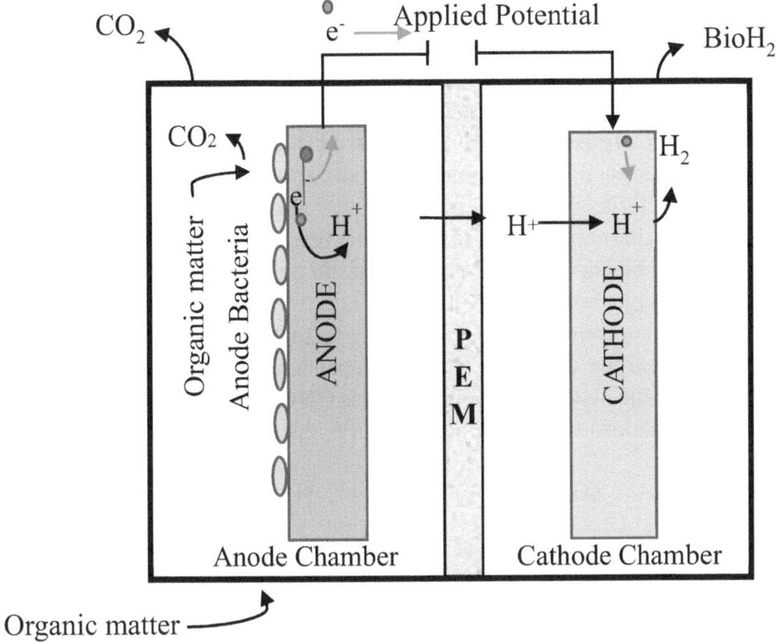

FIGURE 11.5 Dual-chamber MEC.

in the anode chamber, whereas in the cathode chamber, a variety of solutions, for example, salt solutions and buffers, can be added (Parvanova-Mancheva et al., 2022). Under anaerobic conditions, the electrogenic bacteria present in the anode chamber oxidize OM to produce carbon dioxide, protons, and electrons. Protons flow through the solution, arrive at the cathode, and are converted to H_2 when an external voltage is applied, whereas these electrons move from the anode to the cathode (Kumar et al., 2019b). The reactions that take place in a MEC when acetate is used as a substrate are presented in equations (11.2) and (11.3):

$$\text{At anode: } C_2H_4O_2 + 2H_2O \rightarrow 2CO_2 + 8H^+ + 8e^- \tag{11.2}$$

$$\text{At cathode: } 8H^+ + 8\,e^- \rightarrow 4H_2 \tag{11.3}$$

H_2 yield in MEC also depends on certain operational parameters like pH, temperature, applied voltage, substrate, microbe, electrode, and hydraulic retention time (HRT) (Chaurasia and Mondal, 2022). Bio-cathode MECs have three basic types: (i) MEC with one biotic and other abiotic cathodic chambers, (ii) dual MEC where both biotic chambers are (iii) single-chambered MEC with both biotic chambers. In type (ii) with dual-chambered MEC, both cathodes were biotic, i.e., bio-anode and bio-cathode which resulted in higher H_2 yield (Hasany et al., 2016).

11.3.3.2 Light-Dependent Pathway for BioH$_2$ Production

11.3.3.2.1 Biophotolysis Process

Biophotolysis involves utilizing biological systems to produce H$_2$ from water using sunlight energy. Various photosensitive MOs are used as biochemical conversion devices in biophotolysis. These MOs are introduced in photobioreactors that have been specially designed. Autotrophic organisms are those that can photosynthesize, such as cyanobacteria and microalgae. However, microalgae are the most appropriate and efficient alternative MOs studied in the literature for biophotolysis (Ghirardi et al., 2014). The biophotolysis process is classified into two types: indirect biophotolysis and direct biophotolysis.

Indirect biophotolysis occurs in two stages. To form CH biomass, autotrophic organisms take carbon dioxide from the atmosphere and water in the presence of solar radiation in the first stage. The DF of these CHs is performed in the second stage to create H$_2$. Equations (11.4–11.6) illustrate the reaction of indirect biophotolysis.

$$6H_2O + 6CO_2 + Light \rightarrow C_6H_{12}O_6 + 6O_2 \tag{11.4}$$

$$2H_2O + C_6H_{12}O_6 \rightarrow 4H_2 + 2CO_2 + 2CH_3COOH \tag{11.5}$$

$$4H_2 + 2CH_3COOH + Light \rightarrow 8H_2 + 4CO_2 \tag{11.6}$$

On the other side, in direct biophotolysis, there is no intermediate stage; however, in the presence of sun radiation and photosynthetic microorganisms (MOs), water is instantly divided into H$_2$ and oxygen, as indicated in equation (11.7; Kumar et al., 2019a).

$$2H_2O \xrightarrow[\text{Photosynthetic (MOs)}]{\text{Light}} 2H_2 + O_2 \tag{11.7}$$

11.3.3.2.2 PF Process

BioH$_2$ generation through fermentation using biomass substrates is a successful method since it is environment friendly, reduces the need for FF, and reduces pollutants. MOs, including bacteria, fungi, and algae, utilize the substrate in conditions like pH, temperature, and nutrients and produce H$_2$. The fermentation process is divided into two stages: PF (light-dependent) and DF (light-independent; Ahmed et al., 2021). PF is the fermentation of organic substrates using light energy by photosynthetic MOs. It is an anoxic process in which organic substances break down and emit H$_2$ and CO$_2$ in the presence of light. PF is further explained in Section 11.3.1.

11.3.3.3 Light-Independent Pathway for BioH$_2$ Production

11.3.3.3.1 Dark Fermentation

In the absence of light and oxygen, obligate anaerobes and facultative anaerobes perform DF. MOs utilize substrate to produce H$_2$ in DF. DF is further explained in Section 11.3.2.

11.4 FERMENTATION PROCESS FOR BIOMASS CONVERSION TO HYDROGEN

BioH$_2$ is H$_2$ that is conveniently generated from many biomass sources. The fermentation method of producing bioH$_2$ is effective since it is safe for the environment and reduces the need for FF usage and generating pollutants. This method has plenty of opportunity for H$_2$ production to replace FFs. Fermentation can be classified into three types: microbial PEC cells, PF, and DF.

11.4.1 PHOTOFERMENTATION

PF is the fermentation of organic substrates with sunlight by photosynthetic bacteria or algae. In the presence of light, organic molecules such as butyrate, acetate, and lactate break down and release H$_2$ in an anoxic process. Photosynthetic bacteria, including Chlorobium, Chromatium, Halobacterium, Rhodobater, Rhodopseudomonas, and Rhodospirillum, create H$_2$ by PF. The oxygen (O$_2$)-sensitive nitrogenase enzyme, which can generate H$_2$ and decrease nitrogen (N$_2$) at the same time, degrades a wide range of organic substrates, including fructose, glucose, and succinate to organic acids such as acetic and malic acids, into H$_2$ (Das and Basak, 2022). PF is single-stage and two-stage.

Equation (11.8) describes H$_2$ production through PF from adenosine triphosphate (ATP) to adenosine diphosphate (Baeyens et al., 2020);

$$16\text{ATP} + N_2 + 16H_2O + 10H^+ + 8e^- \rightarrow 16\text{ADP} + 2NH_{+4} + 16\text{pi} + H_2 \quad (11.8)$$

PF has been demonstrated to be the most successful strategy due to its high H$_2$ percentage (maximum 58.90%) and high conversion rate for energy (10.12%). PF was also discovered to be superior to the other two methods regarding H$_2$ output. BioH$_2$ production through the PF method could be a viable option. However, its effectiveness in converting sunlight into bioH$_2$ and yielding is low compared to other light-assisted processes, such as biophotolysis. PF is illustrated in Figure 11.6.

PF process depends on many factors for the efficient production of H$_2$. Some of the prominent factors are discussed here.

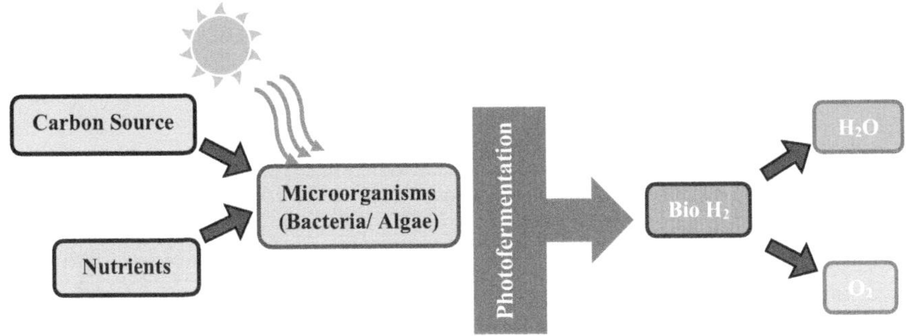

FIGURE 11.6 PF process.

11.4.1.1 Photofermentative MOs

In the PF process, photosynthetic MO metabolizes organic substrates under anoxic or anaerobic conditions into smaller molecules and produces H_2 and CO_2. The photosynthetic MOs use light as energy and organic matter as a source of carbon and electrons. During this fermentation process, $bioH_2$ is produced. PF is becoming increasingly popular due to the wide range of raw materials that can be employed, its high efficiency, environmental friendliness, and potential for large-scale H_2 generation at ambient temperature and pressure (Hitam and Jalil, 2020). Diverse types of photofermentative MOs can be used, for example, purple non-sulfur bacteria (PNSB), *Rhodobacter capsulatus*, *Rhodobacter sphaeroides*, *Lactobacillus amylovorus*, *Rhodopseudomonas palustris* (Adessi et al., 2018), *Rhodopseudomonas palustris* (Batista et al., 2015). These species are extensively used in $bioH_2$ production using waste materials.

11.4.1.2 Bacteria-Assisted Systems (Bacterial Photobioreactors)

Bacterial photobioreactors (PBRs) are the reactors where the bacteria carry PF of carbon-rich substrates. PNSB are a significant group of photofermentative MOs frequently used for photofermentative $bioH_2$ generation. The metabolic flexibility of PNS photosynthetic bacteria is quite strong, and they can function as photoheterotrophs, photoautotrophs, or chemoheterotrophs. *Rhodopseudomonas faecalis*, *Rhodopseudomonas palustris*, *Rhodospirillum rubrum*, *Rhodopseudomonas capsulatus*, *Rhodobacter sulfidophilus*, *Rhodobacter capsulatus*, and *Rhodobacter sphaeroides* are the prominent PNSB which have been shown tremendous results in H_2 production (Mahata et al., 2020). Although using simpler, pure materials as substrates has shown enhanced H_2 production in a shorter period, these substrates are known to raise the cost of the PF and, ultimately, H_2 production. Therefore, researchers are now focusing on using WW and lignocellulosic wastes as substrates in PF.

Combining dark fermentative and photofermentative bacteria to produce $bioH_2$ is another strategy that has been recently evaluated and is gaining importance. DF justifies this strategy by generating several unusable volatile fatty acids (VFAs) and fermentative by-products. With the addition of photofermentative organisms such as the PNSB, these VFAs may easily be used to improve H_2 production. This process of DF and PF can be carried out sequentially or simultaneously. Although the PNSB have been the main participants of PF, both extreme anaerobes (like *Clostridium* and *Desulfovibrio*) and facultative bacteria (like *Enterobacter, Klebsiella, and Escherichia*) have been employed for the DF (Mishra et al., 2019). Consortia and mixed cultures have also been used in a few studies.

11.4.1.3 Algae-Assisted Systems (Algal Photobioreactors)

Algal photobioreactors (APBRs) are the reactors where the algae carry PF of carbon-rich substrates. Many species of algae producing $bioH_2$ exist; however, Chlorella, *Chlamydomonas*, and *Scendesmus* have gained attention in recent years. Algal multiplication pauses in a nutrient-deficient environment, and a fermentative activity takes over, producing H_2. *Chlamydomonas* sp. produced H_2 under a sulfur-deficient anaerobic environment. Under a phosphorous-deprived climate, the marine green algae *Platymonas helgolandica*, which belongs to Chlorella sp., enhanced H_2 production.

Under different light qualities, inactive Chlorella vulgaris and Scenedesmus obliquus were studied for H_2 generation using synthetic WW, and the highest output of H_2 was 128 mL H_2/L, attained by inactive *S. obliquus* in purple light (Putatunda et al., 2022).

11.4.1.4 Enzymes as Biocatalysts for BioH$_2$ Production

The two main kinds of enzymes involved in the production of bioH$_2$ are nitrogenases and hydrogenases. These two enzymes also transport bioH$_2$ in algae-assisted systems.

Hydrogenase (metalloenzyme) is categorized into three types: [NiFe], [FeFe], and [Fe]. The metal present at the active site determines these kinds. [Fe] hydrogenase is found only in bacteria (*Archea*); however, [FeFe] hydrogenase mostly catalyzes the reduction of H^+ ions and creates H_2. In contrast, [NiFe] hydrogenase catalyzes processes that consume H_2. When the consumed H_2 is oxidized to H+, some energy lost during N_2 fixation is restored. Because [FeFe] hydrogenase is particularly oxygen-sensitive, it is inhibited when exposed to aerobic conditions.

Nitrogenase converts molecular nitrogen (N_2) to an ammonium ion that may be used. This ATP-consuming activity has two functions: First, it meets the bacteria's nitrogen need, and second, it maintains the nitrogen cycle in the environment. The process of reduction is linked to the H_2 ion reduction process. As a result, H_2 is produced as a by-product of nitrogen fixation, which is catalyzed by nitrogenase (Putatunda et al., 2022).

11.4.1.5 Factors Affecting BioH$_2$ Production through PF

The type of organism utilized to produce H_2, as well as parameters like light intensity, pH, temperature, substrate quantity, and conversion ratio, all substantially impact the efficiency of bioH$_2$ synthesis.

11.4.1.5.1 Light

An adequate amount of light is required for microbial growth in photobioreactors and the subsequent photosynthetic synthesis of H_2. Light intensity is an important factor during PF since increasing light intensity enhances H_2 synthesis on one hand, while excessive light causes cell damage and death on the other. Therefore, PBRs must be designed so the organisms are subjected to optimal light environments (Kumar et al., 2018). Because of the light saturation effect, light conversion efficiency is frequently low at higher levels. High photon absorption results in the release of surplus photons as heat. The high photon absorption rate can cause photoinhibition in the upper layers of PBRs (Gu et al., 2017).

Although the amount of visible light (400–950 nm) affects how much substrate bacteria consume, too much light can also have the opposite effect, called photoinhibition (Cordara et al., 2018). PF and DF phases were made possible by alternating light and darkness cycles, which resulted in increased rates of H_2 synthesis. Zhang et al. (Zhang et al., 2012) achieved the highest yield of 84.7 mL of H_2/g of total solids and the highest efficiency in light conversion employing dynamic light intensities (4,000, 7,000, 4,000 Lux) and mixing speeds (50, 150, 50 RPM). Because different phases of biological H_2 synthesis necessitate varied light intensities and mixing rates, a dynamic technique is desirable (Ruiz-Marin et al., 2020).

11.4.1.5.2   pH

The pH of the PF processes is crucial for $bioH_2$ generation and subsequent by-products. It influences the activity of the enzymes participating, the hydrolysis of complicated substrates, the variety of MOs, and their metabolic pathways (Srivastava et al., 2020). pH is determined by the type of acids present in the substrate. The ultimate pH decreases because CHs are converted to organic acids, which hinders H_2 production. In batch cultures, pH also impacts the length of the lag phase and, as a result, H_2 gas generation. The culture's starting pH influences the ethanol concentration of anaerobic fermentation products and the organic acid composition. The ideal pH range for H_2 production is determined by the kind of bioprocess and microbe used and the anticipated products and by-products. pH range for the earliest phases of $bioH_2$ generation is between 4.5 and 8.0, with 4.5 being the best value for thermophilic cultures, as reported by the researchers. Nevertheless, the synthesis of organic acids causes the final pH during anaerobic fermentation to drop to 4–4.8, which inhibits hydrogenase action. By adjusting the starting pH values to 5–7.5, scientists reported the effectiveness of photofermentative $bioH_2$ production (Putatunda et al., 2022).

11.4.1.5.3   Temperature

Temperature has the greatest influence on the variety and composition of microbes, the production of organic acids, and the following metabolic pathway in fermentation. The bioreactor's operating temperature influences how well MOs use the substrate and how quickly they develop, which impacts the generation of $bioH_2$ and other metabolic by-products. PF occurs at 30°C–35°C (Chandrasekhar et al., 2015). The temperature ranges studied for H_2 generation include mesophilic (35°C), thermophilic (>50°C), and hyperthermophilic (>65°C). The sequence of maximum H_2 generation is mesophilic > hyperthermophilic > thermophilic (Kumar et al., 2018).

11.4.1.5.4   Hydraulic Retention Time

The bioreactor's HRT affects microbial communities' composition and H_2 generation. At 48-hour HRT, greater specific $bioH_2$ generation rates (0.42 mmol of H_2/g of VS/day) and an increased proportion of H_2 in the generated biogas (23%) were seen (Yue et al., 2021). Moreover, 48-hour HRT increased the amount of *Clostridium sp.* in the bioreactor, while 8- to 24-hour HRT favored *Lactobacillus* and *Enterobacter* predominance (Santiago et al., 2019). Another critical concern for HRT (2–10 hours) is a metabolic shift from acidogenesis to methanogenesis, which is detrimental to H_2 production and can be avoided with shorter HRTs. Nonetheless, for optimal $bioH_2$ generation, HRT must be adjusted by the kind of substrate, the MOs present, and other physicochemical properties of the bioreactor (Chandrasekhar et al., 2015).

11.4.1.5.5   Carbon Source

CHs are an essential component in fermentative H_2 production. Several mono- and polysaccharides have been found as possible $bioH_2$ substrates. The effectiveness of the procedure may vary depending on the carbon source selected. Galactose, sucrose, and glucose are the most common carbon sources, but glucose is preferable as it requires fewer enzymes for fermentation. A mix of sources has been demonstrated to be more useful in some cases (Kumar et al., 2018).

Production of H_2 can be increased by using sewage sludge. DF, PF, and combined DF and PF methods may generate 4, 8, or 12 mol of H_2/mol of hexose or glucose, respectively. Given the energy and substrate used during cellular synthesis, actual yields could be lower than predicted (Sivaramakrishnan et al., 2021). Lower yields might be produced by incomplete glucose breakdown, cellular synthesis substrate usage, or butyrate production rather than acetate. The most productive substrates were water hyacinth, sweet potato starch, WWs, algal biomass, cassava starch, sugar beet molasses, and corncob. Some of the WW used as a carbon source and the H_2 yield are discussed in Table 11.1 (Patel et al., 2018; Putatunda et al., 2022). A maximum amount of H_2 production and COD removal was observed when PNSB was used and the substrate was leachate.

11.4.1.5.6   Macro- and Micronutrients

Macro- and micronutrients also play a crucial role in PF. Proteins are required to develop microbes to maintain the carbon-to-nitrogen (C/N) ratio. The optimal protein content is yet unknown. Nonetheless, the best documented C/N range falls between 10 and 90. On the other hand, high protein concentrations in algal biomass increase ammonium content, lowering the internal pH of H_2-producing bacteria cultures, increasing the energy cost for cell maintenance, and blocking certain enzymes (nitrogenases) essential for bioH_2 generation (Kumar et al., 2016).

The C: N ratio impacts H_2 production; for example, if the ratio is higher than 2/3, H_2 output would decline. Optimum H_2 production (0.27 mol H_2/L daily; 4.8 mol H_2/mol sucrose) was found at a C:N ratio 4:7 (Budiman and Wu, 2018). Ammonia (nitrogen) addition at 0.1 g N/L resulted in a typical maximum mean of 8.5 mL H_2/h. Maintaining the C/N ratio, however, is costly since too much N_2 restrains nitrogenase activity, alters pH to favor microbial growth, and promotes ammonification, which is undesirable for H_2 production. Polypeptone, corn-steep liquor, ammonium carbonate, urea, and nitrogen salts may be more efficient nitrogen sources (Chandrasekhar et al., 2015).

Additional nutrients and inorganic trace elements, including iron, sodium, zinc, nickel, phosphorous, and magnesium, are essential nutritional components for optimum microbial activity and subsequent H_2 output. The addition of iron (10–100 mg/L) alters the activity of hydrogenase and, as a result, the synthesis of H_2, lag phase by-products, and VFA composition (Hay et al., 2013). Fe activates hydrogenase. Fe increases H_2 generation at the optimum dose (25–100 mg/L), while larger quantities are harmful. Similarly, divalent cations (magnesium, manganese, nickel, and zinc) increase H_2 production when present at the ideal amount but become hazardous when present in more significant amounts. Supplementing with molybdenum has some advantages, but only up to a point (Soltan et al., 2017). Phosphorus, also known as phosphate, is essential for microbial energy generation (ATP) and medium buffering, but it also drives the formation of VFAs, which lowers pH and reduces total H_2 output. Adding solid organic wastes rich in CHs to solid waste feedstocks significantly boosts H_2 yield compared to fats and proteins (Putatunda et al., 2022).

11.4.1.5.7   Oxygen Concentration

Since PF must be carried out under strictly anoxygenic conditions, oxygen must either be physically removed by ultrasonication or chemically removed by reducing agents such as 2-mercaptoethanol, sodium sulfite, and L-cystine or by reducing

TABLE 11.1

H_2 Production through PF Using Different WWs

Substrate	MO	Reactor Type	Light Intensity (W/m²)	COD (mg/L)	BioH₂ Yield (L H₂/L)	COD Removal (%)	Reference
Winery WW	PNSB mixed consortium	Laboratory-scale PF reactor	-	1,500	0.468	74	Policastro et al. (2020)
Brewery WW and pulp and paper mill effluent (PPME) (10% BW + 90% PPME)	*Rhodobacter sphaeroides*	100-mL Schott bottles	55.3	2,925 ± 289.91	0.69 mol H₂/L medium	36.7[a]	Hay et al. (2017)
Landfill leachate	PNSB	Batch anaerobic bioreactor	31.6–47.4	43,128	5.754	77	Barghash et al. (2021)
Sugar refinery WW	*Rhodobacter sphaeroides*	Column photobioreactor	200	-	0.005 /h	-	Hay et al. (2013)
Dairy WW	*Rhodobacter sphaeroides*	Small vials (25 mL)	71.1 and 102.7	46,300	3.2	21	Hay et al. (2013)
Olive mill wastewater (OMWW)	*Rhodobacter sphaeroides*	Glass-column photobioreactors	200	1,100	0.01431/h	34	Budiman et al. (2015)
Pretreated brewery WW with banana peels	*Rhodobacter sphaeroides*	Clear glass bioreactor (120 mL)	126	2,675	408.3 mL/L	-	Al-Mohammedawi et al. (2019)
Pretreated brewery (30%) and restaurant (70%) effluents	*Rhodobacter sphaeroides*	Clear glass bioreactor (120 mL)	126		83	46	Al-Mohammedawi et al. (2019)
Pretreated olive mill WW	*Rhodopseudomonas palustris*	Flat glass photobioreactor (1.05 L)	200 µE/m²/s		310		Pintucci et al. (2015)
Combined effluent of palm oil and pulp and paper mills	*R. sphaeroides*	Schott bottle (100 mL)	55.3		8.72	36.9	Budiman and Wu (2016)

[a] Soluble COD removal

gases such as Ar, H_2, and N_2. Although hydrochloric acid (HCl) may be used to remove trace oxygen, this method is not cost-effective on bigger scales. Instead, during DF, microbial strains having a more oxygen-resistant hydrogenase enzyme system, such as the facultative anaerobe *Enterobacter aerogenes*, may be used (Rai and Singh, 2016; Liu et al., 2015). However, other kinds of PBRs use bubble or tube columns, and air-lift flat-panel bioreactors improve mixing and reduce light path while producing a significant amount of biomass and, consequently, H_2 gas (Liu et al., 2015).

11.4.1.5.8 Contaminants

Substances from industrial WWs, colored effluents, high ammonia concentration effluents, inorganic pollutants, heavy metals, and polyaromatic hydrocarbons must be removed or subjected to pretreatment because they can inhibit PF capacity by changing physical parameters like pH or by being harmful to the microbial cultures (Mabutyana and Pott, 2021, Tian et al., 2019). Competition for the substrate will result from contamination by other bacteria. Hence, pretreating industrial effluents makes the bioreactor environment safer. Immobilized microalgae are helpful because they may be used for bioremediation, removing macronutrients, and being a biomass source (Sharma and Arya, 2017).

11.4.1.6 Methods to Enhance Hydrogen Production through PF

Compared to other technologies, $bioH_2$ generation produces less H_2; hence, serious efforts are needed to get beyond these restrictions. The sustainability of a product or process is determined by its impact on society, the market, economics, and the environment. The limitations that need to be overcome include the following:

- Decreasing the partial pressure of H_2 gas while enhancing its purity.
- Minimizing or preventing shifting in metabolic responses.
- Improving the availability of the enzyme system hydrogenase.
- Creating culture strains that produce H_2 more effectively by introducing new microbes with strong nitrogenase and hydrogenase activity, improved feedstock adaptation, and environmental and economic viability (Zhang et al., 2022).
- Modernizing bioprocessing technology by shifting to automatic and controlled systems with continuous monitoring.
- Employing nanoparticles or introducing nanotechnology in PF-enhanced $bioH_2$ production by controlling the limiting factors (Maroušek, 2022).

11.4.2 Dark Fermentation

DF is a process that is independent of light. MOs, such as bacteria or microalgae, are used in heterotrophic fermentation. It is an intriguing technique since it uses mixed biomass waste as a substrate. Compared to other biotechnological processes, DF is defined by its simple technique, independence from light, and potential to employ renewable feedstocks (Chen et al., 2021). Bacteria digest DF's CHs or other organic material under anaerobic conditions to produce H_2 and other intermediates such as

organic acids (VFAs, butyric acid, acetic acid, propionic acid) and alcohols like ethanol. *Clostridium* and the facultative bacteria Enterobacter sp. are among the MOs
capable of generating H_2 under an anaerobic environment.

H_2 production rate increases in the presence of substrates, sucrose, glucose, cellulose, and starch. The H_2 generation rate rises in the presence of substrates, such
as glucose, sucrose, starch, and cellulose. Through complete conversion, 1 mol of
glucose yields 12 mol H_2 atoms, as shown in equation (11.9).

$$C_6H_{12}O_6 + 6H_2O = 12H_2 + 6CO_2 \qquad (11.9)$$

The DF process involves hydrolyzing complex materials into less complex products, that is, complex CHs into reducing sugars, as shown in Figure 11.7. Enzymes
then initiate acidification in this group of simple molecules. Several kinds of
enzymes are secreted by bacteria that ferment. The process of acetogenesis is used
to convert these acid products further into $bioH_2$ and acetate in the following stage.
To synthesize $bioH_2$ along the same pathway, it is required to cease the methanogenesis, and $bioH_2$ generated in the preceding phases must be collected (Das and
Basak, 2022).

Two pathways contribute to the production of $bioH_2$ during the DF process. The
catabolic conversion of formic acid accomplishes one, while the other is accomplished via the hydrogenase pathway's reoxidation of nicotinamide adenine dinucleotide hydride (NADH) (Ahmed et al., 2021).

$BioH_2$ production through the DF method has more yield than other methods.
Pretreatments and enzymes can be utilized to increase the H_2 production efficiency
of the DF process.

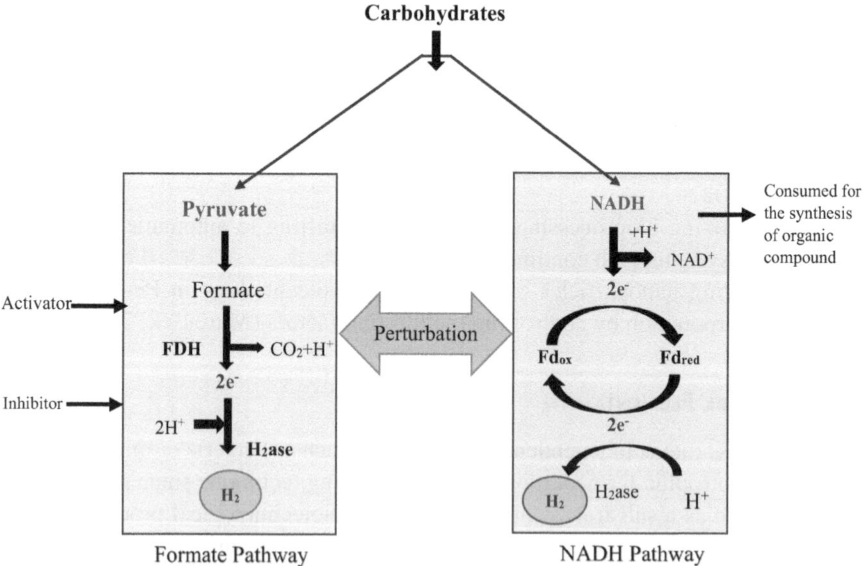

FIGURE 11.7   DF process.

DF has various advantages due to its nondependence on light for H_2 generation, which includes continuous H_2 generation, cheaper energy input, and the convenience of upscaling and managing. Apart from that, DF can use multiple types of waste products as substrates (such as distillery effluent, food wastes), as mentioned in Table 11.2, and hence provide options for both energy production and waste handling (Azwar et al., 2014). The production of H_2 through DF depends on the type of substrate used, the MOs employed, and the operation conditions, that is, pH, temperature, and nutrients.

DF depends on many factors, and hence, the productivity of this fermentation process can be increased by maintaining those factors in a feasible range. These include types of fermentive bacteria, pH, substrate composition, nutrients, VFAs, HRT, temperature, and H_2 partial pressure. Different operational conditions in the DF process yield different quantities of $bioH_2$. These factors have been discussed in the following section.

11.4.2.1 Dark Fermentative MOs

Most of the MOs used for DF to produce biological H_2 are bacteria. Although it has been reported that a number of rumen fungal strains may produce H_2, their suitability for fermentation has not yet been investigated (Yang and Wang, 2019, Stewart et al., 1987). Many fermentative H_2-producing bacteria have so far been identified and partially described. Aerobes, strict anaerobes, and facultative anaerobes are the broad groups into which these bacteria can be divided.

Strict anaerobes grow without oxygen and restrict their growth if oxygen is present. *Clostridium, Ruminococcus, and Thermotoga are the dominant strict anaerobes reported to produce H_2 through DF in the amounts of 2.0 mol/mol glucose, 2.01 mol/ mol glucose, and 2.2 mol/mol glucose,* respectively. *Clostridia* can produce spores, making them simple to manage for industrial purposes. High temperatures cause the *Clostridia* to produce their spores while also killing bacteria. If the conditions are available, the high-temperature spores can be stimulated to create H_2 (Huang et al., 2014). Moreover, *Clostridium* species can be grown in a fermenter-controlled environment. During the logarithmic stage, they make H_2; during the stationary stage, they switch to producing solvents. For H_2 production, various substrates, including glucose, xylose, lactose, and sucrose, are used by *Clostridia*. These strains have high H_2 yields (>2 mol/mol substrates). As many anaerobic thermophilic bacteria have shown they can produce H_2, archaea are generally regarded as that group (Shi et al., 1997, Cao et al., 2022). The benefits of thermophilic H_2 production at high temperatures include increased mass transfer efficiency, lower feedstock contamination risk, lower cooling water use, and higher gas production rates (Taikhao et al., 2015).

Anaerobic and aerobic environments both allow facultative anaerobes to grow. Compared to stringent anaerobes, which cannot live in the presence of oxygen, facultative anaerobes are less sensitive to oxygen. In aerobic circumstances, facultative anaerobes often develop extremely fast before switching to anaerobic metabolism when oxygen levels drop (Ruggeri et al., 2015). Facultative anaerobes are preferred to strict anaerobes because they can be grown to a high cell density with oxygen and produce H_2 at a high rate when the oxygen supply is shut off. Facultative anaerobes like *Bacillus, Enterobacter, and Escherichia are likely to have H_2 through DF in*

TABLE 11.2

H₂ Production through DF Using Different Wastes

Substrate	Process	MO/Inoculum	Reactor Type/Scale	pH	COD (mg/L)	BioH₂ Yield (ml H₂/g VSS)	% COD Removal	Reference
Swine wastewater (WW)	DF with salt pretreatment	*Bacillus* sp.	Lab scale (fermentation flask)	6.0	Soluble COD (sCOD) 4,673.8	7.0	–	Li et al. (2020b)
Starch processing WW	Microbial electrolysis and thermophilic DF	*Thermoanaerobacterium thermosaccharolyticum*	Upflow anaerobic sludge blanket (USAB)	6.5	21,600	465 mL H₂ gCOD⁻¹	58	Khongkliang et al. (2017)
Waste AS	Chitosan-assisted DF	Municipal WW treatment plant	Serum bottles	5.5	–	3.94 ± 0.07	–	Wang et al. (2019a)
Beverage industry WW	DF	Heat-treated AS	Thoroughly stirred tank reactor (CSTR)	6.7	73,0000–76,0000[1]	20	–	Lin et al. (2016)
Beverage industry WW	DF	Heat-treated AS	Internal circulation reactor	5.5	19,203	–	30.35	Alvarez et al. (2022)
Lactate WW	DF	*Clostridium* at 35°C *Sporanaerobacter, Clostridium* and *Pseudomonas* 95% at 45°C	1-L media bottles	8.5	20,300[a]	0.85 mol H₂/ mol lactate	21–30 (35°C) 54–95 (45°C)	Ziara et al. (2019)

(Continued)

TABLE 11.2 (*Continued*)

H$_2$ Production through DF Using Different Wastes

Substrate	Process	MO/Inoculum	Reactor Type/Scale	pH	COD (mg/L)	BioH$_2$ Yield (ml H$_2$/g VSS)	% COD Removal	Reference
Municipal solid wastes 70:30 (organic fraction of municipal solid waste: food waste)	DF	Effluent of stirred tank reactor (SSTR)	Batch bioreactors	5.5	835 g/kg[a]	42.9	–	Angeriz-Campoy et al. (2023)
Cassava pulp and cassava processing WW	DF	Anaerobic granules from the anaerobic digester of a brewery WW treatment plant	2-L bioreactor	6	–	13.72 ± 0.22 L-H$_2$	29.7	Chantawan et al. (2022)
Synthetic swine WW	DF with fruit peel crude enzymes (orange peel, mandarin peel, banana peel)	Heat-treated sludge from an anaerobic bioreactor	500-mL serum bottle	6	6,000	2.01 ± 0.00 (mandarin peel) mol H$_2$/ mol glucose	–	Feng et al. (2023)
Dairy WW	Nickel ferrite nanoparticles catalyzed DF	*Escherichia coli* bacteria	20 Vials (50 mL)	6.5	4.5	241.3	83.6	Fahoul et al. (2022)
Synthetic WW	DF	Sludge from the bottom of a lake	500-mL serum bottle	5	10,000	70.94 ± 4.7	37.13 ± 1.86	Tang et al. (2022)

[a] Soluble COD.

amounts of 2.28 mol/mol glucose, 2.23 mol/mol glucose, and 2 mol/mol glucose, respectively.

Most of the recognized hydrogenases are very oxygen-sensitive. These enzymes are inactive in oxygen, mainly when producing H_2. As a result, aerobes are typically thought to be unable to produce H_2. The only type of aerobic bacteria evolving H_2 is *Alcaligenes,* which are reported to generate 100 µmol/(g Protein. hour) H_2 through the process of DF (Cao et al., 2022).

11.4.2.2 Factors Affecting DF

11.4.2.2.1 Temperature

The metabolic process of the bacteria that produce H_2 is greatly influenced by temperature. Most H_2-producing bacteria are mesophilic; most bacteria that produce H_2 prefer temperatures between 20°C and 45°C (Pradhan et al., 2015). H_2 output increased as temperature increased between 33°C and 39°C but declined with further improvement in temperature. When the temperature is maintained between 35°C and 38°C, the reactor's anaerobic activated sludge (AS) has the most active metabolism and the highest methane and H_2 production rates. Because high temperatures can decrease H_2's solubility in the liquid phase and thermophiles in fermenting MOs have a greater tolerance limit for high temperatures, it has been reported that anaerobic fermentation for the generation of H_2 benefits from high temperatures. However, fermentation at higher temperatures needs more significant thermal energy input, resulting in energy costs. In light of the authentic WW and substrate conditions, the fermentation temperature should be maintained at $(36+1)$°C (Qu et al., 2022).

The influence of temperature variations will decrease the reaction system's biomass and H_2 production rate. As a result, the temperature variation in the daily operation of an anaerobic fermentation reactor is kept to a maximum of 2°C–3°C. Also, as the susceptibility of fermentation MOs to temperature variations is strongly connected with organic load, it is essential to focus more on maintaining stable temperature management during high-load operation (Qu et al., 2022).

11.4.2.2.2 pH

A correct pH is crucial to ensuring the development of H_2-producing MOs and the activity of enzymes that produce H_2. The metabolic pathways of fermentation are also impacted by pH, which results in several forms of H_2-producing fermentations. The suitable pH for fermentative H_2 production ranges from 4.5 to 6.5. However, the optimum pH for DF is 5.5, as reported by many researchers (Table 11.2). The mixed culture of bacteria in the reactor may undergo a community succession process by keeping the pH value within a defined range. It can achieve the desired level of fermentation and make the desired flora prevail to become the dominant community (Li et al., 2017).

11.4.2.2.3 Micro- and Macronutrients

The nutrients needed by the H_2-producing bacteria include C, N, P, and inorganic metal elements. The fermentation substrate typically serves as the carbon source. A diverse range of substrates, organic wastes, and solid OM can be used in the DF

H_2 generation. Several nitrogen sources include protein, nitrate, nitrite, and ammonium salt. Ammonium is a buffer for organic acids and a nitrogen source (Hay et al., 2013). Phosphates are often the source of phosphorus. The development of bacteria that produce H_2 and the production of enzymes both depend heavily on metal ions.

Small concentrations of manganese, iron, sodium, zinc, potassium, iodine, cobalt, ammonium, nickel, copper, molybdenum, and calcium are required in the medium for maximum H_2 production. These metal ions sometimes improve H_2 production by controlling the other competitive MOs group. The large concentration of nutrients and metal ions inhibits the DF process. Qu et al. (2022) suggested some optimum values of the metal ions concentration for maximum H_2 production.

Substrate type, temperature, pH, macro- and micronutrients, and the type of MO employed play a significant role in the production of H_2 (Table 11.2). These conditions vary with the kind of substrate. Therefore, standard operation conditions cannot be defined.

11.5 COMPARISON OF DIFFERENT BIOH$_2$ PRODUCTION METHODS

Different methods for bioH$_2$ production have been discussed in detail. Table 11.3 compares these methods based on the merits and demerits of biological processes for bioH$_2$ production.

11.6 FACTORS CAUSING INHIBITION IN FERMENTATION PROCESS

Despite improvements in H_2-generating methods, the different limiting factors in diverse fermentative H_2 production systems continue to go unrecognized and undocumented.

These elements, including inorganic, organic, and biological inhibitors, may have been produced during the pretreatment and fermentation processes or have always been contained in the substrate. As a result, there is an urgent need to manage inhibition-causing factors in fermentative H_2 generation, which is the root cause of low H_2 yield and poor H_2 production.

Managing these inhibiting elements will improve fermentative H_2 production and make its commercialization more likely. Some of the prominent inhibition-causing factors are discussed below;

11.6.1 INORGANIC INHIBITORS

Specific inorganic components, which include heavy metal ions, have an inhibitory effect in the fermentation process, mainly in DF (Duruibe et al., 2007). The source of these inorganic components is feedstocks or substrates such as WW, waste AS, FW, and complex materials (Ji et al., 2013).

The concentrations of inorganic components define the toxicity or inhibitory effect (Bao et al., 2013).

TABLE 11.3

Merits and Demerits of Bio H_2 Production Methods

Bio H_2 Method	Merits	Demerits
PF	• Photosynthetic bacteria can utilize varieties of spectral energy • Diverse types of substrates can be processed • Can process DF effluent • Bioremediation • Efficient conversion of organic acid wastes to H_2 and CO_2	• Lower conversion efficiency of light • Bio H_2 production through photosynthetic bacteria stays low • Nonhomogeneous distribution of light affects the whole fermentation process • Light-dependent process and continuous bioH_2 production require a light source, especially at night • Costly PBRs
DF	• Several metabolites can be generated, and substrates can be applied • Treats different types of substrates • No dependency on light • No issues related to limited O_2 concentrations • Bioremediation • Continuous H_2 production, day and night	• Low H_2 production efficiency and not thermodynamically feasible • O_2 accumulation causes inhibition • Low bioH_2 production
Direct biophotolysis	• A comprehensive strategy to produce H_2 from water and sunlight • Requires simple cultivation • Simple substrate of H_2O • CO_2 consumption	• The process is inhibited by high light intensity and O_2 • High purity O_2 and H_2 streams cannot be separated
Indirect biophotolysis	• H_2 production from water • Separate O_2 production • N_2 fixation • Metabolite by-product transformation to H_2	• There is less absorption of hydrogenate • Hydrogenates absorbed can be eliminated • The enzyme hydrogenase produces CO_2 while producing less H_2
MEC	• A nonpolluting technology that generates power in an eco-friendly way • Bioremediation • Significant reduction in COD • Capable of processing black fermentation waste • Excellent recovery of H_2 • Substrate deterioration is high	• Solar systems are costlier but can provide a more effective solution • Higher capital costs • Loss of H_2 • Issues with scaling, power supply, stability, and mode of operation

Thiosulfinate has a solid potential to suppress the generation of dark fermentative H_2 when food waste is used as substrate. Thiosulfinate at a concentration of 2.5 g/L has been shown to reduce the generation of methane and H_2 by affecting the metabolic activity, reproduction, and growth of microbes (Tao et al., 2020; Zhao et al., 2019).

Zinc and copper are the most harmful heavy metals in the dark fermentative H_2 generation process, while Cr and Cd have less inhibitory impact. Therefore, it is challenging to find out the order of heavy metal toxicity universally. Also, because batch experiments were used to study the toxic effects of heavy metals, inhibiting other factors was not considered (Chen et al., 2021).

11.6.2 LIGHT METALS

Ca, Na, and Mg are the light metals most studied in fermentative H_2 production. They can help bacteria grow, but large quantities of these soft metals can prevent DF from producing H_2 (Bundhoo and Mohee, 2016).

11.6.3 HYDROGEN GAS

The partial pressure of H_2 rises as more H_2 accumulates in the reactor's headspace, increasing the concentration of dissolved H_2 gas in the liquid phase. The accumulation of H_2 gas can interfere with H_2 production as a product.

Le Chatlier's principle states that as the amount of H_2 in a liquid increases, the bio-reaction that produces H_2 will be hindered (Sivagurunathan et al., 2016).

11.6.4 ORGANIC INHIBITORS

VFAs, furan derivatives, and phenolic chemicals are examples of organic inhibitors. These organic components are produced during the fermentation process or the processing of specific substrates.

In the dark fermentative H_2 production process, the concentration of VFAs needs to be maintained to conduct the fermentation process smoothly (Bundhoo and Mohee, 2016).

The inoculum used significantly impacted the furan inhibitory action because the anaerobic digester sludge inoculum was entirely degraded and did not prevent the creation of H_2 (Palmqvist and Hahn-Hägerdal, 2000).

The source of phenolic inhibitors is lignocellulose pretreatment and common pollutants in industrial WW (Sivagurunathan et al., 2017; Pradeep et al., 2015). Phenolic chemicals hinder cell communication and harm cell membranes. Because smaller molecular weight chemicals may more easily infiltrate cell membranes, they are thought to be more hazardous than higher molecular weight phenolic compounds (Zhang and Chen, 2020).

11.6.5 BIO-INHIBITORS

A class of peptides known as bacteriocins may remain active across a wide pH range and are temperature-tolerant. Bacteriocins can kill other bacteria irreversibly, even at low concentrations (Wang et al., 2019b). Bio-inhibitors are usually observed in the DF process, released either by lactic acid bacteria (LAB) or in the substrate (Alvarez-Sieiro et al., 2016). LAB has inhibited batch tests and even continuous systems (Woraprayote et al., 2016).

11.7 PROSPECTIVE WAYS TO MITIGATE INHIBITION

11.7.1 THROUGH DILUTION

Dilution is an efficient approach for lowering various inhibitors in the production of fermentative H_2. The purpose and impact of the dilution technique is to drop the concentration of inhibitors below the inhibitory threshold by adding additional broth or even ordinary tap water, etc. Dilution has been studied and utilized to reduce bacteriocins and VFAs in producing dark fermentative H_2 (Zhao et al., 2019).

11.7.2 ELIMINATION

Removing the inhibitors from the DF H_2 generation process will immediately reduce the inhibitor concentrations below the inhibitory threshold. Although little research has been conducted, removal effectively lowers the inhibition caused by heavy metal ions.

Although the quantities of heavy metals in H_2 production substrates are often not high, biosorbents have been explored to remove heavy metals and have proven to be successful at removing heavy metals at low concentrations (Qin et al., 2020).

During the processing of lignocellulosic biomass, lignin, hemicellulose, and sugar degradation produce microbial inhibitors. *Coniochaeta ligniaria* is a fungus with the innate capacity to metabolize various substances, including furan and aromatic aldehydes, known to operate as important fermenting MO inhibitors. *C. ligniaria* was employed in a study to break down and eliminate inhibitory substances from pretreatment rice hulls, a commonly available agricultural residue full of glucose. In contrast to uninterrupted fermentations, which either failed or generated less biofuel, bio-based liquors converted glucose to 0.58% (w/v) biofuel (Nichols et al., 2014).

In addition to biosorbents, heavy metals may be eliminated from substrates utilized in the synthesis of H_2 via bioleaching and electro-kinetic remediation techniques (Xu et al., 2017).

Sound H_2 generation may be accomplished from heavy metal-contaminated substrates by co-cultivating microalgae and H_2 producers. Unfortunately, there is little study on how to reduce light metal inhibition.

11.7.3 DEACTIVATION

In addition to dilution and removal, inactivating the inhibitors is another way to lessen their inhibition. The inactivation approach has been investigated in inhibiting bacteriocins, thiosulfate, furan, and phenolic chemicals.

Furan and phenolic chemicals have inhibitory effects that could be lessened by detoxification processes such as pretreatments, as discussed in Section 11.3.2. These pretreatment procedures are frequently substrate-focused, time-consuming, or expensive (Jönsson et al., 2013).

The structural stability of thiosulfate can be compromised, and the control (Tao et al., 2020).

The addition of proteases like trypsin can successfully inactivate bacteriocins that induce. Combining heat pretreatment at 100°C and base pretreatment at pH 9 can undermine and inactivate thiosulfinate structural integrity (Li et al., 2020a).

11.7.4 Regulating the Operational Parameters

In producing fermentative H_2, several operational variables exist, such as pH, temperature, C/N. Different H_2 production outcomes may be obtained by changing a few operating factors, which can also aid in reducing inhibition (Policastro et al., 2022).

11.8 CONCLUSION

$BioH_2$ is one of the most influential alternative energy sources to replace FFs. DF and PF by stringent anaerobes, facultative anaerobe microalgae, cyanobacteria, and PNSB are the most frequent methods of producing $bioH_2$. The MO's type, oxygen content, fermentation materials, pH, temperature, and enzyme concentrations all affect the activity of H_2-producing enzymes during DF and PF and their function in H_2 production. Lignocellulosic waste biomass has been regarded as one of the best feedstock sources for the effective generation of $bioH_2$ among the waste biomass sources that are readily accessible.

DF has more potential and is more straightforward than PF for industrialized and organized H_2 generation. Low H_2 production, however, is the main obstacle to this method's continued development. Several researchers worked hard to enhance the capabilities of DF.

The H_2 generation rate might be significantly boosted when the concentration of the substrate (glucose and starch) substantially improves. Still, it should not exceed the acceptable limit due to the buildup of suppressive VFAs throughout the process.

Due to the complex composition of biomass, pretreatment is necessary before the fermentation process can begin. Acid and enzymatic hydrolysis and hybrid pretreatment technologies are attractive because they improve decreasing sugar production from waste biomass sources. As metabolic pathway engineering and synthetic biology have matured, more improvements are on the horizon. Developing novel modified strains and advances in combined/hybrid fermentation methods bring up new avenues for increasing $bioH_2$ production efficiency. Concurrent synthesis of biomethane and other value-added chemicals is another excellent alternative for improving the technological and economical aspects of $bioH_2$ production. Overall, current $bioH_2$-generating advancements provide a viable future alternative to standard FFs.

REFERENCES

Acar, C. & Dincer, I. 2019. Review and evaluation of hydrogen production options for a better environment. *Journal of Cleaner Production*, 218, 835–849.

Acar, C., Dincer, I. & Naterer, G. F. 2016. Review of photocatalytic water-splitting methods for sustainable hydrogen production. *International Journal of Energy Research*, 40, 1449–1473.

Adessi, A., Venturi, M., Candeliere, F., Galli, V., Granchi, L. & De Philippis, R. 2018. Bread wastes to energy: Sequential lactic and photo-fermentation for hydrogen production. *International Journal of Hydrogen Energy*, 43, 9569–9576.

Ahmed, S. F., Rafa, N., Mofijur, M., Badruddin, I. A., Inayat, A., Ali, M. S., Farrok, O. & Yunus Khan, T. 2021. Biohydrogen production from biomass sources: Metabolic pathways and economic analysis. *Frontiers in Energy Research*, 9, 753878.

Ajanovic, A., Sayer, M. & Haas, R. 2022. The economics and the environmental benignity of different colors of hydrogen. *International Journal of Hydrogen Energy*, 47, 24136–24154.

Al-Mohammedawi, H. H., Znad, H. & Eroglu, E. 2019. Improvement of homofermentative biohydrogen production using pre-treated brewery wastewater with banana peel waste. *International Journal of Hydrogen Energy*, 44, 2560–2568.

Alvarez, A. J., Fuentes, K. L., Arias, C. A. & Chaparro, T. R. 2022. Production of hydrogen from beverage wastewater by dark fermentation in an internal circulation reactor: Effect on pH and hydraulic retention time. *Energy Conversion and Management: X*, 15, 100232.

Alvarez -Sieiro, P., Montalbán-López, M., Mu, D. & Kuipers, O. P. 2016. Bacteriocins of lactic acid bacteria: Extending the family. *Applied Microbiology and Biotechnology*, 100, 2939–2951.

Angeriz -Campoy, R., Fdez-Güelfo, L. A., Álvarez-Gallego, C. J. & Romero-García, L. I. 2023. Pre-composting of municipal solid wastes as enhancer of bio-hydrogen production through dark fermentation process. *Fuel*, 333, 126575.

Arun, J., Sasipraba, T., Gopinath, K., Priyadharsini, P., Nachiappan, S., Nirmala, N., Dawn, S., Chi, N. T. L. & Pugazhendhi, A. 2022. Influence of biomass and nanoadditives in dark fermentation for enriched bio-hydrogen production: A detailed mechanistic review on pathway and commercialization challenges. *Fuel*, 327, 125112.

Azwar, M., Hussain, M. & Abdul-Wahab, A. 2014. Development of biohydrogen production by photobiological, fermentation and electrochemical processes: A review. *Renewable and Sustainable Energy Reviews*, 31, 158–173.

Baeyens, J., Zhang, H., Nie, J., Appels, L., Dewil, R., Ansart, R. & Deng, Y. 2020. Reviewing the potential of bio-hydrogen production by fermentation. *Renewable and Sustainable Energy Reviews,* 131, 110023.

Bakhtyari, A., Makarem, M. A. & Rahimpour, M. R. 2019. Hydrogen production through pyrolysis. In Fuel *Cells* and *Hydrogen Production*, pp. 1–28. Springer.

Banu, J. R., Merrylin, J., Usman, T. M., Kannah, R. Y., Gunasekaran, M., Kim, S.-H. & Kumar, G. 2020. Impact of pretreatment on food waste for biohydrogen production: A review. *International Journal of Hydrogen Energy*, 45, 18211–18225.

Bao, M., Su, H. & Tan, T. 2013. Dark fermentative bio-hydrogen production: Effects of substrate pre-treatment and addition of metal ions or L-cysteine. *Fuel*, 112, 38–44.

Barghash, H., Okedu, K. E. & Al Balushi, A. 2021. Bio-hydrogen production using landfill leachate considering different photo-fermentation processes. *Frontiers in Bioengineering and Biotechnology,* 9, 644065.

Batista, A. P., Ambrosano, L., Graça, S., Sousa, C., Marques, P. A., Ribeiro, B., Botrel, E. P., Neto, P. C. & Gouveia, L. 2015. Combining urban wastewater treatment with biohydrogen production-an integrated microalgae-based approach. *Bioresource Technology*, 184, 230–235.

Bhagwat, S. & Olczak, M. 2020. *Green hydrogen: Bridging the energy transition in Africa and Europe,* European University Institute.

Budiman, P. M. & Wu, T. Y. 2016. Ultrasonication pre-treatment of combined effluents from palm oil, pulp and paper mills for improving photofermentative biohydrogen production. *Energy Conversion and Management,* 119, 142–150.

Budiman, P. M. & Wu, T. Y. 2018. Role of chemicals addition in affecting biohydrogen production through photofermentation. *Energy Conversion and Management*, 165, 509–527.

Budiman, P. M., Wu, T. Y., Ramanan, R. N. & MD. Jahim, J. 2015. Improvement of biohydrogen production through combined reuses of palm oil mill effluent together with pulp and paper mill effluent in photofermentation. *Energy & Fuels*, 29, 5816–5824.

Bundhoo, M. Z. & Mohee, R. 2016. Inhibition of dark fermentative bio-hydrogen production: A review. *International Journal of Hydrogen Energy*, 41, 6713–6733.

Burton, N., Padilla, R., Rose, A. & Habibullah, H. 2021. Increasing the efficiency of hydrogen production from solar powered water electrolysis. *Renewable and Sustainable Energy Reviews*, 135, 110255.

Cao, Y., Liu, H., Liu, W., Guo, J. & Xian, M. 2022. Debottlenecking the biological hydrogen production pathway of dark fermentation: Insight into the impact of strain improvement. *Microbial Cell Factories,* 21, 166.

Chandrasekhar, K., Lee, Y.-J. & Lee, D.-W. 2015. Biohydrogen production: strategies to improve process efficiency through microbial routes. *International Journal of Molecular Sciences,* 16, 8266–8293.

Chantawan, N., Moungprayoon, A., Lunprom, S., Reungsang, A. & Salakkam, A. 2022. High-solid dark fermentation of cassava pulp and cassava processing wastewater for hydrogen production. *International Journal of Hydrogen Energy*, 47, 40672–40682.

Chaurasia, A. K. & Mondal, P. 2022. Enhancing biohydrogen production from sugar industry wastewater using Ni, Ni-Co and Ni-Co-P electrodeposits as cathodes in microbial electrolysis cells. *Chemosphere*, 286, 131728.

Chen, Y., Yin, Y. & Wang, J. 2021. Recent advance in inhibition of dark fermentative hydrogen production. *International Journal of Hydrogen Energy*, 46, 5053–5073.

Chozhavendhan, S., Rajamehala, M., Karthigadevi, G., Praveenkumar, R. & Bharathiraja, B. 2020. A review on feedstock, pretreatment methods, influencing factors, production and purification processes of bio-hydrogen production. *Case Studies in Chemical and Environmental Engineering,* 2, 100038.

Cordara, A., Re, A., Pagliano, C., Van Alphen, P., Pirone, R., Saracco, G., Dos Santos, F. B., Hellingwerf, K. & Vasile, N. 2018. Analysis of the light intensity dependence of the growth of Synechocystis and of the light distribution in a photobioreactor energized by 635 nm light. *PeerJ*, 6, e5256.

Correa, C. R. & Kruse, A. 2018. Supercritical water gasification of biomass for hydrogen production—Review. *The Journal of Supercritical Fluids*, 133, 573–590.

Dai, B.-L., Guo, X.-J., Yuan, D.-H. & Xu, J.-M. 2018. Comparison of different pretreatments of rice straw substrate to improve biogas production. *Waste and Biomass Valorization*, 9, 1503–1512.

Das, S. R. & Basak, N. 2022. Optimization of process parameters for enhanced biohydrogen production using potato waste as substrate by combined dark and photo fermentation. *Biomass Conversion and Biorefinery*, 14, 1–21.

Dawood, F., Anda, M. & Shafiullah, G. 2020. Hydrogen production for energy: An overview. *International Journal of Hydrogen Energy*, 45, 3847–3869.

Deivayanai, V., Yaashikaa, P., Kumar, P. S., & Rangasamy, G. 2022. A comprehensive review on the biological conversion of lignocellulosic biomass into hydrogen: Pretreatment strategy, technology advances and perspectives. *Bioresource Technology,* 128166. doi: 10.1016/j.biortech.2022.128166.

Dinesh, G. K., Chauhan, R. & Chakma, S. 2018. Influence and strategies for enhanced biohydrogen production from food waste. *Renewable and Sustainable Energy Reviews*, 92, 807–822.

Dong, L., Cao, G., Zhao, L., Liu, B. & Ren, N. 2018. Alkali/urea pretreatment of rice straw at low temperature for enhanced biological hydrogen production. *Bioresource Technology*, 267, 71–76.

Duruibe, J. O., Ogwuegbu, M. & Egwurugwu, J. 2007. Heavy metal pollution and human biotoxic effects. *International Journal of physical sciences*, 2, 112–118.

Elbari, H., Lahboubi, N., Habchi, S., Rachidi, S., Bayssi, O., Nabil, N., Mortezaei, Y. & Villa, R. 2022. Biohydrogen production from fermentation of organic waste, storage and applications. *Cleaner Waste Systems,* 100043.

El-Shafie, M., Kambara, S. & Hayakawa, Y. 2019. Hydrogen production technologies overview. doi: 10.4236/jpee.2019.71007.

Elzinga, D., Baritaud, M., Bennett, S., Burnard, K., Pales, A., Philibert, C., Cuenot, F., D' Ambrosio, D., Dulac, J. & Heinen, S. 2014. *Energy technology perspectives 2014: Harnessing electricity's potential.* International Energy Agency (IEA), Paris, France.

Eskandari, A. 2019. *A preliminary theoretical and experimental study of a photo-electrochemical cell for solar hydrogen production.* Université Clermont Auvergne (UCA).

Fahoul, N., Sayadi, M. H., Rezaei, M. R. & Homaeigohar, S. 2022. Nickel ferrite nanoparticles catalyzed dark fermentation of dairy wastewater for biohydrogen production. *Bioresource Technology Reports,* 19, 101153.

Feng, S., Ngo, H. H., guo, W., Khan, M. A., Zhang, S., Luo, G., Liu, Y., An, D. & Zhang, X. 2023. Fruit peel crude enzymes for enhancement of biohydrogen production from synthetic swine wastewater by improving biohydrogen-formation processes of dark fermentation. *Bioresource Technology,* 372, 128670.

Gautam, R., Nayak, J. K., Ress, N. V., Steinberger-Wilckens, R. & Ghosh, U. K. 2023. Bio-hydrogen production through microbial electrolysis cell: Structural components and influencing factors. *Chemical Engineering Journal,* 455, 140535.

Ghirardi, M. L., King, P. W., Mulder, D. W., Eckert, C., Dubini, A., Maness, P.-C. & Yu, J. 2014. Hydrogen production by water biophotolysis. *Microbial bioenergy: Hydrogen production,* pp. 101–135. Springer.

Gu, J., Zhou, Z., Li, Z., Chen, Y., Wang, Z., Zhang, H. & Yang, J. 2017. Photosynthetic properties and potentials for improvement of photosynthesis in pale green leaf rice under high light conditions. *Frontiers in Plant Science,* 8, 1082.

Hasany, M., Mardanpour, M. M. & Yaghmaei, S. 2016. Biocatalysts in microbial electrolysis cells: A review. *International Journal of Hydrogen Energy,* 41, 1477–1493.

Hay, J. X. W., Wu, T. Y., Juan, J. C. & MD. Jahim, J. 2013. Biohydrogen production through photo fermentation or dark fermentation using waste as a substrate: Overview, economics, and future prospects of hydrogen usage. *Biofuels, Bioproducts and Biorefining,* 7, 334–352.

Hay, J. X. W., Wu, T. Y., Juan, J. C. & MD. Jahim, J. 2017. Effect of adding brewery wastewater to pulp and paper mill effluent to enhance the photofermentation process: Wastewater characteristics, biohydrogen production, overall performance, and kinetic modeling. *Environmental Science and Pollution Research,* 24, 10354–10363.

Hitam, C. & Jalil, A. 2020. A review on biohydrogen production through photo-fermentation of lignocellulosic biomass. *Biomass Conversion and Biorefinery,* 13, 1–19.

Huang, H.-W., Lung, H.-M., Yang, B. B. & Wang, C.-Y. 2014. Responses of microorganisms to high hydrostatic pressure processing. *Food Control,* 40, 250–259.

Huang, J., Feng, H., Huang, L., Ying, X., Shen, D., Chen, T., Shen, X., Zhou, Y. & Xu, Y. 2020. Continuous hydrogen production from food waste by anaerobic digestion (AD) coupled single-chamber microbial electrolysis cell (MEC) under negative pressure. *Waste Management,* 103, 61–66.

IEA2014. *Energy technology perspectives 2014-harnessing electricity's potential,* p. 382. IEA, Paris.

Ji, M., Jiang, X. & Wang, F. 2013. Feasibility of anaerobic co-digestion of waste activated sludge and corn straw to produce methane-batch experiment. *Asian Journal of Chemistry,* 25, 8793–8796.

Jönsson, L. J., Alriksson, B. & Nilvebrant, N.-O. 2013. Bioconversion of lignocellulose: Inhibitors and detoxification. *Biotechnology for Biofuels,* 6, 1–10.

Karthikeyan, O. P., Trably, E., Mehariya, S., Bernet, N., Wong, J. W. & Carrere, H. 2018. Pretreatment of food waste for methane and hydrogen recovery: A review. *Bioresource Technology,* 249, 1025–1039.

K Keçebaş, A. and M. Kayfeci, Hydrogen production Solar hydrogen production: processes systems and technologies. 2019.

Khongkliang, P., Kongjan, P., Utarapichat, B., Reungsang, A. & Sompong, O. 2017. Continuous hydrogen production from cassava starch processing wastewater by two-stage thermophilic dark fermentation and microbial electrolysis. *International Journal of Hydrogen Energy*, 42, 27584–27592.

Kim, D., Kim, G., Oh, D. Y., Seong, K.-W. & Park, K. Y. 2022. Enhanced hydrogen production from anaerobically digested sludge using microwave assisted pyrolysis. *Fuel*, 314, 123091.

Kondusamy, D. & Kalamdhad, A. S. 2014. Pre-treatment and anaerobic digestion of food waste for high rate methane production—A review. *Journal of Environmental Chemical Engineering*, 2, 1821–1830.

Kumar, A. N., Min, B. & Mohan, S. V. 2018. Defatted algal biomass as feedstock for short chain carboxylic acids and biohydrogen production in the biorefinery format. *Bioresource Technology*, 269, 408–416.

Kumar, G., Zhen, G., Kobayashi, T., Sivagurunathan, P., Kim, S. H. & Xu, K. Q. 2016. Impact of pH control and heat pre-treatment of seed inoculum in dark H_2 fermentation: A feasibility report using mixed microalgae biomass as feedstock. *International Journal of Hydrogen Energy*, 41, 4382–4392.

Kumar, M., Oyedun, A. O. & Kumar, A. 2019a. A comparative analysis of hydrogen production from the thermochemical conversion of algal biomass. *International Journal of Hydrogen Energy*, 44, 10384–10397.

Kumar, S., Sharma, S., Thakur, S., Mishra, T., Negi, P., Mishra, S., Hesham, A. E. L., Rastegari, A. A., Yadav, N. & Yadav, A. N. 2019b. Bioprospecting of microbes for biohydrogen production: Current status and future challenges. Bioprocessing for *biomolecules production*, 443–471. doi: 10.1002/9781119434436.ch22.

routesLamb, J. J., Hillestad, M., Rytter, E., Bock, R., Nordgård, A. S., Lien, K. M., ... & Pollet, B. G. (2020). Traditional routes for hydrogen production and carbon conversion. In *Hydrogen, biomass and bioenergy* (pp. 21-53). Academic Press.

Li, M., Guo, C., Luo, B., Chen, C., Wang, S. & Min, D. 2020a. Comparing impacts of physicochemical properties and hydrolytic inhibitors on enzymatic hydrolysis of sugarcane bagasse. *Bioprocess and Biosystems Engineering*, 43, 111–122.

Li, P., Cai, D., Luo, Z., Qin, P., Chen, C., Wang, Y., Zhang, C., Wang, Z. & Tan, T. 2016. Effect of acid pretreatment on different parts of corn stalk for second generation ethanol production. *Bioresource Technology*, 206, 86–92.

Li, W., Guo, J., Cheng, H., Wang, W. & Dong, R. 2017. Two-phase anaerobic digestion of municipal solid wastes enhanced by hydrothermal pretreatment: Viability, performance and microbial community evaluation. *Applied Energy*, 180, 613–622.

Li, X., Guo, L., Liu, Y., Wang, Y., She, Z., Gao, M. & Zhao, Y. 2020b. Effect of salinity and pH on dark fermentation with thermophilic bacteria pretreated swine wastewater. *Journal of Environmental Management*, 271, 111023.

Lim, D., Lee, B., Lee, H., Byun, M., Cho, H.-S., Cho, W., Kim, C.-H., Brigljević, B. & Lim, H. 2021. Impact of voltage degradation in water electrolyzers on sustainability of synthetic natural gas production: Energy, economic, and environmental analysis. *Energy Conversion and Management*, 245, 114516.

Lin, C.-Y., Leu, H.-J. & Lee, K.-H. 2016. Hydrogen production from beverage wastewater via dark fermentation and room-temperature methane reforming. *International Journal of Hydrogen Energy*, 41, 21736–21746.

Liu, B.-F., Jin, Y.-R., Cui, Q.-F., Xie, G.-J., Wu, Y.-N. & Ren, N.-Q. 2015. Photo-fermentation hydrogen production by Rhodopseudomonas sp. nov. strain A7 isolated from the sludge in a bioreactor. *International Journal of Hydrogen Energy*, 40, 8661–8668.

Ma, C., Liu, J., Ye, M., Zou, L., Qian, G. & Li, Y.-Y. 2018. Towards utmost bioenergy conversion efficiency of food waste: Pretreatment, co-digestion, and reactor type. *Renewable and Sustainable Energy Reviews*, 90, 700–709.

Mabutyana, L. & Pott, R. W. 2021. Photo-fermentative hydrogen production by *Rhodopseudomonas palustris* CGA009 in the presence of inhibitory compounds. *International Journal of Hydrogen Energy*, 46, 29088–29099.

Mahata, C., Ray, S. & Das, D. 2020. Optimization of dark fermentative hydrogen production from organic wastes using acidogenic mixed consortia. *Energy Conversion and Management*, 219, 113047.

Maroušek, J. 2022. Nanoparticles can change (bio) hydrogen competitiveness. *Fuel*, 328, 125318.

Martínez -Patiño, J. C., Romero, I., Ruiz, E. N., Cara, C., Romero- García, J. M. & Castro, E. 2017. Design and optimization of sulfuric acid pretreatment of extracted olive tree biomass using response surface methodology. *BioResources*, 12, 1779–1797.

Mikhaylov, A., Moiseev, N., Aleshin, K. & Burkhardt, T. 2020. Global climate change and greenhouse effect. *Entrepreneurship and Sustainability Issues*, 7, 2897.

Milani, D., Kiani, A. & Mcnaughton, R. 2020. Renewable-powered hydrogen economy from Australia's perspective. *International Journal of Hydrogen Energy*, 45, 24125–24145.

Mishra, P., Krishnan, S., Rana, S., Singh, L., Sakinah, M. & Ab Wahid, Z. 2019. Outlook of fermentative hydrogen production techniques: An overview of dark, photo and integrated dark-photo fermentative approach to biomass. *Energy Strategy Reviews*, 24, 27–37.

Mishra, P., Singh, D., Sonvane, Y. & Ahuja, R. 2020. Enhancement of hydrogen storage capacity on co-functionalized GaS monolayer under external electric field. *International Journal of Hydrogen Energy*, 45, 12384–12393.

Mosca, L., Jimenez, J. A. M., Wassie, S. A., Gallucci, F., Palo, E., Colozzi, M., Taraschi, S. & Galdieri, G. 2020. Process design for green hydrogen production. *International Journal of Hydrogen Energy*, 45, 7266–7277.

Nichols, N. N., Hector, R. E., Saha, B. C., Frazer, S. E. & Kennedy, G. J. 2014. Biological abatement of inhibitors in rice hull hydrolyzate and fermentation to ethanol using conventional and engineered microbes. *Biomass and Bioenergy*, 67, 79–88.

Noussan, M., Raimondi, P. P., Scita, R. & Hafner, M. 2020. The role of green and blue hydrogen in the energy transition—A technological and geopolitical perspective. *Sustainability*, 13, 298.

Pagliaccia, P., Gallipoli, A., Gianico, A., Montecchio, D. & Braguglia, C. 2016. Single stage anaerobic bioconversion of food waste in mono and co-digestion with olive husks: Impact of thermal pretreatment on hydrogen and methane production. *International Journal of Hydrogen Energy*, 41, 905–915.

Palmqvist, E. & Hahn-Hägerdal, B. 2000. Fermentation of lignocellulosic hydrolysates. II: Inhibitors and mechanisms of inhibition. *Bioresource Technology*, 74, 25–33.

Parvanova-Mancheva, T., Vasileva, E. & Beschkov, V. 2022. Bio-hydrogen production through microbial electrolysis cells. Journal of Chemical Technology & Metallurgy, 57, 140535.

Patel, S. K., Lee, J.-K. & Kalia, V. C. 2018. Nanoparticles in biological hydrogen production: An overview. *Indian Journal of Microbiology*, 58, 8–18.

Pintucci, C., Padovani, G., Giovannelli, A., Traversi, M. L., Ena, A., Pushparaj, B. & Carlozzi, P. 2015. Hydrogen photo-evolution by *Rhodopseudomonas palustris* 6A using pre-treated olive mill wastewater and a synthetic medium containing sugars. *Energy Conversion and Management*, 90, 499–505.

Policastro, G., Giugliano, M., Luongo, V., Napolitano, R. & Fabbricino, M. 2022. Enhancing photo fermentative hydrogen production using ethanol rich dark fermentation effluents. *International Journal of Hydrogen Energy*, 47, 117–126.

Policastro, G., Luongo, V. & Fabbricino, M. 2020. Biohydrogen and poly-β-hydroxybutyrate production by winery wastewater photofermentation: Effect of substrate concentration and nitrogen source. *Journal of Environmental Management*, 271, 111006.

Pradeep, N., Anupama, S., Navya, K., Shalini, H., Idris, M. & Hampannavar, U. 2015. Biological removal of phenol from wastewaters: A mini review. *Applied Water Science*, 5, 105–112.

Pradhan, N., Dipasquale, L., D' Ippolito, G., Panico, A., Lens, P. N., Esposito, G. & Fontana, A. 2015. Hydrogen production by the thermophilic bacterium Thermotoga neapolitana. *International Journal of Molecular Sciences,* 16, 12578–12600.

Putatunda, C., Behl, M., Solanki, P., Sharma, S., Bhatia, S. K., Walia, A. & Bhatia, R. K. 2022. Current challenges and future technology in photofermentation-driven biohydrogen production by utilizing algae and bacteria. *International Journal of Hydrogen Energy,* 48, 21088–21109.

Qin, H., Hu, T., Zhai, Y., Lu, N. & Aliyeva, J. 2020. The improved methods of heavy metals removal by biosorbents: A review. *Environmental Pollution,* 258, 113777.

Qu, X., Zeng, H., Gao, Y., Mo, T. & Li, Y. 2022. Bio-hydrogen production by dark anaerobic fermentation of organic wastewater. *Frontiers in Chemistry,* 10. doi: 10.3389/fchem.2022.978907.

Qureshi, F., Yusuf, M., Kamyab, H., Vo, D.-V. N., Chelliapan, S., Joo, S.-W. & Vasseghian, Y. 2022. Latest eco-friendly avenues on hydrogen production towards a circular bioeconomy: Currents challenges, innovative insights, and future perspectives. *Renewable and Sustainable Energy Reviews,* 168, 112916.

Rai, P. K. & Singh, S. 2016. Integrated dark-and photo-fermentation: Recent advances and provisions for improvement. *International Journal of Hydrogen Energy,* 41, 19957–19971.

Ravindran, R. & Jaiswal, A. K. 2016. A comprehensive review on pre-treatment strategy for lignocellulosic food industry waste: Challenges and opportunities. *Bioresource Technology,* 199, 92–102.

Roy, R., Rahman, M. & Raynie, D. 2020. Recent advances of greener pretreatment technologies of lignocellulose. *Current Research in Green and Sustainable Chemistry,* 3, 100035.

Ruggeri, B., Tommasi, T., Sanfilippo, S., Ruggeri, B., Tommasi, T. & Sanfilippo, S. 2015. Ecological mechanisms of dark H2 production by a mixed microbial community. In *BioH$_2$ & BioCH$_4$ through anaerobic digestion: From research to full-scale applications,* pp. 1–24. doi: 10.1007/978-1-4471-6431-9_1.

Ruiz-Marin, A., Canedo-López, Y. & Chávez-Fuentes, P. 2020. Biohydrogen production by Chlorella vulgaris and Scenedesmus obliquus immobilized cultivated in artificial wastewater under different light quality. *Amb Express,* 10, 1–7.

Santiago, S. G., Trably, E., Latrille, E., Buitrón, G. & Moreno- Andrade, I. 2019. The hydraulic retention time influences the abundance of Enterobacter, Clostridium and Lactobacillus during the hydrogen production from food waste. *Letters in Applied Microbiology,* 69, 138–147.

Sharma, A. & Arya, S. K. 2017. Hydrogen from algal biomass: A review of production process. *Biotechnology Reports,* 15, 63–69.

Shetty, D. J., Kshirsagar, P., Tapadia-Maheshwari, S., Lanjekar, V., Singh, S. K. & Dhakephalkar, P. K. 2017. Alkali pretreatment at ambient temperature: A promising method to enhance biomethanation of rice straw. *Bioresource Technology,* 226, 80–88.

Shi, Y., Weimer, P. & Ralph, J. 1997. Formation of formate and hydrogen, and flux of reducing equivalents and carbon in Ruminococcus flavefaciens FD-1. *Antonie Van Leeuwenhoek,* 72, 101–109.

Show, K.-Y., Yan, Y., Zong, C., Guo, N., Chang, J.-S. & Lee, D.-J. 2019. State of the art and challenges of biohydrogen from microalgae. *Bioresource Technology,* 289, 121747.

Shukla, A., Kumar, D., Girdhar, M., Kumar, A., Goyal, A., Malik, T. & Mohan, A. 2023. Strategies of pretreatment of feedstocks for optimized bioethanol production: Distinct and integrated approaches. *Biotechnology for Biofuels and Bioproducts,* 16, 44.

Sivabalan, K., Hassan, S., Ya, H. & Pasupuleti, J. 2021. J., 2021. A review on the characteristic of biomass and classification of bioenergy through direct combustion and gasification as an alternative power supply. In *Journal of physics: conference series* (Vol. 1831, No. 1, p. 012033). IOP Publishing. *Journal of Physics: Conference Series,* 012033.

Sivagurunathan, P., Kumar, G., Bakonyi, P., Kim, S.-H., Kobayashi, T., Xu, K. Q., Lakner, G., Tóth, G., Nemestóthy, N. & Bélafi-Bakó, K. 2016. A critical review on issues and overcoming strategies for the enhancement of dark fermentative hydrogen production in continuous systems. *International Journal of Hydrogen Energy*, 41, 3820–3836.

Sivagurunathan, P., Kumar, G., Mudhoo, A., Rene, E. R., Saratale, G. D., Kobayashi, T., Xu, K., Kim, S.-H. & Kim, D.-H. 2017. Fermentative hydrogen production using lignocellulose biomass: An overview of pre-treatment methods, inhibitor effects and detoxification experiences. *Renewable and Sustainable Energy Reviews*, 77, 28–42.

Sivaramakrishnan, R., Shanmugam, S., Sekar, M., Mathimani, T., Incharoensakdi, A., Kim, S.-H., Parthiban, A., Geo, V. E., Brindhadevi, K. & Pugazhendhi, A. 2021. Insights on biological hydrogen production routes and potential microorganisms for high hydrogen yield. *Fuel*, 291, 120136.

Solarte-Toro, J. C., Romero-García, J. M., Martínez-Patiño, J. C., Ruiz-Ramos, E., Castro-Galiano, E. & Cardona-Alzate, C. A. 2019. Acid pretreatment of lignocellulosic biomass for energy vectors production: A review focused on operational conditions and techno-economic assessment for bioethanol production. *Renewable and Sustainable Energy Reviews*, 107, 587–601.

Soltan, M., Elsamadony, M. & Tawfik, A. 2017. Biological hydrogen promotion via integrated fermentation of complex agro-industrial wastes. *Applied Energy*, 185, 929–938.

Srivastava, N., Srivastava, M., Abd_Allah, E. F., Singh, R., Hashem, A. & Gupta, V. K. 2021. Biohydrogen production using kitchen waste as the potential substrate: A sustainable approach. *Chemosphere*, 271, 129537.

Srivastava, N., Srivastava, M., Kushwaha, D., Gupta, V. K., Manikanta, A., Ramteke, P. & Mishra, P. 2017. Efficient dark fermentative hydrogen production from enzyme hydrolyzed rice straw by Clostridium pasteurianum (MTCC116). *Bioresource Technology*, 238, 552–558.

Srivastava, N., Srivastava, M., Mishra, P., Kausar, M. A., Saeed, M., Gupta, V. K., Singh, R. & Ramteke, P. 2020. Advances in nanomaterials induced biohydrogen production using waste biomass. *Bioresource Technology*, 307, 123094.

Stewart, C., Mcpherson, C. A. & Cansunar, E. 1987. The effect of lasalocid on glucose uptake, hydrogen production and the solubilization of straw by the anaerobic rumen fungus Neocallimastix frontalis. *Letters in Applied Microbiology*, 5, 5–7.

Taikhao, S., Incharoensakdi, A. & Phunpruch, S. 2015. Dark fermentative hydrogen production by the unicellular halotolerant cyanobacterium *Aphanothece halophytica* grown in seawater. *Journal of Applied Phycology*, 27, 187–196.

Tang, T., Chen, Y., Liu, M., Du, Y. & Tan, Y. 2022. Effect of pH on the performance of hydrogen production by dark fermentation coupled denitrification. *Environmental Research*, 208, 112663.

Tao, Z., Wang, D., Yao, F., Huang, X., Wu, Y., Du, M., Chen, Z., An, H., Li, X. & Yang, Q. 2020. The effects of thiosulfinates on methane production from anaerobic co-digestion of waste activated sludge and food waste and mitigate method. *Journal of Hazardous Materials*, 384, 121363.

Tian, H., Li, J., Yan, M., Tong, Y. W., Wang, C.-H. & Wang, X. 2019. Organic waste to biohydrogen: A critical review from technological development and environmental impact analysis perspective. *Applied Energy*, 256, 113961.

Unni, R., Reshmy, R., Madhavan, A., Binod, P., Pandey, A. & Sindhu, R. 2022.. In *Organic waste to biohydrogen,* pp. 163–179. Springer.

CRC Press.

Vavouraki, A. I., Angelis, E. M. & Kornaros, M. 2013. Optimization of thermo-chemical hydrolysis of kitchen wastes. *Waste Management*, 33, 740–745.

Vickers, N. J. 2017. Animal communication: When i'm calling you, will you answer too? *Current Biology*, 27, R713–R715.

Wang, D., Zhang, D., Xu, Q., Liu, Y., Wang, Q., Ni, B.-J., Yang, Q., Li, X. & Yang, F. 2019a. Calcium peroxide promotes hydrogen production from dark fermentation of waste activated sludge. *Chemical Engineering Journal*, 355, 22–32.

Wang, G., Cheng, Q., Zhao, W., Liao, Q. & Zhang, H. 2022. Review on the transport capacity management of oil and gas pipeline network: Challenges and opportunities of future pipeline transport. *Energy Strategy Reviews*, 43, 100933.

Wang, J. & Yin, Y. 2018. Fermentative hydrogen production using various biomass-based materials as feedstock. *Renewable and Sustainable Energy Reviews*, 92, 284–306.

Wang, J., Zhang, S., Ouyang, Y. & Li, R. 2019b. Current developments of bacteriocins, screening methods and their application in aquaculture and aquatic products. *Biocatalysis and Agricultural Biotechnology*, 22, 101395.

Wang, S., Lu, A. & Zhong, C.-J. 2021. Hydrogen production from water electrolysis: Role of catalysts. *Nano Convergence*, 8, 1–23.

Wei, H., Junhong, T. & Yongfeng, L. 2016. Utilization of food waste for fermentative hydrogen production. *Physical Sciences Reviews*, 1. doi: 10.1515/9783110342420-006.

Woraprayote, W., Malila, Y., Sorapukdee, S., Swetwiwathana, A., Benjakul, S. & Visessanguan, W. 2016. Bacteriocins from lactic acid bacteria and their applications in meat and meat products. *Meat Science*, 120, 118–132.

Xu, F., Li, Y., Ge, X., Yang, L. & Li, Y. 2018. Anaerobic digestion of food waste—Challenges and opportunities. *Bioresource Technology*, 247, 1047–1058.

Xu, Y., Zhang, C., Zhao, M., Rong, H., Zhang, K. & Chen, Q. 2017. Comparison of bioleaching and electrokinetic remediation processes for removal of heavy metals from wastewater treatment sludge. *Chemosphere*, 168, 1152–1157.

Yadav, V. S., Vinoth, R., & Yadav, D. (2018). Bio-hydrogen production from waste materials: a review. In *MATEC Web of Conferences* (Vol. 192, p. 02020). EDP Sciences.02020.

Yakaboylu, G. A., Jiang, C., Yumak, T., Zondlo, J. W., Wang, J. & Sabolsky, E. M. 2021. Engineered hierarchical porous carbons for supercapacitor applications through chemical pretreatment and activation of biomass precursors. *Renewable Energy*, 163, 276–287.

Yang, G. & Wang, J. 2019. Changes in microbial community structure during dark fermentative hydrogen production. *International Journal of Hydrogen Energy*, 44, 25542–25550.

Yin, Y. & Wang, J. 2018. Pretreatment of macroalgal Laminaria japonica by combined microwave-acid method for biohydrogen production. *Bioresource Technology*, 268, 52–59.

Yin, Y., Zhuang, S. & Wang, J. 2018. Enhanced fermentative hydrogen production using gamma irradiated sludge immobilized in polyvinyl alcohol (PVA) gels. *Environmental Progress & Sustainable Energy*, 37, 1183–1190.

Yodwong, B., Guilbert, D., Phattanasak, M., Kaewmanee, W., Hinaje, M. & Vitale, G. 2020. AC-DC converters for electrolyzer applications: State of the art and future challenges. *Electronics*, 9, 912.

Yue, T., Jiang, D., Zhang, Z., Zhang, Y., Li, Y., Zhang, T. & Zhang, Q. 2021. Recycling of shrub landscaping waste: Exploration of bio-hydrogen production potential and optimization of photo-fermentation bio-hydrogen production process. *Bioresource Technology*, 331, 125048.

Zhang, Q., Jin, P., Li, Y., Zhang, Z., Zhang, H., Ru, G., Jiang, D., Jing, Y. & Zhang, X. 2022. Analysis of the characteristics of paulownia lignocellulose and hydrogen production potential via photo fermentation. *Bioresource Technology*, 344, 126361.

Zhang, Y., Fan, X., Yang, Z., Wang, H., Yang, D. & Guo, R. 2012. Characterization of H_2 photoproduction by a new marine green alga, *Platymonas helgolandica* var. tsingtaoensis. *Applied Energy*, 92, 38–43.

Zhang, Z. & Chen, Y. 2020. Effects of microplastics on wastewater and sewage sludge treatment and their removal: A review. *Chemical Engineering Journal*, 382, 122955.

Zhao, X., Ye, S., Qi, N., Li, X., Bao, N., Xing, D. & Ren, N. 2019. Mechanisms of enhanced bio-H_2 production in Ethanoligenens harbinense by l-cysteine supplementation: Analyses at growth and gene transcription levels. *Fuel*, 252, 143–147.

Ziara, R. M., Miller, D. N., Subbiah, J. & Dvorak, B. I. 2019. Lactate wastewater dark fermentation: The effect of temperature and initial pH on biohydrogen production and microbial community. *International Journal of Hydrogen Energy*, 44, 661–673.

Section IV

Other Renewable Resources for Hydrogen Production

12 Application of Solar Energy in Hydrogen Production

Fatemeh Zarei-Jelyani, Mohammad Zarei-Jelyani, Fatemeh Salahi, and Mohammad Reza Rahimpour

12.1 INTRODUCTION

At present, the patterns observed in the consumption and production of energy are markedly unsustainable from an environmental, economic, and social perspective. Without a clear choice, the growing usage of fossil fuels would raise worries about the security of energy reserves, and by 2050, human greenhouse gas (GHG) emissions will have more than double (Hosseini et al., 2013). It is imperative to alter the present trajectory and give due consideration to carbon capture and storage (CCS) within energy systems, while also extensively implementing renewable and sustainable energy sources (Hwang et al., 2013). In order for governments and industrial sectors to take the necessary steps toward implementing environmentally friendly energy generating technologies, a clear road map is needed (Yi et al., 2019). The utilization of hydrogen fuel as an energy carrier is deemed to be a sustainable solution that is poised to play an important role in the global energy landscape. It is expected to contribute significantly to the reduction of carbon emissions in both the industrial and transportation sectors. Hydrogen fuel is scarce in the free state, and its production is predominantly artificial. Currently, the primary source of hydrogen extraction involves hydrocarbons, which results in a significant release of carbon dioxide emissions (Hosseini and Wahid, 2016). The process of electrolyzing water is a widely used method for producing hydrogen, whereby the application of electrical energy causes the separation of water molecules into hydrogen, which serves as a clean energy carrier, and oxygen, which is a safe by-product of the reaction (Hosseini and Wahid, 2020). Similar to gasoline, hydrogen is a combustible fuel. Nevertheless, in contrast to gasoline flames that propagate at ground level, hydrogen flames ascend vertically due to its buoyancy, as it is 14.4 times less dense than air (Sulaiman et al., 2015). Hydrogen is a multifaceted energy source that can be derived from a diverse array of resources and utilized in various applications throughout the entirety of the energy industry. Hydrogen can be classified into three distinct categories, namely "gray hydrogen," which is produced without carbon sequestration from natural gas (NG), the "blue hydrogen," which is produced from NG with CCS, and "green hydrogen," which is produced from renewable energy sources and does not produce any carbon emissions. This classification is based on information obtained from various

DOI: 10.1201/9781003382270-16

sources. The cost-effectiveness of gray hydrogen in comparison with blue and green hydrogen is attributed to the predominant influence of NG prices. Since the cost of NG is the primary factor, gray hydrogen has low cost than blue and green varieties. There are multiple factors that exert an influence on the pricing of green hydrogen. Assessing the cost of green hydrogen necessitates consideration of the pricing of environmentally sustainable electrical power utilized in the electrolysis process (Balcombe et al., 2018). Additionally, the cost of electrolysis, the technology used to create hydrogen from water, should be addressed (Glenk and Reichelstein, 2019). The price of renewable energy has dropped dramatically in recent years and is expected to drop much more in the near future. Hence, there is a genuine potential for the production of green hydrogen to cater to the needs of transportation systems and domestic usage (Poompavai and Kowsalya, 2019). At present, water electrolysis accounts for roughly 5% of the total hydrogen supply, while the majority is derived from fossil fuels (Ni et al., 2007). By connecting energy demand and supply in both distributed and centralized systems, hydrogen may increase overall low-carbon energy system flexibility. While the potential benefits of hydrogen fuel for energy and environmental security in end-use applications are encouraging, the implementation of hydrogen generation technologies and its transmission, distribution, and retail infrastructure pose significant challenges.

Renewable energy sources provide a means of producing hydrogen without emitting harmful substances. Conversely, hydrogen serves as a viable option for utilizing renewable energy in the production of transportation fuel and ensuring a consistent supply of power. Renewable technologies possess a remarkable adaptability that enables their utilization in addressing a wide range of energy requirements. While the costs associated with certain renewable technologies may exceed those of conventional energy sources, the progress in technology and the growing demand within the energy market have resulted in a decrease in prices. Furthermore, the environmental advantages assist to offset the greater expenditures. The synergistic combination of hydrogen and renewable energy technology holds great promise for the future. Solar energy is extensively recognized as a highly abundant renewable energy source, and the utilization of solar energy for hydrogen production is regarded as a paramount approach toward achieving sustainable energy solutions. Many researchers have worked on studying the various solar hydrogen generation technologies using energy and exergy analysis. The study conducted by Wang et al. (2012) examined the underlying mechanisms involved in solar-to-hydrogen reactions. These reactions encompass thermochemical cycles, which utilize heat to separate water molecules; electrolysis, which employs electric potential for splitting water molecules; and photochemical reactions that use photon-activated electrons from auxiliary reagents to trigger and separate water molecules. Erickson and Goswami (2001) conducted a comprehensive examination of the theoretical foundations and present technological advancements pertaining to hydrogen derived from solar energy. Joshi et al. (2010) conducted an investigation on the exergetic evaluation of methods for producing hydrogen through solar energy. The solar hydrogen production system has been categorized according to various factors, including the energy input and solar thermal, as well as the type of chemical reactants used. Additionally, various hydrogen generation processes such as reforming electrolysis, cracking, and gasification have been considered in this classification. The sustainability of a solar photovoltaic-based hydrogen system has been examined in a case study using exergy efficiency and sustainability index as analytical tools. Subsequently, Joshi et al. (2010) conducted a comparative analysis of the sustainability index between solar thermal and photovoltaic hydrogen generation systems. Among the several solar hydrogen generation systems found, some have achieved industrial and commercial maturity, while others are still under investigation. Beyond their variety, these technologies are divided into three families: photochemical, thermochemical, and electrochemical technologies that provide a potential option for solar energy storage.

The aim of this chapter is to compile an extensive discussion of solar energy-based technologies for hydrogen synthesis, specifically focusing on fixtures. The methods used for hydrogen production from solar energy are classified into three distinct categories: photochemical processes, thermochemical processes, and electrochemical processes, which are discussed in this chapter. This chapter facilitates the contemplation of the feasibility and future prospects for the generation and use of solar hydrogen as an energy carrier, particularly in places that possess enough solar radiation.

12.2 HYDROGEN GENERATION FROM NEW ENERGY

The use of finite fossil fuels for hydrogen generation adds significantly to the growth in the greenhouse effect. The technical and financial aspects of addressing this issue would involve implementing measures such as CO_2 harnessing. These measures have the potential to alter the parameters of alternative economic solutions, specifically in the realm of renewable energy (Short et al., 2005). The development of hydrogen generation systems based on renewable energy sources occurs as much as feasible without emitting GHGs (Ngoh and Njomo, 2012). These methods offer alternative approaches to hydrogen production using fossil fuels. The link between the generation of renewable power and the demands for fixed and portable energy is provided by hydrogen. When power generated by solar photovoltaics, ocean, geothermal, wind, and hydro-technologies is utilized to manufacture and store hydrogen, the renewable source gains value and may cover a wide range of demands. In the context of transportation applications, hydrogen can be used to transform renewable resources into car fuel. One of the most common hydrogen economies aims is to use renewable hydrogen for transportation fuel, which can be generated domestically and emissions-free. In a recent study conducted by Koroneos and Dimitrios (Christopher and Dimitrios, 2012), the authors examined and compared different hydrogen production processes that rely on renewable energy sources, with a specific focus on exergy efficiency. Three factors must be considered in all hydrogen production techniques: the raw material, the energy required for synthesis, and the manufacturing process. Most processes have a large number of significant variations (Agbossou et al., 2001). Figure 12.1 provides a comprehensive overview of the various techniques employed for hydrogen generation utilizing renewable energy sources, including wind power, geothermal and hydroelectricity energy.

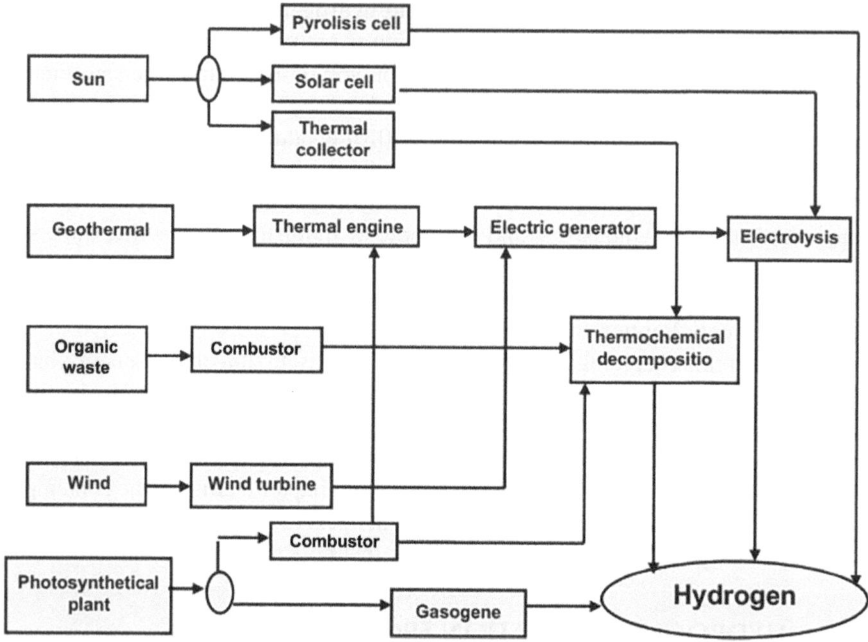

FIGURE 12.1 Techniques of hydrogen production from renewable energies (Ngoh and Njomo, 2012).

12.3   HYDROGEN GENERATION FROM SOLAR ENERGY

12.3.1   Photochemical Processes

The photochemical processes utilize solar radiation to facilitate the hydrolysis of water. Currently, there exist two distinct procedures, namely the photoelectrochemical and the photobiological.

12.3.1.1   The Bio-Production of H_2 from Photosynthetic Microorganisms

Photobiological processes rely on the ability of certain organisms, including cyanobacteria, photosynthetic bacteria, and green algae, to serve as biological catalysts for the generation of hydrogen through the utilization of water and various enzymes, like hydrogenase and nitrogenase. The existence of diverse microbial physiology and metabolism has resulted in a multitude of pathways through which microorganisms can generate H_2, each exhibiting apparent benefits and potential drawbacks. The initial observation of the H_2 metabolism in green algae dates back to the early 1940s and was made by Hans Gaffron. The author noted that green algae have the ability to utilize H_2 as an electron donor during CO_2 fixation or produce H_2 under both dark and light conditions in anaerobic environments (Benemann, 1996). The physiological importance of hydrogen metabolism in algae remains a topic of fundamental research. However, the production of H_2 gas through photohydrogen generation by green algae is a subject of interest due to its ability to utilize the abundant resources of light and water (Miao et al., 2004, Miyake et al., 1999).

Microbial conversions can be conducted under ambient conditions, although the primary limitations are the reduced rate of hydrogen production and lower yield. The enzymatic regulation of various processes is facilitated by hydrogen-producing enzymes, including hydrogenase and nitrogenase. The majority of photosynthetic microorganisms possess hydrogenases, which can be categorized into two distinct types including uptake hydrogenases and reversible hydrogenases. Uptake hydrogenases, namely Ni-Fe hydrogenases and Ni-Fe-Se hydrogenases, serve as significant catalysts for hydrogen utilization. Reversible hydrogenases possess the capacity to both generate and consume hydrogen contingent upon the prevailing reaction conditions, as suggested by their nomenclature. Nitrogenase is primarily composed of two major components, namely MoFe protein and Fe protein. The enzyme nitrogenase exhibits the capability to employ magnesium adenosine triphosphate (MgATP) and electrons for the purpose of reducing diverse substrates, which may include protons.

12.3.1.2 Photoelectrolysis of Water

The process of water photoelectrolysis involves the semiconducting photocatalyzer dissociation through the application of an electric current via illumination. Photoelectrochemical cells (PECs) are devices that utilize photoactive electrodes. These electrodes are immersed in an aqueous electrolyte or water and are exposed to solar radiation, which can facilitate the decomposition of water into oxygen and hydrogen. Thermodynamic cycles offer a means of hydrogen production that does not rely on the use of carbonized fossils. This method involves the use of exhaustion and has the potential to produce significant amounts of CO_2, which is a major contributor to the greenhouse impact. The process of water decomposition involves the implementation of either a thermos electrochemical or a thermal decomposition mechanism, which is facilitated by the provision and subsequent storage of solar energy. This is achieved through a series of reactions, the cumulative effect of which is equivalent to the overall process of water decomposition (equation (12.1)).

$$H_2O \rightarrow H + \left(\frac{1}{2}\right)O \qquad \Delta H = 284 \text{ kJ/mol} \qquad (12.1)$$

The investigation of photoelectrochemical processes is still ongoing. The implementation is anticipated to occur solely in the long term.

12.3.2 THERMOCHEMICAL PROCESSES

The use of solar energy for hydrogen generation via thermochemical methods includes a variety of processes such as thermochemical cycles; direct thermolysis of water; and hydrocarbon cracking, reforming, and gasification. The techniques described above employ concentrated solar radiation as a heat source at high temperatures to help the endothermic reaction. High sunlight concentration ratios may be achieved using a variety of ways, including parabolic disks, tower systems, and solar furnaces.

12.3.2.1 Hydrocarbon Solar Cracking

Hydrocarbon cracking involves the cogeneration of black carbon and hydrogen. Matovitch (Matovich, 1978) proposed the utilization of methane decomposition as a method of generating hydrogen. The reactor that was developed enabled the full dissociation of CH_4 within a matter of seconds, achieved at 2,100 K. In 1993, Steinberg (Steinberg, 1986) and Muradov (Muradov, 1993) provided further endorsement for this approach. The following is the presentation of an endothermic reaction.

$$CH_4 \rightarrow C + H_2 \tag{12.2}$$

Theoretically, the thermodynamic process of methane decomposition achieves full completion at approximately 1,300 K. In reality, the stability of methane is high and the kinetic decomposition process is not favorable. On the other hand, the features of black carbon are also influenced by the temperature at which decomposition occurs. In reality, the stability of methane is high and the kinetic decomposition process is not favorable. On the other hand, the properties of black carbon are also influenced by the temperature at which decomposition occurs. The aforementioned factors have resulted in a proclivity toward utilizing reaction temperatures in excess of 1,500 K.

Research on the catalytic and/or thermal decomposition of hydrocarbons using solar channels has been conducted by various scholars and institutions. Steinfeld et al. (1999) and Hirsch and Steinfeld (2004) carried out studies in Switzerland at PSI, while the National Renewable Energy Laboratory (NREL; Dahl et al., 2004) conducted research in the United States. Dahl et al. (2001) achieved a peak methane conversion rate of 90% at a temperature of 2,133 K using a suspended particle reactor (aerosol reactor) with a mixture of Ar/CH_4. Hirsch and Steinfeld (2004) conducted experimental investigations on methane cracking in a reactor exposed to a 2.8 MW/m^2 solar flux intensity, while being direct irradiation with a 5 KW power, at extremely low temperatures.

The effluent stream of methane contained carbon particulates that functioned as both a radiation absorber and a reactive surface for the cracking reaction. In domains characterized by temperatures ranging from 1,510 to 1,680 K, the conversion of methane was found to be between 44% and 66%. Additionally, a conversion rate of 67% was measured in a vortex-type solar reactor. The generation of H_2 and black carbon was investigated by Rodat et al. (2011) using a prototype solar reactor, as illustrated in the experimental diagram previously described.

The solar method has recently been utilized to produce hydrogen through the most commonly utilized pathway of methane cracking, as evidenced by prior research. Regrettably, this mode of production entails the consumption of nonrenewable fossil resources and the emission of polluting waste materials.

12.3.2.2 The Solar Steam Hydrocarbon Reforming

The simplified reaction can be employed to represent the process of steam gasification of solid carbonaceous materials, as well as the steam reforming of oil, NG, and other hydrocarbons (Hosseini and Wahid, 2016).

$$C_xH_y + xH_2O \rightarrow \frac{y}{2}xH_2 + xCO \tag{12.3}$$

This reaction occurs within a significantly elevated temperature between 840°C and 950°C and under a moderate pressure of approximately 20–30 bars. The production of pure hydrogen, which is required for various applications, involves a relatively intricate production process. The composition utilized in the steam reforming unit for production purposes encompasses a range of substances, including but not limited to NG, propane, naphthalene, and methane.

The principal component of NG is methane, which typically requires desulfurization prior to being fed into the steam reforming process.

$$CH_4 + H_2O \rightarrow CO + 3H_2 \tag{12.4}$$

$$CO + H_2O \rightarrow CO_2 + H_2 \tag{12.5}$$

$$CH_4 + 2H_2O \rightarrow CO_2 + 4H_2 \tag{12.6}$$

The gas sample acquired consisted primarily of hydrogen (H_2), carbon dioxide (CO_2) at a volume percentage of 16%–20%, water vapor (H_2O), and trace amounts of carbon monoxide (CO) and methane (CH_4) (Pen et al., 1996; Beghi, 1986). In general, the outcome of the two antecedent reactions exhibits an endothermic nature. The parasitic reactions involved in the hydrocarbon decomposition exhibit equal propensity and typically culminate in the production of soot. The subsequent phases entail the segregation of carbon dioxide and hydrogen, as well as the eradication of residual impurities, primarily methane and by-products of carbon dioxide. Numerous studies and initiatives have been conducted globally regarding steam reforming.

Numerous experimental studies have been conducted to investigate the feasibility of solar hydrogen production through solar steam gasification and solar steam reforming reactors (Puig-Arnavat et al., 2013). Certain systems have been utilized in the industrial sector to generate process heat by burning fossil fuels. However, this approach leads to a decrease in efficiency due to the irreversible nature of external combustion. Nevertheless, through the utilization of solar energy's heat, it is possible to not only mitigate the emission of pollutants but also enhance the fuel's calorific value. Given the relatively high energy loss associated with steam reforming and steam gasification decomposition methods in the carbon sequestration process, solar cracking may be considered as a more favorable option for hydrogen production (Jiang et al., 2016).

In a study conducted by Pen et al. (1996), noncatalytic steam propane reforming was investigated at high temperatures using a solar reactor. Also, Böhmer et al. (1991) conducted experimental research on catalytic steam methane reforming in a directly irradiated solar reactor and got a rate of H_2 gas generation that oscillated between 45% and 53%. Under identical conditions, Raner Tamme and colleagues (2001) achieved an 87% methane conversion rate. The reactor's output products include a molar proportion of hydrogen of 49.3%.

12.3.2.3 Biomass Thermochemical Transformation

Biomass is made up of all the plants that grow on the earth's surface. It is produced by the photosynthesis of H_2O, CO_2, and sunshine, which results in the formation of molecules, cellulose, lignocellulose, and lignin with the same chemical formula as $C_6H_9O_4$. It is feasible to recover stored energy in a combustible form (Goyal et al., 2008) through a transformation process that is more or less efficient on both energetic and economic grounds. This is accomplished by the processes of alcoholic fermentation, methanation, combustion, and thermochemical reactions (Boutin et al., 2002).

This approach is especially useful for valuing lignocellulosic materials like wood or straw. The gasification process of organic substances (Albertazzi et al., 2005) involves a series of interconnected operations that require the concurrent transfer of physical quantities and the regulation of reaction times and proportions at a specific moment. Gasification operations presently account for 430 million Nm^3 of the world's daily hydrogen production (Shoko et al., 2006).

The translation process can typically be represented by the following reaction:

$$C_6H_9O_4 + 2H_2O \rightarrow 6CO + 6.5H_2 \tag{12.7}$$

Additional hydrogen is obtained through the "gas shift" process in the event that we want to promote hydrogen production:

$$6CO + 6H_2O \rightarrow 6CO_2 + 6H_2 \tag{12.8}$$

Biomass has the potential to be a dependable energy source for hydrogen generation. Biomass is plentiful, renewable, and simple to utilize. Because of the photosynthesis of green plants, net CO_2 emissions are almost zero across the life cycle. The thermochemical pyrolysis and hydrogen gasification methods are economically viable and will rival the conventional way of NG reforming. Biological dark fermentation is a potential process for creating hydrogen that might be used commercially in the future. The advancement of these technologies will include the use of biomass in order to develop a sustainable hydrogen economy.

12.3.3 Electrochemical Processes

Aside from gas reformation, the process of electrolysis of water is extensively used in industrial environments for the creation of hydrogen. Water electrolysis is an electrochemical phenomenon that facilitates the separation of water into its fundamental components, hydrogen and oxygen, through the utilization of electrical energy (Padin et al., 2000). This process involves two distinct chemical reactions occurring at the anode and the cathode:

$$\text{At the anode } H_2O + \text{electricity} \rightarrow 2H^+ + \frac{1}{2}O_2 + 2e^- \tag{12.9}$$

$$\text{At the cathode } H_2O + \text{electricity} \rightarrow 2H^+ + \frac{1}{2}O_2 + 2e^- \tag{12.10}$$

$$\text{The general electrolysis reaction } H_2O + \text{electricity} \rightarrow H_2 + \frac{1}{2}O_2 \quad (12.11)$$

Electrolysis can be conducted under two distinct temperature conditions: ambient and high temperature.

The electrodes are separated spatially by an electrolytic conductor made up of ions, which allows for effective ionic species exchange between them. During the process, the system's electrical energy is transformed to chemical energy in the form of hydrogen. While water electrolysis is well understood, it is not economically feasible unless electricity is provided from a renewable source. The most important technologies in the electrolysis of water process are ceramic oxide electrolytes, polymer membranes, alkaline electrolytes, and steam electrolysis, which have current electrolytic efficiencies of 65%–85%. Proton exchange membrane (PEM) electrolyzers are one of the most practicable methods for producing high-quality hydrogen. Grigoriev et al. (2006) performed a review of PEM technology electrolyzers.

12.4 PHOTOVOLTAIC ELECTROLYSIS SYSTEMS

The primary economic benefits of hydrogen generation using electrolysis are its capacity to be scaled up and its emission-free nature when generated using renewable energy sources.

Photovoltaic hydrogen is a popular method for producing hydrogen using the energy generated by photovoltaic cells. This approach involves the use of a photovoltaic power plant to generate energy for the purpose of electrolyzing water. The use of a photovoltaic power plant system with enhanced efficiency results in an elevation of the electrolyzer's potential, hence facilitating the production of hydrogen. The use of photovoltaic power plants for hydrogen production is a sustainable and ecologically conscious approach. Consequently, there has been significant scholarly investigation into hydrogen manufacturing systems that rely on solar energy, particularly photovoltaic technology (Fereidooni et al., 2018).

Figure 12.2 illustrates a schematic representation of a hydrogen production technique utilizing solar thermal energy. This method comprises three primary components, namely the concentrating collector, electrical generator, and electrolyzer. In order to consistently supply the necessary thermal energy to a heat engine, it is possible to incorporate a heat storage unit into the system. When solar radiation is directed toward the absorber connected to the heat engine, a portion of the absorbed energy is transformed into mechanical work, while the remaining energy is wasted. The utilization of an electrical generator facilitates the conversion of mechanical energy into electrical power. Then, the electricity that is produced is employed for the purpose of electrolyzing water. The aforementioned procedure involves the relocation of components such as the generator and heat engine, resulting in a higher maintenance demand when compared to alternative solar-to-hydrogen systems. The high-temperature electrolysis (HTE) of steam is anticipated to exhibit a lower electrical energy consumption in comparison with electrolysis conducted at low temperatures, as depicted. This can be attributed to the more advantageous thermodynamic

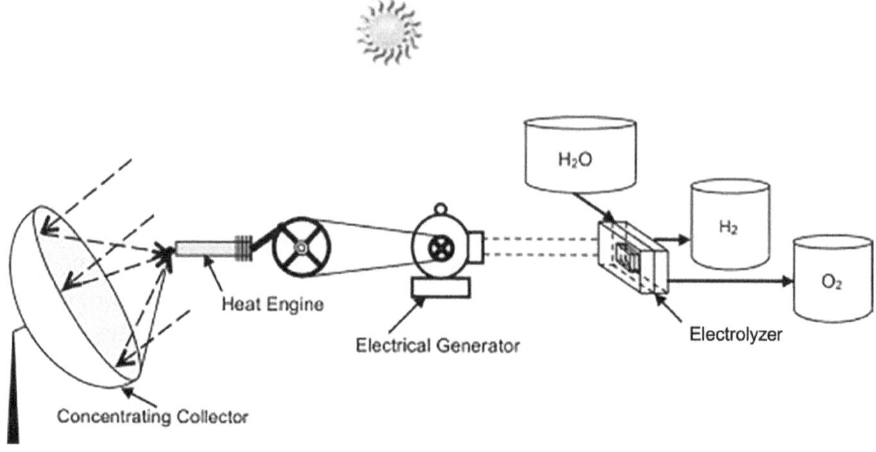

FIGURE 12.2   Schematic of solar thermal process for converting solar energy into hydrogen (Joshi et al., 2016).

and electrochemical kinetic conditions that facilitate the reaction (Dutta, 1990). The thermodynamic conditions exhibit greater favorability as indicated by the decrease in molar Gibbs energy of the reaction (ΔG) from 1.23 eV (237 kJ/mol) at room temperature to 0.95 eV at 900°C (183 kJ/mol). Conversely, the molar enthalpy of the reaction (ΔH) remains relatively constant, with a value of 1.3 eV or 249 kJ/mol at 900°C [95]. Heat may therefore contribute a considerable portion of the energy necessary for a perfect (loss-free) HTE. The utilization of solid oxide steam electrolysis (SOSE) in combination with renewable energy sources holds significant promise as a viable technology (Shin et al., 2007; Zhang et al., 2010).

The use of photovoltaic systems in conjunction with electrolyzers is a cost-effective approach for solar-to-hydrogen conversion, which is currently being researched in the fields of electrolysis and solar energy. The use of photovoltaic systems in conjunction with water electrolysis for hydrogen generation started in the early 1970s (Bilgen, 2001). Given the ecological and economic concerns at hand, a potentially cost-effective approach for solar hydrogen generation involves using a photovoltaic current source. This technique entails the direct conversion of solar energy into electricity, so offering a viable solution. Lodhi (Lodhi, 1995) developed the idea of converting solar photovoltaic cells into hydrogen and tested the viability of utilizing hydrogen as a clean fuel to replace liquid and gaseous fossil fuels. If we assume that the efficiency of photovoltaic cells is roughly 20% and the efficiency of the electrolysis system is approximately 80% (Rzayeva et al., 2001), then the overall efficiency of a solar hydrogen manufacturing plant (Yilanci et al., 2009) would be estimated to be around 16%. The exergy and energy efficiency of a photovoltaic-based solar-to-hydrogen unit ranges from 3.68% to 4.84%, according to estimates by Joshi et al. (2010). The low efficiency was ascribed to the photovoltaic system's poor efficiency (Joshi et al., 2009). While photovoltaic-based hydrogen generation is not a particularly economical technology, it is an ecologically benign one that produces no GHGs or noise pollution while in operation, and since it has no moving parts, maintenance is simple. The efficiency of photovoltaic systems is reduced at high temperatures. Consequently, it is advantageous to transfer the heat generated by the photovoltaic system to an electrolyzer, since higher temperatures may enhance the performance of the whole system. The primary photovoltaic systems function by using a single junction inside the semiconductor material to convert a fraction of the solar irradiation into electrical power. The remaining solar spectrum is then transformed into heat (Green et al., 2021). In order to enhance the efficiency of photovoltaic systems, it is essential that solar cells exhibit a broader spectral responsiveness to include a greater range of the solar spectrum. Hence, the third generation of photovoltaics encompasses several technologies such as multijunction solar cells (MJSCs), hot carrier solar cells, multiband and thermal photovoltaics, Gratzel solar cells, and the introduction of organic polymer-based solar cells.

12.5 CONCLUSION

This chapter investigated several approaches to using solar energy for the purpose of hydrogen synthesis. Although cracking and steam reforming of hydrocarbons offer additional benefits, they are nonetheless constrained by their nonrenewable fossil energy character and environmental consequences. Water electrolysis is an intriguing approach that has great promise for development with the commercialization of high-temperature electrolyzers in tropical regions with abundant solar radiation. In comparison, solar-powered electrolysis and photochemical technologies provide greater advantages for hydrogen filling stations. These technologies need fewer procedures, eliminate the requirement for external power sources, and eliminate the need for additional hydrogen distribution systems. Further material development for high-temperature processes is required, with an emphasis on high-temperature heat exchangers and membranes for solar thermal processes. Hence, the predominant solar hydrogen uses systems worldwide mostly include photovoltaic hydrogen systems designed for both transportation and stationary purposes. Future research needs to be done to harness solar energy to produce hydrogen.

REFERENCES

Agbossou, K., Chahine, R., Hamelin, J., Laurencelle, F., Anouar, A., St-Arnaud, J.-M. & Bose, T. 2001. Renewable energy systems based on hydrogen for remote applications. *Journal of Power Sources*, 96, 168–172.

Albertazzi, S., Basile, F., Brandin, J., Einvall, J., Hulteberg, C., Fornasari, G., Rosetti, V., Sanati, M., Trifirò, F. & Vaccari, A. 2005. The technical feasibility of biomass gasification for hydrogen production. *Catalysis Today*, 106, 297–300.

Balcombe, P., Speirs, J., Johnson, E., Martin, J., Brandon, N. & Hawkes, A. 2018. The carbon credentials of hydrogen gas networks and supply chains. *Renewable and Sustainable Energy Reviews*, 91, 1077–1088.

Beghi, G. 1986. A decade of research on thermochemical hydrogen at the Joint Research Centre-Ispra. *Hydrogen Systems*, 153–171.

Benemann, J. 1996. Hydrogen biotechnology: Progress and prospects. *Nature Biotechnology*, 14, 1101–1103.

Bilgen, E. 2001. Solar hydrogen from photovoltaic-electrolyzer systems. *Energy Conversion and Management*, 42, 1047–1057.

Böhmer, M., Langnickel, U. & Sanchez, M. 1991. Solar steam reforming of methane. *Solar Energy Materials*, 24, 441–448.

Boutin, O., Ferrer, M. & lédé, J. 2002. Flash pyrolysis of cellulose pellets submitted to a concentrated radiation: Experiments and modelling. *Chemical Engineering Science*, 57, 15–25.

Christopher, K. & Dimitrios, R. 2012. A review on exergy comparison of hydrogen production methods from renewable energy sources. *Energy & Environmental Science*, 5, 6640–6651.

Dahl, J. K., Buechler, K. J., Finley, R., Stanislaus, T., Weimer, A. W., Lewandowski, A., Bingham, C., Smeets, A. & Schneider, A. 2004. Rapid solar-thermal dissociation of natural gas in an aerosol flow reactor. *Energy*, 29, 715–725.

Dahl, J. K., Tamburini, J., Weimer, A. W., Lewandowski, A., Pitts, R. & Bingham, C. 2001. Solar-thermal processing of methane to produce hydrogen and syngas. *Energy & Fuels*, 15, 1227–1232.

Dutta, S. 1990. Technology assessment of advanced electrolytic hydrogen production. *International Journal of Hydrogen Energy*, 15, 379–386.

Erickson, P. & Goswami, D. Y. 2001. Hydrogen from solar energy—An overview of theory and current technological status. *IECEC—36th Intersociety Energy Conversion Engineering Conference,* Savannah, Georgia, USA: IEEE. pp. 573–580.

Fereidooni, M., Mostafaeipour, A., Kalantar, V. & Goudarzi, H. 2018. A comprehensive evaluation of hydrogen production from photovoltaic power station. *Renewable and Sustainable Energy Reviews*, 82, 415–423.

Glenk, G. & Reichelstein, S. 2019. Economics of converting renewable power to hydrogen. *Nature Energy*, 4, 216–222.

Goyal, H. B., Seal, D. & Saxena, R. C. 2008. Bio-fuels from thermochemical conversion of renewable resources: A review. *Renewable and Sustainable Energy Reviews*, 12, 504–517.

Green, M., Dunlop, E., Hohl- Ebinger, J., Yoshita, M., Kopidakis, N. & Hao, X. 2021. Solar cell efficiency tables (version 57). *Progress in photovoltaics: Research and applications*, 29, 3–15.

Grigoriev, S. A., Porembsky, V. I. & Fateev, V. N. 2006. Pure hydrogen production by PEM electrolysis for hydrogen energy. *International Journal of Hydrogen Energy*, 31, 171–175.

Hirsch, D. & Steinfeld, A. 2004. Solar hydrogen production by thermal decomposition of natural gas using a vortex-flow reactor. *International Journal of Hydrogen Energy*, 29, 47–55.

Hosseini, S. E. & Wahid, M. A. 2016. Hydrogen production from renewable and sustainable energy resources: Promising green energy carrier for clean development. *Renewable and Sustainable Energy Reviews*, 57, 850–866.

Hosseini, S. E. & Wahid, M. A. 2020. Hydrogen from solar energy, a clean energy carrier from a sustainable source of energy. *International Journal of Energy Research*, 44, 4110–4131.

Hosseini, S. E., Wahid, M. A. & Aghili, N. 2013. The scenario of greenhouse gases reduction in Malaysia. *Renewable and Sustainable Energy Reviews*, 28, 400–409.

Hwang, J. Y., Shi, S., Sun, X., Zhang, Z. & Wen, C. 2013. Electric charge and hydrogen storage. *International Journal of Energy Research*, 37, 741–745.

Jiang, D., Yang, W. & Tang, A. 2016. A refractory selective solar absorber for high performance thermochemical steam reforming. *Applied Energy*, 170, 286–292.

Joshi, A. S., Dincer, I. & Reddy, B. V. 2009. Performance analysis of photovoltaic systems: A review. *Renewable and Sustainable Energy Reviews*, 13, 1884–1897.

Joshi, A. S., Dincer, I. & Reddy, B. V. 2010. Exergetic assessment of solar hydrogen production methods. *International Journal of Hydrogen Energy*, 35, 4901–4908.

Joshi, A. S., Dincer, I. & Reddy, B. V. 2016. Effects of various parameters on energy and exergy efficiencies of a solar thermal hydrogen production system. *International Journal of Hydrogen Energy*, 41, 7997–8007.

Lodhi, M. 1995. A hybrid system of solar photovoltaic, thermal and hydrogen: A future trend. *International Journal of Hydrogen Energy*, 20, 471–484.

Matovich, E. 1978. High temperature chemical reaction processes utilizing fluid-wall reactors. Google Patents.

Miao, X., Wu, Q. & Yang, C. 2004. Fast pyrolysis of microalgae to produce renewable fuels. *Journal of Analytical and Applied Pyrolysis*, 71, 855–863.

Miyake, J., Wakayama, T., Schnackenberg, J., Arai, T. & Asada, Y. 1999. Simulation of the daily sunlight illumination pattern for bacterial photo-hydrogen production. *Journal of Bioscience and Bioengineering*, 88, 659–663.

Muradov, N. 1993. How to produce hydrogen from fossil fuels without CO_2 emission. *International Journal of Hydrogen Energy*, 18, 211–215.

Ngoh, S. K. & Njomo, D. 2012. An overview of hydrogen gas production from solar energy. *Renewable and Sustainable Energy Reviews*, 16, 6782–6792.

Ni, M., Leung, M. K., Leung, D. Y. & Sumathy, K. 2007. A review and recent developments in photocatalytic water-splitting using TiO_2 for hydrogen production. *Renewable and Sustainable Energy Reviews*, 11, 401–425.

Padin, J., Veziroglu, T. N. & Shahin, A. 2000. Hybrid solar high-temperature hydrogen production system. *International Journal of Hydrogen Energy*, 25, 295–317.

Pen, M., Gomez, J. & Fierro, J. G. A. 1996. New catalytic routes for syngas and hydrogen production. *Applied Catalysis A: General*, 144, 7–57.

Poompavai, T. & Kowsalya, M. 2019. Control and energy management strategies applied for solar photovoltaic and wind energy fed water pumping system: A review. *Renewable and Sustainable Energy Reviews*, 107, 108–122.

Puig-Arnavat, M., Tora, E., Bruno, J. & Coronas, A. 2013. State of the art on reactor designs for solar gasification of carbonaceous feedstock. *Solar Energy*, 97, 67–84.

Rodat, S., Abanades, S. & Flamant, G. 2011. Co-production of hydrogen and carbon black from solar thermal methane splitting in a tubular reactor prototype. *Solar Energy*, 85, 645–652.

Rzayeva, M., Salamov, O. & Kerimov, M. 2001. Modeling to get hydrogen and oxygen by solar water electrolysis. *International Journal of Hydrogen Energy*, 26, 195–201.

Shin, Y., Park, W., Chang, J. & Park, J. 2007. Evaluation of the high temperature electrolysis of steam to produce hydrogen. *International Journal of Hydrogen Energy*, 32, 1486–1491.

Shoko, E., Mclellan, B., Dicks, A. L. & Da Costa, J. C. D. 2006. Hydrogen from coal: Production and utilisation technologies. *International Journal of Coal Geology*, 65, 213–222.

Short, W., Blair, N. & Heimiller, D. 2005. Modeling the Market Potential of Hydrogen from Wind and Competing Sources. National Renewable Energy Lab, Golden, CO (US).

Steinberg, M. 1986. The direct use of natural gas for conversion of carbonaceous raw materials to fuels and chemical feedstocks. *International Journal of Hydrogen Energy*, 11, 715–720.

Steinfeld, A., Sanders, S. & Palumbo, R. 1999. Design aspects of solar thermochemical engineering-a case study: Two-step water-splitting cycle using the Fe3O4/FeO redox system. *Solar Energy*, 65, 43–53.

Sulaiman, A., Inambao, F. & Bright, G. 2015. Solar-hydrogen energy: An effective combination of two alternative energy sources that can meet the energy requirements of mobile robots. In *Robot Intelligence Technology and Applications 3: Results from the 3rd International Conference on Robot Intelligence Technology and Applications*, pp. 819–832. Fuzhou, China: Springer.

Tamme, R., Buck, R., Epstein, M., Fisher, U. & Sugarmen, C. 2001. Solar upgrading of fuels for generation of electricity. *Journal of Solar Energy Engineering*, 123, 160–163.

Wang, Z., Roberts, R., Naterer, G. & Gabriel, K. 2012. Comparison of thermochemical, electrolytic, photoelectrolytic and photochemical solar-to-hydrogen production technologies. *International Journal of Hydrogen Energy*, 37, 16287–16301.

Yi, H., Yan, M., Huang, D., Zeng, G., Lai, C., Li, M., Huo, X., Qin, L., Liu, S. & Liu, X. 2019. Synergistic effect of artificial enzyme and 2D nano-structured Bi2WO6 for eco-friendly and efficient biomimetic photocatalysis. *Applied Catalysis B: Environmental*, 250, 52–62.

Yilanci, A., Dincer, I. & Ozturk, H. K. 2009. A review on solar-hydrogen/fuel cell hybrid energy systems for stationary applications. Progress *in Energy and Combustion Science*, 35, 231–244.

Zhang, X.-R., Yamaguchi, H. & Cao, Y. 2010. Hydrogen production from solar energy powered supercritical cycle using carbon dioxide. *International Journal of Hydrogen Energy*, 35, 4925–4932.

13 Application of Hydropower Energy in Hydrogen Production
Tide, Wave, Water Flow, and Fall

*Zeynab Farzizada, Agshin Garashli,
and Rasoul Moradi*

13.1 INTRODUCTION

Our reliance on fossil fuels is being reduced because of the use of renewable energy sources. Renewable energy sources like solar, wind, and hydropower have grown in significance due to the rising global energy demand and the negative environmental effects of the use of fossil fuels. The force of water flowing through turbines, which turns the kinetic energy of the water into electricity, produces hydropower energy. When compared to nonrenewable energy sources, hydropower energy is considered to be one of the most environmentally friendly renewable energy sources. Hydropower plants, however, have the potential to negatively affect aquatic ecosystems, as well as to lead to the displacement of communities living in affected areas. The planning and implementation phases of hydropower plant development must carefully evaluate the effects that the plants will have on the environment and society.

A notable and promising renewable energy source in the fight against climate change and the advancement of sustainable development, is hydropower energy despite existing obstacles.

Since hydropower is a clean, renewable energy source that does not emit any harmful pollutants or greenhouse gases, it is a good replacement for fossil fuels (Abolhosseini et al. 2014). This chapter will discuss producing hydrogen using hydropower energy. Hydropower energy, a renewable and sustainable energy source that has existed for centuries, is produced by the movement of water.

Tidal energy is a type of hydroelectric power that generates electricity from the kinetic energy of ocean tides. When tidal power plants are used, the energy from the tides can be used to split water molecules into hydrogen and oxygen, a process known as electrolysis, which produces hydrogen. Due to its great degree of predictability and stability, tidal energy is a reliable fuel for large-scale hydrogen generation.

DOI: 10.1201/9781003382270-17

Another type of hydroelectric power source called wave energy uses the kinetic energy of ocean waves to produce electricity. Wave power plants, like tidal power plants, can also produce hydrogen by converting wave energy into electrical power suitable for electrolysis. Wave energy is a highly scalable energy source due to its abundance in many regions worldwide, but is less predictable than tidal energy.

The most common kind of hydropower that uses the energy of moving water in rivers or streams to produce electricity is called water flow, also referred to as hydro-electric power. Hydroelectric power plants have the capability of producing hydrogen through the utilization of water flow energy to generate electricity for the aim of electrolysis. Hydroelectric power is a great option for producing hydrogen because of its excellent predictability and dependability.

Another hydropower source is waterfalls, which can be used to produce hydrogen by utilizing the energy of the falling water to power hydroelectric power plants. The electricity produced can then be used for electrolysis. In certain places, waterfalls can be a useful source of electricity even though they are not as accessible as other hydropower sources.

Hydropower technologies that can be used to generate hydrogen primarily fall into four categories: tide, wave, water flow, and hydropower. As a dependable, scalable, and low-carbon electricity source, hydropower energy has a significant potential for producing hydrogen (Rourke et al. 2009). Utilizing clean and sustainable electricity from renewable sources, like hydropower, is one way to produce the significant amount of hydrogen needed for the electrolysis of water. Reliability, scalability, and low-carbon footprint are a few benefits of producing hydrogen with hydropower energy.

It is possible to produce electricity using any hydropower technology, and this electricity can then be used to electrolyze water to separate its molecules into hydrogen and oxygen (Armaroli and Balzani 2011).

Using hydropower energy to produce hydrogen is an efficient method to produce fuel that is both ecologically beneficial and renewable. Hydropower sources such as water flow, falls, waves, and tides provide the potential for hydrogen production via electrolysis. These energy sources are highly consistent and reliable, which makes them ideal for producing hydrogen on a big scale. However, hydropower plant construction and maintenance can be expensive, and it is important to take the effects of the plant on the environment into account.

13.2 TIDAL ENERGY

Tidal energy is a type of hydropower that converts the energy obtained from the natural rise and fall of tides into electrical energy. Tides result from the combined effects of gravitational kinematic forces due to the motions of the moon, the sun, and the rotation of the earth. Tidal is believed to be one of the major renewable energy sources besides solar, wind, and hydro energy and is only implemented along coastlines and can be produced using a variety of technologies (Kazim 2010). As a result, coastal countries are in a unique situation both to prevent global warming and have energy security using clean energy (Chowdhury et al. 2021). Compared to other renewable energy sources, tidal power has the distinct advantage of being very predictable (Pelc and Fujita 2002).

Tidal fences, tidal turbines, and tidal barrages are the main methods used to produce tidal energy.

13.2.1 TIDAL FENCES

Tidal fences are a form of turbine that functions as a big turnstile. They produce power by harnessing the strong tidal currents close to the coast. With its barrage-like architecture, which includes horizontal tidal stream turbines as a bridge over a watercourse, the project might combine infrastructure and electricity generation (Waters and Aggidis 2016). Tidal fences consist of turbines that span an entire channel where tidal flow creates relatively fast currents. Tidal fences could be installed wherever tidal flows and topographic constraints create predictable currents of 2 m/s or greater (Pelc and Fujita 2002).

13.2.3 TIDAL TURBINES

Tidal current turbines use the kinetic energy of moving water to generate electricity. Tidal current technology is similar to wind energy technology (Rourke et al. 2009). However, the operating conditions differ significantly. Tidal current turbines must be able to generate electricity during both flood and ebb tides, as well as withstand structural loads when not producing electricity (Rourke et al. 2010). Most common methods of tidal current energy extraction are horizontal-axis tidal current turbines (HATCTs) and vertical-axis tidal current turbines (VATCTs).

HATCT: The HATCT has grown in popularity as a commonly used device for harnessing tidal energy around the world, owing to its exceptional hydrodynamic stability and excellent coefficient for energy gain. A typical representation of a HATCT is shown in Figure 13.1. The rotation axis is in parallel to the current stream direction, which helps to capture more energy from current stream (Qian et al. 2019).

VATCT: The VATCT is a device that uses the kinetic energy of tidal currents to generate electricity. Figure 13.2 illustrates that regardless of the incoming flow direction, the rotating direction of the VATCT stays constant. Furthermore, positioning the turbine's generator on top of the waterline shaft can significantly reduce the complexity and cost of the underwater seal (Qian et al. 2019).

Figure 13.3 depicts a comparison between the HATCTs and VATCTs. Although the VATCT has many advantages over the HATCT, few of them are currently being used commercially because of the lower level of research and shorter time spent in development. The HATCT is easier to design because of its straightforward structure and uses the blade element momentum theory (BEMT) to analyze the pouches, in contrast to the VATCT, which has more intricate movements (Qin et al. 2022).

The current obstacles limiting the development of tidal current turbines include the complexity of their installation and maintenance, the logistics of electricity transmission, loading conditions, and possible environmental consequences (Rourke et al. 2010).

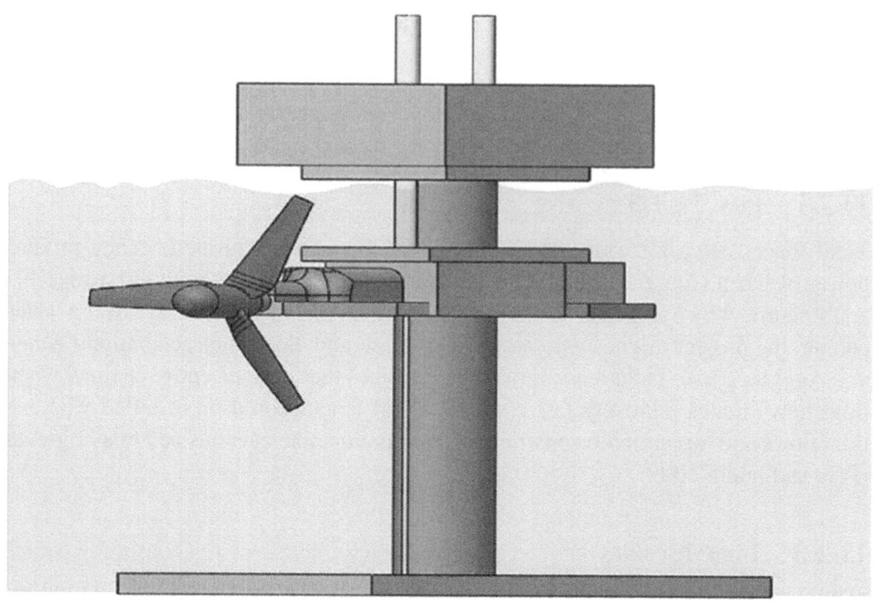

FIGURE 13.1 HATCT. Adapted from Qian et al. 2019.

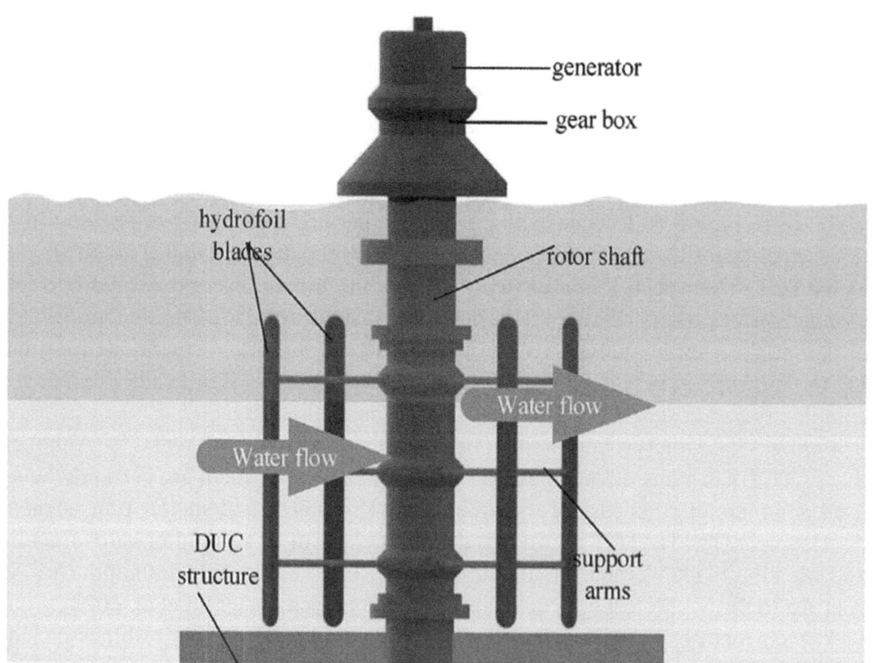

FIGURE 13.2 VATCT (Qian et al. 2019).

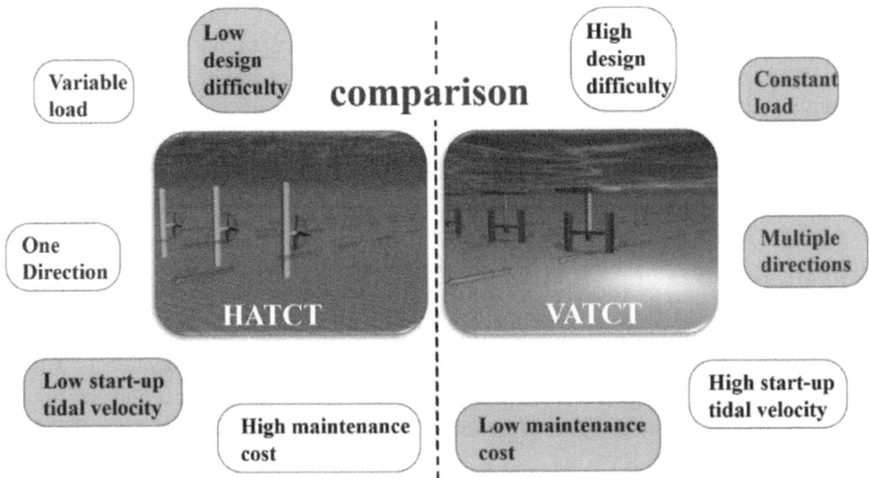

FIGURE 13.3   HATCT and VATCT comparison (Qin et al. 2022).

13.2.3 TIDAL BARRAGES

The most effective way to harness the power of the tides is through tidal barrages, which are typically built as a dam across an estuary or bay with a tidal range of at least 5 m (Rourke et al. 2010). Using this method, the gravitational potential energy that results from the difference in height between high and low tides is used to compress air or turn turbines to create electricity, which is then used to power the dam. Production of energy from tidal barrages is a proven and stable technology, with several tidal locales across the world considered suitable for development (Rourke et al. 2010).

The primary obstacles limiting the development of tidal barrage systems are the high costs of construction and the possible environmental effects, as there are no major technological limitations that need to be addressed. Building a tidal barrier requires a large quantity of materials to withstand the powerful forces produced by the dammed water (Rourke et al. 2010).

Moreover, tidal energy is a far cleaner alternative for other renewable energy sources and may provide enormous amounts of electricity. The placement of the turbine in relation to the tidal currents determines the possibility for high power production. Tidal energy has a negligible ecological impact because it is a sustainable and pollution-free energy source (Chowdhury et al. 2021). Tidal energy as a renewable energy source has the potential to spur advancements in social, technical, industrial, and governmental fields as well as energy security and sustainable development (Midilli 2016). Although it has the potential to generate electricity through tidal energy, there are several challenges related to its development and dissemination of information about ocean energy resources. While increasing affordability can drive innovation, incentives, and cost reductions to other sustainable energy sources, it can also result in an underestimation of the existing potential of tidal energy (Chowdhury et al. 2021).

The production of hydrogen could benefit greatly from the use of tidal power because it is a renewable energy source that emits no greenhouse gases directly and has a minimal environmental impact. Notably, tidal power can be used to generate electricity or lessen the load on an already-operating power plant. Tidal power can be used to focus its electrical output toward the electrolysis of saltwater, which will help produce hydrogen (Dincer and Joshi 2013).

In conclusion, tidal energy has enormous potential to serve as a reliable and sustainable source of energy for the production of hydrogen. Reducing reliance on fossil fuels and greenhouse gas emissions can be accomplished by using tidal energy to produce hydrogen. Additionally, because of how predictable and consistent it is, tidal energy is a highly reliable and consistent source of energy that can be used to produce hydrogen on a large scale (Almoghayer et al. 2022).

But there are still some issues that need to be resolved as tidal energy technology for producing hydrogen is still in its infancy. The high initial investment costs of tidal energy projects, the requirement for efficient energy storage options, and the potential effects of tidal energy systems on marine ecosystems are some of the difficulties (Pelc and Fujita 2002).

However, with continued technological development and increased funding for tidal energy research and development, these difficulties are likely to be outweighed by the potential advantages of tidal energy for hydrogen production. The long-term provision of a sustainable and dependable source of hydrogen, which can power a variety of applications, including transportation, heating, and power generation, can be greatly aided by tidal energy (Orhan et al. 2012).

13.3 WAVE ENERGY

Generally, a wave can be described as a disturbance that travels through a medium from one location to another location (Chenari et al., 2014).

Ocean wave power is a renewable energy source that harnesses the kinetic energy of the waves generated by the wind's effect on the surface of the ocean (Falcão 2010; Blackledge et al. 2013). Figure 13.4 shows how waves are created. The differential heating of the earth by the sun leads to the creation of air circulation, commonly known as wind, which in turn blows over the surface of the oceans. As the wind blows across open bodies of water, it creates a pushing force on the surface water particles, causing them to travel in a circular and vertical path. Waves emerge in the water as a result of this rolling motion (Chenari et al. 2014; Blackledge et al. 2013).

As wind blows over open bodies of water, it exerts a propulsive force on water particles on the surface, causing them to move in a vertical and circular path. As a result of this rolling action, waves are created in the water (Chenari et al. 2014).

In the revolutionary era of Napoleonic Paris, a visionary man named Monsieur Girard dared to imagine harnessing the untamed power of the sea. In 1799, he obtained a patent for a machine designed with his son, a device that would capture the raw energy of ocean waves and transform it into mechanical power. Their invention promised to revolutionize the industry, from powering pumps and sawmills to driving entire cities. Girard's bold vision was an early glimpse into the vast potential of wave energy and a reminder of humanity's unquenchable thirst for innovation (Falcão 2010; Fitterman 2021).

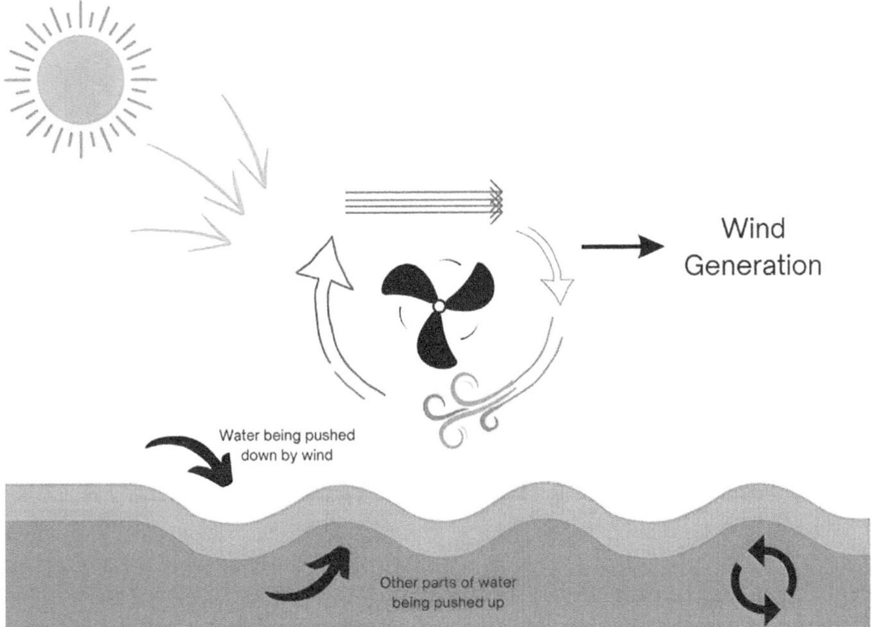

FIGURE 13.4 Wave creation steps (Chenari et al. 2014).

The history of wave power dates back several decades, with numerous reviews and publications documenting its evolution. Among the pioneering works is McCormick's book from 1981, followed by the books of Shaw, Charlier, Justus, Ross, Brooke, and Cruz. In addition, reports commissioned by the UK Department of Energy and the European Thematic Network on Wave Energy offer valuable insights into the state-of-the-art in the field. Yoshio Masuda is known as the creator of modern wave energy technology. Thanks to his extensive studies on the subject starting in the 1940s, the first navigation buoy powered by wave energy was developed in Japan. Eventually, Masuda proposed the creation of a larger apparatus known as the Kaimei barge, which functioned as a testing ground for several oscillating water columns powered by various kinds of air turbines. Despite limited success, the Kaimei testing program played a crucial role in advancing the theoretical understanding of wave energy absorption (Falcão 2010).

13.3.1 Potential

Ocean wave energy is gaining significant attention and recognition as a significant and promising source of energy in numerous countries (Falcão 2010). Wave energy has been identified to have the greatest potential for generating energy among all renewable sources, making it a promising area of study. According to Figure 13.5, the worldwide wave energy potential is substantial enough in order to provide electricity for the entire planet (Chenari et al. 2014).

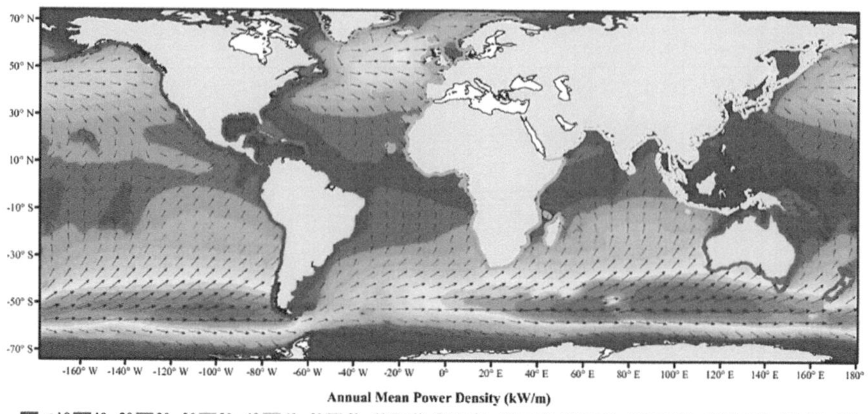

FIGURE 13.5 Wave energy potential (Gunn and Stock-Williams 2012).

There is an estimated theoretical yearly energy potential of up to 2.64 trillion kilowatt hours in waves off the coast of the United States alone. This amounts to 64% of utility-scale energy generation in the United States in 2021. The coasts of the USA, Europe, Japan, and New Zealand are seen to be viable places for wave-based energy production (US Energy Information Administration, n.d.). The World Energy Council has reported that harnessing wave power could produce around 2 terawatts of energy, which is double the current worldwide electricity production (Blackledge et al. 2013). Despite this, the challenge lies in effectively extracting such a vast amount of energy. Particularly during the oil crisis of the 1970s, there has been a rise in the studies and research on wave energy. This association between changes in wave energy technology and oil prices is depicted in Figure 13.6.

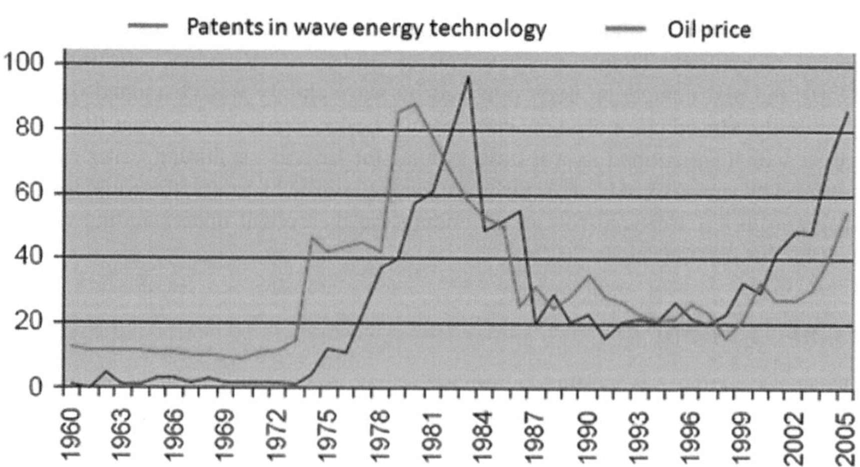

FIGURE 13.6 Relationship between oil price and advancements in wave energy technology (Blackledge et al. 2013).

Wave energy is affected by various factors such as bottom effects and sheltering from land. In favorable offshore settings, power per unit length (kW/m) ranges from 20 to 70 kW/m which is a common unit of measurement for wave energy. Because there are minimal seasonal changes, the southern beaches of South America, Africa, and Australia show promise for wave energy (Falcão 2010).

The energy contained in waves can be determined using the equation (13.1):

$$P = 0.49H^2T \tag{13.1}$$

where "H" is the wave height, "T" is the wave period, and "P" is the power stored in the wave. Other variables, such as wind speed, the period duration on the surface of water, and the degree of wind–water interaction, affect how much energy is contained in waves (Chenari et al. 2014).

Wave energy is harnessed through the use of wave energy converters (WECs) which transform wave energy into electrical power. This is done by transforming wave energy into energy in working fluids, which is then converted into mechanical energy using a motor or turbine. Lastly, a generator that produces electrical energy is rotated using the mechanical energy. Unlike other renewable energy sources, wave energy devices are not all the same in their design. The fact that there are over a thousand patented wave energy conversion systems in Europe, North America, and Japan illustrates the variety of methods employed in this industry (Chenari et al. 2014).

There are many designs for these devices, and they can be categorized based on their location and type. Wave energy converters generally installed at three main locations: onshore, nearshore, and offshore. Onshore devices have the advantage of being closer to the utility network, but they have less energy to exploit due to the shallow waters close to the shoreline. Nearshore devices typically have a direct connection to the bottom and have a considerable power potential, although their movement may be restricted by the direction of the waves. Although the installation and maintenance of offshore wave energy devices are more difficult and the expenses of connecting them to the utility grid are greater, these devices are located in ocean areas with large water depth, beyond the nearshore regions, and have the potential to harness the most energy. The energy from waves is converted into mechanical energy using a motor or a turbine, which then rotates a generator to produce electrical energy (Blackledge et al. 2013). By distinguishing between the power that might be generated from these three sites, respectively, and depths, Figures 13.7 and 13.8 illustrate the variations regarding the location of the WECs.

Wave energy converters (WECs) are devices used to convert the energy from ocean waves into electrical energy. Various types of WECs are currently being developed and tested, classified based on their design and location. The early research on WECs focused on floating devices and resulted in three classifications: point absorbers, terminators, and attenuators.

Point absorbers are tiny devices that produce energy by utilizing the water's vertical displacement at a particular location. These devices might be fixed underwater constructions or buoyant structures that float on the water's surface. They do not depend on the direction of the waves and instead utilize changes in water pressure to generate electricity. Figure 13.9 displays "The Wavebob device" which is an example of a point absorber device.

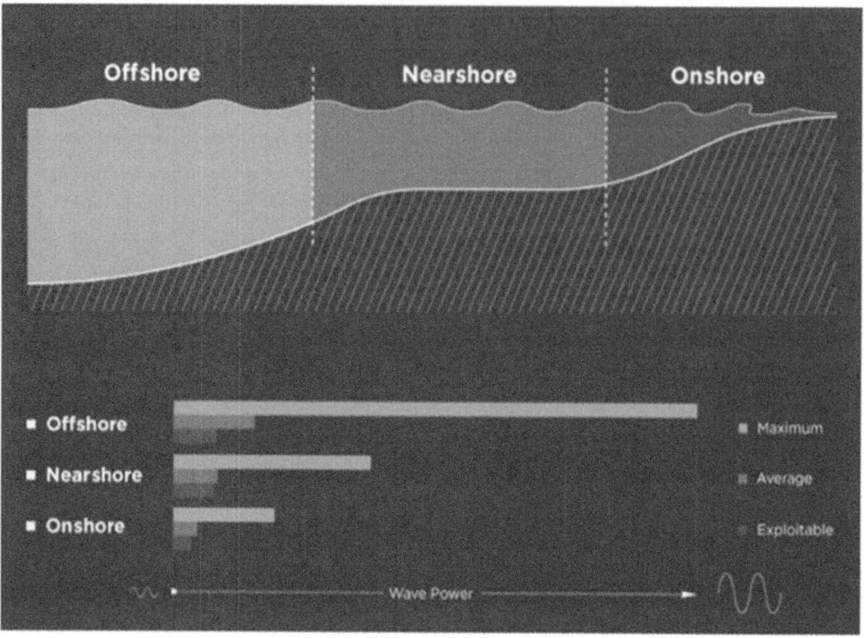

FIGURE 13.7 Relationship between the location and power output of wave energy converters (Blackledge et al. 2013).

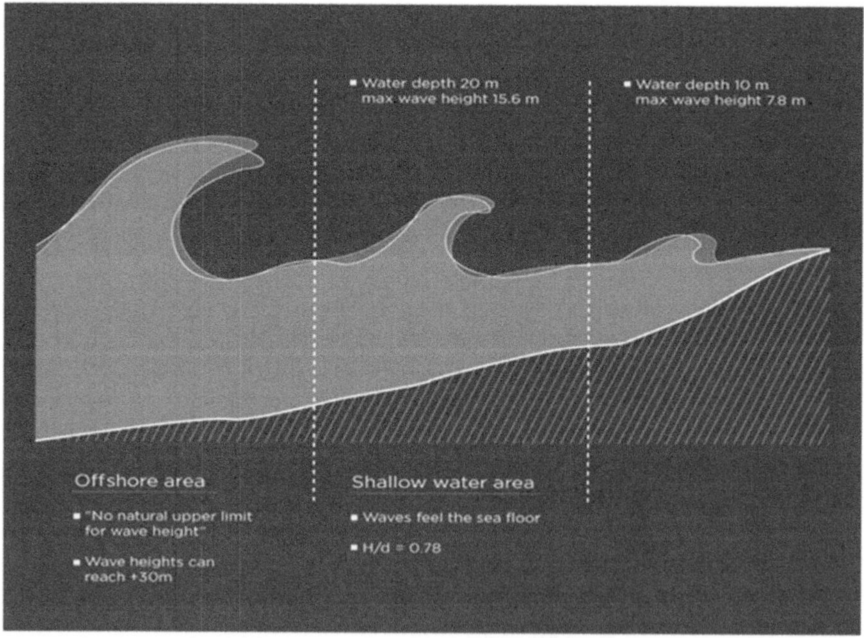

FIGURE 13.8 Classification of wave energy locations with respect to depths (Blackledge et al. 2013).

FIGURE 13.9 The Wavebob device (Tarrant and Meskell 2016).

According to Blackledge et al. (2013), attenuators are long, floating devices with several segments that are oriented to match the direction of wave propagation. The different heights and forces of oncoming waves cause a flexing motion, which is connected to hydraulic pumps or other converters. Because attenuators have a smaller surface area and interact with the oncoming waves less, the effects of hydrodynamic forces including drag, slamming, and inertia are lessened. Due to this design, these forces are less likely to seriously harm offshore devices (Chenari et al. 2014; Blackledge et al. 2013). Figure 13.10 displays "The Pelamis" which is an example of an attenuator device.

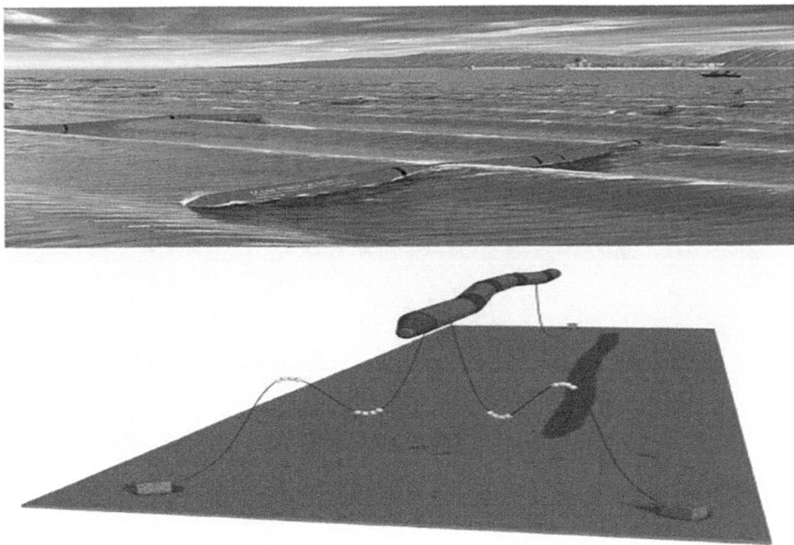

FIGURE 13.10 Pelamis by Ocean Power Delivery Ltd. (Blackledge et al. 2013).

Devices known as oscillating wave surge converters are made up of a deflector that is perpendicular to the path of the incoming waves and hinged at a fixed location. By oscillating back and forth, the deflector generates power by utilizing the waves' horizontal velocity. This method, which produces electricity, is based on the conversion of wave energy (Blackledge et al. 2013). Figure 13.11 displays "The Oyster device" which is an example of oscillating wave surge converters.

The design of wave energy converters has seen some creative advancements recently, which has led to the introduction of new models, such as the overtopping device and the oscillating water column, broadening the classification spectrum.

Overtopping devices have reservoirs that use the force of impinging waves to raise the level of saltwater over the surrounding sea level, producing electricity. The water is then recycled back into the sea after being used to power turbines, which generate electricity. According to Blackledge et al. (2013), overtopping devices can be created and tested for both onshore and floating offshore applications. The working principle of "The Wavedragon device," an example of an overtopping device, is shown in Figure 13.12.

FIGURE 13.11   The Oyster device by Aquamarine Power (Blackledge et al. 2013).

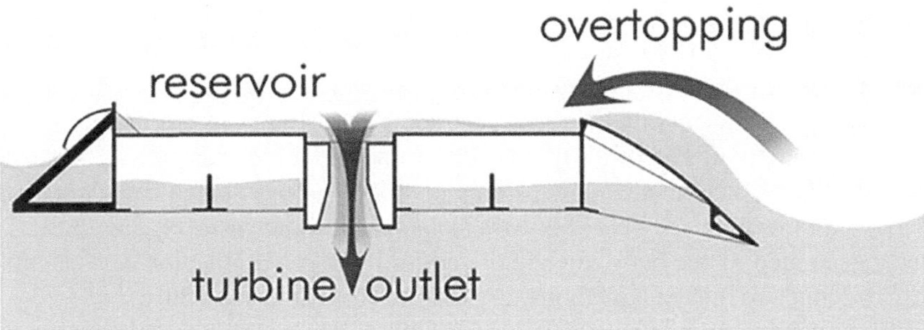

FIGURE 13.12   The Wavedragon device (Blackledge et al. 2013).

Wave power is a promising renewable energy source that has several advantages and disadvantages. One of the main advantages of wave energy is its relatively smaller environmental impact compared to other renewable energy sources. Wave energy requires no construction of access roads and no land use. However, wave power can have some impacts on marine life, including fish and marine mammals (Chenari et al. 2014).

Another advantage of wave energy is its high energy density. It carries the highest energy density of all renewable energy sources, which makes it a highly efficient and cost-effective energy source. One of the advantages of wave energy converters over wind turbines is their reduced visual intrusiveness, as well as their low requirements for infrastructure and land use. In addition, wave energy offers a favorable chance to reduce carbon emissions and significantly boost energy output.

Wave energy also has a few unique features that set it apart from other renewable energy sources. For instance, the demand for power in areas with moderate climates is reflected in the natural fluctuation of wave energy in accordance with seasonal fluctuations. Wave energy devices may produce power for longer periods than other renewable energy sources like solar and wind energy, which increases their dependability as a source of energy (Blackledge et al. 2013).

Unpredictable fluctuations at various time intervals are one of the negative aspects of wave power. The consistent and dependable generation of power may face difficulties as a result of this variability. Because it depends on variables like the condition of the sea and seasonal variations, wave power can be highly variable and unpredictable. This variability makes it challenging to rely on wave power as a consistent and reliable source of energy (Falcão 2010).

In conclusion, wave power is a promising renewable energy source that has many advantages, including its high energy density, negligible environmental impact, and potential for significant contributions to energy production. However, its variability remains a major disadvantage that needs to be addressed before it can become a reliable source of energy.

The production of hydrogen through wave energy has been a topic of interest in the renewable energy industry for the past few decades. With the increasing demand for sustainable energy sources and the need to reduce greenhouse gas emissions, hydrogen production by using wave energy is a promising solution. The main method of producing hydrogen by wave energy is through the process of electrolysis, where the energy produced from wave power is used to power the chemical reaction. This chapter section will discuss the different ways in which hydrogen production can be achieved by wave energy and will examine some case studies in this field.

The University of Palermo's Energy Department (DEIM) has created a buoyant point absorber system that can produce up to 80 kW of power when it converts wave energy into electrical power. This new technology is shown in Figure 13.13. The power generated by the buoy can be utilized for hydrogen production, and scientists are exploring two different methods for doing so (Franzitta et al. 2016).

The first strategy uses an isolated setup with an electrolyzer, a desalination system, and a hydrogen storage system located on the internal buoy. Hydrogen is produced and stored using this technology exclusively using wave energy. The second approach is to connect multiple DEIM point absorber buoys to a central plant on the mainland, where the desalination and hydrogen-producing systems are located. The energy produced by each buoy's power unit is transferred to the central plant for hydrogen production (Franzitta et al. 2016).

Another system proposed for hydrogen production by wave power is composed of four main blocks: wave converters, a reverse osmosis (RO) plant, an electrolysis unit, and a compression unit (Franzitta et al. 2016). The wave converters are the primary energy source, which provides electricity to power the reverse osmosis plant, where seawater is desalinated to obtain water for the electrolysis unit. The electrolysis unit then converts water into hydrogen and oxygen using proton exchange membrane (PEM) electrolyzers. Following production, the hydrogen is compressed before being transported by barges or ships to the final consumers.

The four main elements of this concept, as shown in the process diagram, are represented in Figure 13.14: the compression unit, the wave converters, the electrolysis unit, and the reverse osmosis plant.

FIGURE 13.13   Graphic representation of DEIM point absorber (Franzitta et al. 2016).

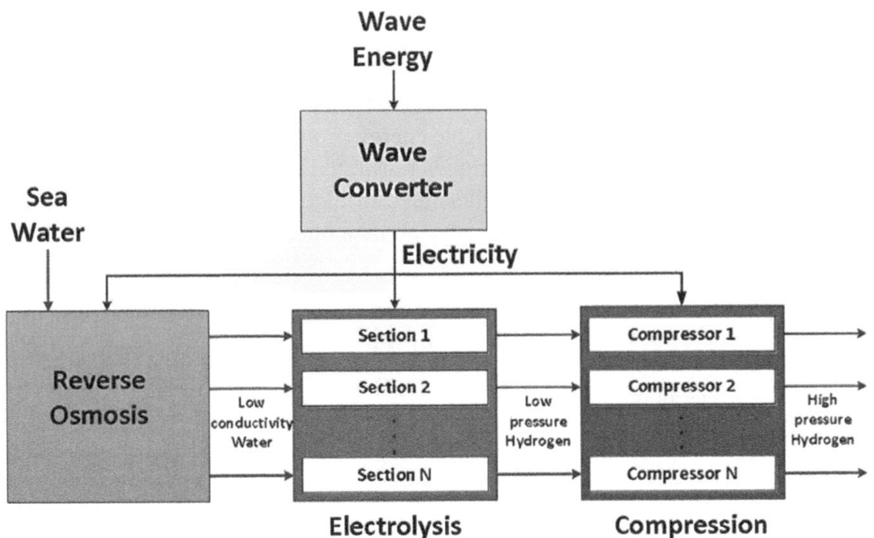

FIGURE 13.14   Process diagram (Serna and Tadeo 2014).

PEM electrolysis is a well-established and commercially available technique. As long as the electricity used comes from renewable sources, it can operate intermittently and produce significant amounts of hydrogen without emitting greenhouse gases. There can be no interruptions to the process (Serna and Tadeo 2014). Seawater desalination can be used to achieve the specific inlet water quality needed for PEM electrolyzers to function properly over the long term. Because of its ability to be automated and its flexibility in meeting different flow and conductivity requirements, reverse osmosis is a viable desalination technique for marine applications (Serna and Tadeo 2014).

In the context of an acidic proton exchange membrane cell, it is hypothesized that the process of liquid water splitting proceeds in accordance with the following half-cell reactions:

Equation (13.2): Half-cell reaction of liquid water splitting.

$$\text{anode: } H_2O \text{ (liq)} \quad \tfrac{1}{2}O_2(g) + 2H^+ + 2e^-$$

$$\text{cathode: } 2H + 2e^- \quad H_2(g) \quad \text{full reaction: } H_2O \text{ (liq)} \quad H_2(g) + \tfrac{1}{2}O_2(g)$$

The electrolysis application that is being suggested is based on a modified version of a design that was originally meant to be used for automated desalination of water using renewable energy sources in remote areas. The system has an array of batteries as part of its provisional storage mechanism to control energy production and consumption. An important consideration for the plant is the energy cost per cubic meter of water produced; this cost is roughly 2 kWh/m^3. This is a crucial factor to take into account when assessing the economic viability and efficiency of the plant.

Simulation testing of the proposed system shows that, according to buoy measurements, it operates satisfactorily under standard parameters for 165 days. The quantity of hydrogen produced by the system indicates how much energy is available at any given time, and it uses this information to modify production. The simulation results demonstrate the feasibility and potential of hydrogen production by wave power, providing a promising avenue for sustainable energy production in the future. The proposed system's structure for desalination is illustrated in Figure 13.15.

The hydrogen production rate is illustrated in Figure 13.16, which demonstrates that the suggested controller adjusts the production rate to correspond with the amount of available energy.

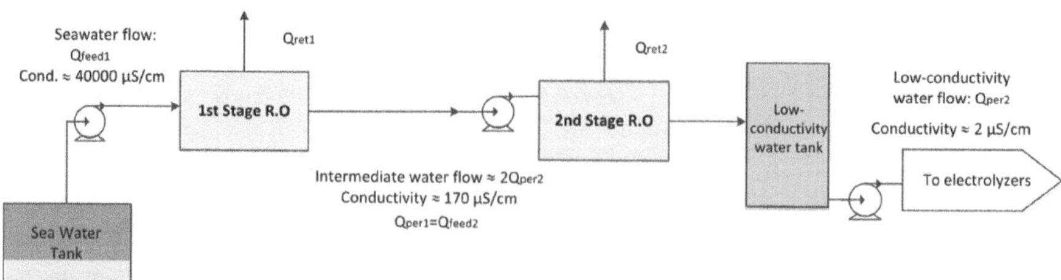

FIGURE 13.15   Proposed structure of the desalination system (Serna and Tadeo 2014).

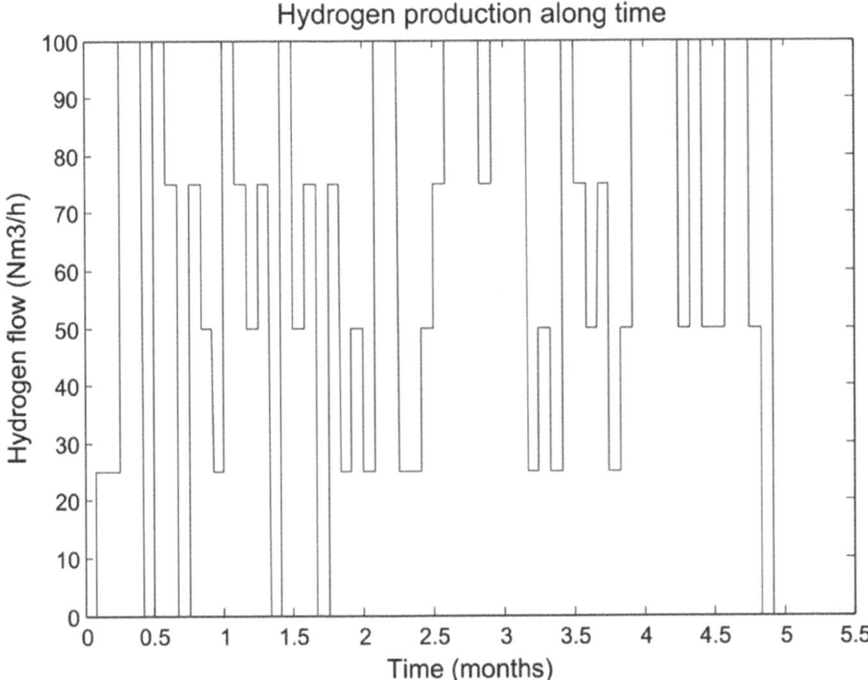

FIGURE 13.16 Hydrogen production over time (Serna and Tadeo 2014).

Another prototype has been developed that can produce both electric energy and hydrogen gas from the power of sea waves. The prototype is composed of four main components which are all installed inside a buoy. These components include an electric power generator, an AC-DC converter, an electrolyzer, and a hydrogen storage unit. A permanent magnet linear generator (PMLG) is the device in the prototype that uses the buoy's oscillatory motion to generate electrical energy. The PMLG is composed of a stator and a translator and has six electrical phases. The generator produces voltages that are directly proportional to the translator's speed. An AC-DC converter is used to convert the AC power generated by the PMLG into DC power which is then used to power the electrolyzer. The electrolyzer separates water molecules into oxygen and hydrogen gas, and the produced hydrogen gas is stored in a hydrogen storage tank. Overall, this prototype demonstrates the potential for generating electricity and hydrogen gas from sea waves (Colucci et al. 2015).

13.4 WATER FLOW ENERGY

The energy derived from the motion of water, such as tidal currents, streams, and rivers, is considered renewable and is referred to as hydrokinetic energy or water flow energy. Unlike traditional hydropower that requires the construction of dams and reservoirs, water flow energy systems use various devices like turbines, oscillating water columns, and piezoelectric energy harvesters to capture the kinetic energy of water flow and convert it into electricity (Center for Climate and Energy Solutions [C2ES], n.d.).

For centuries, hydroelectric power has been employed. The Romans built turbines to grind grains into flour and bread, not to generate electricity. Before the Industrial Revolution, water mills were widely used to generate energy for a range of tasks, such as cutting wood, grinding grains, and producing steel by starting hot fires. Water flow must be controlled in order to extract energy from moving water. In order to do this, sizable reservoirs are built often accomplished through the construction of a dam that creates an artificial lake or reservoir. Next, water is directed through the tunnels of the dam, turning turbines to produce energy (National Geographic Education, n.d.).

Large, fast-flowing rivers like the Columbia River produce the most hydroelectricity. The Bonneville Dam is one of the many dams on the Columbia River. It generates more than a million watts of electricity a year, which is enough to power a large number of residential and commercial properties. With the increasing global demand for clean and sustainable energy, the development and deployment of water flow energy systems have become essential for meeting the energy needs of communities and reducing greenhouse gas emissions (National Geographic Education, n.d.).

An alternative form of hydropower that does not require building dams is hydrokinetic electricity, which includes tidal and wave power. This technology, which generates power from currents or waves, is presently undergoing various stages of research, development, and implementation. Low-head hydropower has been used for hydrokinetic energy for more than a century and is currently a commercially viable source. It has a power output range of 1–250 kW.

Water flow energy is a renewable energy source that can be harnessed in various ways to generate electricity. One way scientists have developed to harness this energy is through a device that uses a special material called piezoelectric film. This material can convert the energy from water flow into electrical energy.

Researchers have built and tested prototypes of the device, finding that the size of the piezoelectric film and the type of material used can affect how much electricity the device can produce. The device has the potential to produce a small amount of electricity when exposed to water flow.

Another example of a mesoscale system for recovering energy from water flow is an electromagnetic energy harvester that runs on a turbine driven by the facility's turbo expanders and throttling valves losing water pressure. This type of harvester can output a power of 150 W. Additionally, there have been investigations into converting the kinetic energy of fluid motion into mechanical energy of a resonator to harvest energy from fluid motion. As an example, a piezoelectric membrane placed behind a bluff body can harvest energy from the von Kármán vortex street (Wang and Liu 2011).

Previous research has suggested that a small graphene sheet immersed in hydrochloric acid can produce voltage from water flow. However, new research has found that larger graphene samples do not produce measurable voltage when isolated from the solution due to the adsorption of H_3O+ cations onto the graphene by strong covalent bonds. When both the graphene and its metal electrodes are exposed to the solution, water flow can induce voltages, but the energy harvested is mainly from the exposed electrodes and not the graphene itself (Yin et al. 2012).

In conclusion, there are various ways to harness the energy from water flow to generate electricity. The development of devices that use piezoelectric film, electromagnetic harvesters, and resonators can help make the use of water flow energy more efficient and accessible.

Hydrogen production from water flow energy is a promising method of producing clean and renewable energy. The process utilizes the kinetic energy of water flow to drive an electrolysis reaction that separates water molecules into hydrogen and oxygen gases. The primary aspect of this process is using a mechanical device, such as a turbine, to transform the flow of water's kinetic energy into electrical energy, which is then used to run the electrolysis apparatus.

The electrolysis unit consists of two electrodes, usually made of platinum or other metals, immersed in water. These electrodes are connected to a source of electrical power, such as a battery or solar panel. Water molecules split into individual oxygen and hydrogen gas molecules through a process known as electrolysis when electrodes are charged with an electrical current. The oxygen atom in the water molecule is separated from its two hydrogen atoms during this process, and the latter are then released as gas. The hydrogen gas is collected and stored for later use as a fuel source.

This method of producing hydrogen is considered to be a green and environmentally friendly way to produce energy because it uses the natural flow of water to produce electricity. Unlike traditional methods of hydrogen production, which rely on nonrenewable fossil fuels, this method produces zero greenhouse gas emissions and has minimal impact on the environment.

While the technology for hydrogen production from water flow energy is currently developing, in its early phases, it holds great promise for the future of clean energy. As more research is conducted and advances are made in the technology, it is likely that this method of hydrogen production will become an increasingly important source of clean energy.

13.5 WATERFALL ENERGY

Utilizing the energy of falling water to create electricity is known as waterfall energy. In the modern era, hydroelectric power has become a major source of energy for the entire world. For thousands of years, sawing and grinding grains have been accomplished with the help of hydroelectric energy. For nations wishing to switch to low-carbon energy sources, waterfall energy is a desirable alternative because it is safe, dependable, and sustainable. In this method of generating energy, falling water is utilized to turn turbines that produce electricity, which can then be used to produce hydrogen (Lee and Wang 2008).

The energy released by falling water can be captured and used to generate electricity. In the simplest terms possible, water falls due to gravity, which transforms kinetic energy into mechanical energy, which can then be transformed into a form of usable electrical energy (Chakraborty et al. 2015).

The amount of power that can be produced once the facility is complete is calculated prior to the development of a hydroelectric power site. The volume of water released—referred to as discharge—and the height at which the waterfalls—referred to as head—determine the actual output of energy at a dam. Therefore, a certain amount of energy will be produced by a given volume of water falling at a given distance. The type of turbine to be used depends on the head and discharge at the power site as well as the desired generator rotational speed (Chakraborty et al. 2015). Within the water's flow, turbines capture the kinetic energy and transform it into mechanical energy. High-speed turbine rotation as a result drives a generator, which transforms mechanical energy into electrical energy. The water flow and vertical distance (also referred to as "head") that the water falls through determine how much hydroelectric power is produced (Shaw 2017).

The gross head (H) is the vertical distance that represents the maximum drop in water between the upstream and downstream water levels. Because of the energy lost during the movement of water to and from the turbine, the actual head that the turbine experiences is slightly lower than the gross head. The term "net head" describes the lowered head (Uhunmwangho and Okedu 2009). In general, more electricity can be produced with higher head and flow rates.

The flow rate (Q) is the amount of water flowing through a river in a unit of time; it is commonly expressed in cubic meters per second (m^3/sec). The flow rate for small-scale systems can also be expressed in liters per second, where one cubic meter is equal to 1,000 L/sec (Uhunmwangho and Okedu 2009).

While waterfalls in their natural state can be used to generate hydroelectricity, most hydroelectric plants get their water from waterfalls that were built by humans. Dams are used to create these waterfalls by limiting a river's natural flow into channels that feed water to turbines. Because the water flow is controlled, a higher pressure is produced in a smaller area, increasing the efficiency of energy collection (Sciencing, n.d.).

The water used to generate electricity is redirected back to its original source. If the body of water being used does not dry up, hydroelectric power can be produced continuously for all of time. It is a completely clean source of energy as well. Once installed, the power plants convert fuel without producing any waste by-products. Dams that have been built have the ability to close their gates and store water for use when demand for electricity is higher (Chakraborty et al. 2015).

Using waterfall energy as a source of electricity has a lot of advantages. It is a renewable source of energy, so unlike fossil fuels, it will not run out. It is also a clean source of energy because it does not emit greenhouse gases or pollute the air. Third, energy from hydropower can be stored so that it can be used later. Unlike solar or wind energy, which depends on the weather, a waterfall power plant can produce electricity on demand like other hydropower plants.

However, it also has some significant disadvantages, like a detrimental effect on the environment: Dam construction can have a significant negative impact on the environment, for example, displacing wildlife and changing water flow patterns. Hydroelectric dams can also be expensive to maintain, even though operating costs are low, particularly if the dam is situated in a remote area. Failure risk is yet another drawback to take into account: The Banqiao Dam collapse in China in 1975, which resulted in the deaths of an estimated 171,000 people, shows that dam failures can have disastrous effects (Muda et al. 2021).

The use of waterfall energy for the production of hydrogen has a number of advantages, including being clean and renewable and having the potential to lower greenhouse gas emissions. It has been studied that using hydroelectric power instead of other methods to produce hydrogen could significantly lower carbon dioxide emissions (Koroneos et al. 2004).

In conclusion, waterfall energy is a promising renewable energy source that has been used by people for a very long time. Hydroelectric power plants can use the energy produced by falling water to generate electricity for homes and businesses (Bagher et al. 2015). While this form of energy does have some drawbacks, such as the need for a reliable water supply and the potential for environmental harm to aquatic habitats, it also has many benefits, such as lower greenhouse gas emissions and lower operating costs than fossil fuels. Waterfall energy has the potential to be a significant contributor to our transition to a more sustainable energy future by helping us meet our energy needs while lowering our carbon footprint.

13.6 ENVIRONMENTAL IMPACTS

Significant environmental effects result from the use of hydropower energy in hydrogen generation through the utilization of tide, wave, water flow, and fall.

Tidal barrage construction, however, can have significant ecological effects, especially on bird feeding grounds, in coastal estuaries or bays. Barrages and tidal fences must be built in coastal areas with specific environmental requirements, and their construction has the ability to change tidal processes over vast areas of land (Serna and Tadeo 2014; Bagher et al. 2015).

The barrage structure may have a design life of more than 100 years, making the long-term effects important even though the life of the turbines may be limited. Although data are scarce, there is little evidence that using underwater tidal stream energy devices will result in higher rates of mortality for pelagic species like fish and marine mammals (Frid et al. 2012).

The alteration of the natural water flow and its impacts on aquatic ecosystems are possible consequences of using waterfalls to generate electricity. Fish, amphibians, and other aquatic invertebrates frequently have important habitats in waterfalls and the streams that support them. The capacity of these organisms to locate food, shelter, and reproduction can be impacted by changing the water's flow. According to a study by Vannote et al., variations in water flow can have an ecosystem-wide cascading impact that alters species composition and reduces biodiversity (Robin et al. 1980).

Wave energy collectors have the ability to change the habitats of the water column and seabed, changing the wave environment for some distances from the installation. The magnitude and location of the development will determine the extent of the effects, and most effects would be reversible if an installation were removed. Overall, even though using renewable energy has many advantages, it is still important to plan carefully and take into account any possible effects if you want to minimize ecological harm (Pelc and Fujita 2002; Frid et al. 2012).

Despite the potential environmental impact, wet renewables are becoming more economical and offshore energy resources are expected to become a significant source of renewable energy in the near future (Frid et al. 2012).

13.7 ECONOMIC CONSIDERATIONS

Hydropower has drawn a lot of attention as the world moves toward more renewable and sustainable energy sources because of its potential to produce hydrogen through a variety of processes, including tides, waves, water flow, and falls. The economic benefits of this use of hydropower energy for hydrogen production are encouraging and merit further investigation, from lowering carbon emissions to addressing issues with energy security.

With low-head and run-of-the-river types, hydropower plants can produce base-load energy by using about 60% of the river flow. Overflowing the spillway, the final 40% of the flow is wasted. Hydrogen production from hydroelectric power plants has been found to be economically and technically viable. By using a gas engine to power this hydrogen, the 40% of extra flow that would otherwise be wasted can be used further (Tarnay 1985).

Tidal energy costs vary by region and the technology being used. But as tidal energy costs have decreased over time, it is now a desirable option for producing hydrogen. The International Energy Agency (IEA) discovered that tidal energy was a practical way to produce hydrogen at a low cost (IEA, 2019).

Due to its variability and the technology needed to capture it, wave energy generally costs more than tidal energy. Although wave energy is expected to become more affordable in the future, it does have the advantage of being accessible in many places (López et al. 2013).

Another hydroelectric source that can be used to produce hydrogen is waterfall energy or run-of-the-river hydropower. By redirecting a part of a river's flow through a turbine, waterfall energy can be captured. Depending on where it is located and the technology being used, waterfall energy costs can vary (Von Sperling 2012).

The use of hydropower energy for the cost-effective production of hydrogen from tide, wave, water flow, and fall has shown promising results. Each type of hydropower energy has a different price based on the location and technology employed, but they all have the potential to generate green hydrogen at a competitive price.

Table 13.1 provides a comparative analysis of four hydropower sources, namely tidal, wave, flow, and fall, in terms of their energy source, hydrogen production potential, advantages, disadvantages, and key challenges.

13.8 CONCLUSION

In conclusion, the use of hydropower energy in the production of hydrogen is a hopeful answer for issues with both energy and the environment. Since it can be used to produce hydrogen, a clean and renewable fuel, hydropower has long been a popular way to produce energy.

Tidal energy is a stable and dependable hydropower source that can be used to run turbines to create hydrogen. Another potential hydropower source is wave energy, which can be converted into hydrogen using wave energy converters. Additionally, the water flow can be used to create energy, which can then be used to electrolyze water to create hydrogen. Finally, hydroelectric dams can capture the power of falling water to make electricity that can be used to create hydrogen.

TABLE 13.1

Comparison of Hydropower Sources in Terms of Hydrogen Production

Hydropower Source	Tidal (Reference)	Wave (Reference)	Water Flow (Reference)	Fall (Reference)
Energy source	Tides	Waves	Flowing water	Gravity
Hydrogen production potential	High	High	Medium	High
Advantages	Predictable, consistent (Almoghayer et al. 2022)	Reliable, smaller environmental impact compared to other renewables (Chenari et al. 2014; Blackledge et al. 2013)	Widely available (Bagher et al. 2015)	Significantly lowers carbon dioxide emissions (Koroneos et al. 2004)
Disadvantages	Potential negative effect on marine ecosystems (Pelc and Fujita 2002)	Unpredictable (Falcão 2010)	Limited power output	Detrimental effect on the environment
Key challenges	High initial investment costs (Pelc and Fujita 2002)	Development of cost-effective technologies (Kassem and Majid 2016)	Achieving a high level of efficiency while avoiding negative environmental impacts	Expensive to maintain (Muda et al. 2021)

Hydrogen generated through hydropower is a clean, renewable energy source that has a variety of uses, including power production and transportation. However, the use of hydropower energy in the production of hydrogen represents a potential answer to our energy and environmental problems, so it should be investigated and developed further.

REFERENCES

Abolhosseini, S., Heshmati, A., & Altmann, J. (2014). A review of renewable energy supply and energy efficiency technologies. *Renewable and Sustainable Energy Reviews*, 39, 748–764.

Almoghayer, M. A., Woolf, D. K., Kerr, S., & Davies, G. (2022). Integration of tidal energy into an island energy system—A case study of Orkney islands. *Energy*, 242, 122547.

Armaroli, N., & Balzani, V. (2011). The hydrogen issue. *ChemSusChem*, 4(1), 21–36.

Bagher, A. M., Vahid, M., Mohsen, M., & Parvin, D. (2015). Hydroelectric energy advantages and disadvantages. *American Journal of Energy Science*, 2, 17–20.

Blackledge, J., Coyle, E., Kearney, D., McGuirk, R., & Norton, B. (2013). Estimation of wave energy from wind velocity. *Engineering Letters*, 21, 158–170.

Center for Climate and Energy Solutions (C2ES). (2024). *Renewable Energy*. https://www.c2es.org/

Chakraborty, S., Ahmad, M. I., Guin, A., Mukherjee, S., Goswami, R. P., & Roy, R. (2015). Hydropower: Its amazing potential—A theoretical perspective. *International Journal of Civil and Environmental Engineering*, 2, 56–60.

Chenari, B., Saadatian, S. S., & Ferreira, A. D. (2014). Wave energy systems: An overview of different wave energy converters and recommendation for future improvements. In *INTED 2014 Proceedings*, Valencia, Spain, pp. 6266–6272.

Chowdhury, M. S., Rahman, K. S., Selvanathan, V., Nuthammachot, N., Suklueng, M., Mostafaeipour, A., Habib, A., Akhtaruzzaman, M., Amin, N., & Techato, K. (2021). Current trends and prospects of tidal energy technology. *Environment, Development and Sustainability*, 23(1), 8179–8194.

Colucci, A., Boscaino, V., Cipriani, G., Curto, D., Di Dio, V., Franzitta, V., Trapanese, M., & Viola, A. (2015). An inertial system for the production of electricity and hydrogen from sea wave energy. In *OCEANS 2015—MTS/IEEE*, Washington, DC, pp. 1–10.

Dincer, I., & Joshi, A. S. (2013). *Hydrogen Production Methods*. SpringerBr, pp. 7–20.

Falcão, A. F. D. O. (2010). Wave energy utilization: A review of the technologies. *Renewable Sustainable Energy Reviews*, 14, 899–918

Fitterman, D. (2021). *Capturing the Ocean's Energy*. E360.

Franzitta, V., Curto, D., Rao, D., & Viola, A. (2016). Hydrogen production from sea wave for alternative energy vehicles for public transport in Trapani (Italy). *Energies*, 9, 850.

Frid, C., Andonegi, E., Depestele, J., Judd, A., Rihan, D., Rogers, S. I., & Kenchington, E. (2012). The environmental interactions of tidal and wave energy generation devices. *Environmental Impact Assessment Review*, 32, 133–139.

Gunn, K., & Stock-Williams, C. (2012). Quantifying the global wave power resource. *Renewable Energy*, 44(1), 296–304.

International Energy Agency. (2019). *The Future of Hydrogen*. Paris, France: International Energy Agency.

Kassem, A. H., & Majid, M. R. (2016). Ocean energy technology development—The challenge of this millennium. *Journal of Ocean Engineering and Technology*, 30(4), 225–231.

Kazim, A. (2010). Strategy for a sustainable development in the UAE through hydrogen energy. *Renewable Energy*, 35(10), 2257–2269.

Koroneos, C., Dompros, A., Roumbas, G., & Moussiopoulos, N. (2004). Life cycle assessment of hydrogen fuel production processes. *International Journal of Hydrogen Energy*, 29, 1443–1450.

Lee, D. J., & Wang, L. (2008). Small-signal stability analysis of an autonomous hybrid renewable energy power generation/energy storage system part I: Time-domain simulations. *IEEE Transactions on Energy Conversion*, 23, 311–320.

López, I., Andreu, J., Ceballos, S., De Alegría, I. M., & Kortabarria, I. (2013). Review of wave energy technologies and the necessary power-equipment. *Renewable and Sustainable Energy Reviews*, 27, 413–423.

Midilli, A. (2016). Green hydrogen energy system: A policy on reducing petroleum-based global unrest. *International Journal of Global Warming*, 10(3), 354–370.

Muda, R. S., Abdullah, F. S., Hussain, M. R., Tukiman, I., & Sani, J. A. (2021). Flood mapping for dam break assessment in Cameron Highlands. *International Journal of Advanced Engineering*, 3, 46–63.

National Geographic Education. (n.d.). *Hydroelectric Energy: Power from Running Water*. National Geographic Society.

Orhan, M. F., Dincer, I., Rosen, M. A., & Kanoglu, M. (2012). Integrated hydrogen production options based on renewable and nuclear energy sources. *Renewable Sustainable Energy Reviews*, 16, 6059–6082.

Pelc, R., & Fujita, R. M. (2002). Renewable energy from the ocean. *Marine Policy*, 26(6), 471–479.

Qian, P., Feng, B., Liu, H., Tian, X., Si, Y., & Zhang, D. (2019). Review on configuration and control methods of tidal current turbines. *Renewable and Sustainable Energy Reviews*, 108, 125–139.

Qin, Z., Tang, X., Wu, Y. T., & Lyu, S. K. (2022). Advancement of tidal current generation technology in recent years: A review. *Energies*, 15(4), 8042.

Robin, L., Vannote, G., Wayne Minshall, Kenneth W., Cummins, James R., Sedell, and Colbert E. Cushing. (1980). The river continuum concept. Canadian Journal of Fisheries and Aquatic Sciences, 37, 130–137.

Rourke, F. O., Boyle, F., & Reynolds, A. (2009). Renewable energy resources and technologies applicable to Ireland. *Renewable and Sustainable Energy Reviews*, 13(8), 1975–1984.

Rourke, F. O., Boyle, F., & Reynolds, A. (2010). Tidal energy update 2009. *Applied Energy*, 87(2), 398–409.

Sciencing. (n.d.). *How Does a Waterfall Generate Power?*

Serna, Á., & Tadeo, F. (2014). Offshore hydrogen production from wave energy. *International Journal of Hydrogen Energy*, 39, 1549–1557.

Shaw, R. (2017). International Energy Agency. (2010). CO_2 Emissions from Fuel Combustion.

Tarnay, D. S. (1985). Hydrogen production at hydro-power plants. *International Journal of Hydrogen Energy*, 10, 577–584.

Tarrant, K., & Meskell, C. (2016). Investigation on parametrically excited motions of point absorbers in regular waves. *Ocean Engineering*, 111, 67–81.

U.S. Energy Information Administration. (2022). Wave Power. Energy Explained.https://www.eia.gov/energyexplained/hydropower/wave-power.php

Uhunmwangho, R., & Okedu, E. K. (2009). Small hydropower for sustainable development. *Pacific Journal of Science and Technology*, 10, 535–543.

Von Sperling, E. (2012). Hydropower in Brazil: Overview of positive and negative environmental aspects. *Energy Procedia*, 18, 110–118.

Wang, D. A., & Liu, N. Z. (2011). A shear mode piezoelectric energy harvester based on a pressurized water flow. *Sensors and Actuators A: Physical*, 167, 449–458.

Waters, S., & Aggidis, G. (2016). Tidal range technologies and state of the art in review. *Renewable and Sustainable Energy Reviews*, 59, 514–529.

Yin, J., Zhang, Z., Li, X., Zhou, J., & Guo, W. (2012). Harvesting energy from water flow over graphene? *Nano Letters*, 12, 1736–1741.

14 Application of Geothermal Energy in Hydrogen Production

Tolga Ayzit, Alper Özmumcu, and Alper Baba

14.1 INTRODUCTION

Hydrogen is the most abundant element in the universe. However, it is found only in small amounts in the Earth's atmosphere. Essentially all of the hydrogen on Earth is in the form of water or hydrocarbons. Because the hydrogen molecule has the highest energy density relative to its mass, it can travel very long distances without weighing much. For decades, liquid hydrogen has been the main fuel of all space programmes. It was hydrogen that put mankind on the moon. To this day, space exploration relies on hydrogen propulsion and fuel cells to power space rockets.

Hydrogen has a higher buoyancy than air. It therefore disperses and dilutes rapidly, provided that adequate ventilation has been provided to prevent gas accumulation. In the event of a leak, the lower explosion limit of hydrogen is similar to that of natural gas but better than that of gasoline. Without any additives, it is odourless and has a less visible flame than gasoline. When hydrogen is burned with air, it produces water, which absorbs energy as it vaporises. Therefore, the radiant heat of a hydrogen flame is less than that of natural gas. The heat of combustion of hydrogen is also higher than that of natural gas due to its higher energy density per mass. However, hydrogen tends to leak due to its small molecular size. All these properties result in a net energy loss comparable to that of natural gas.

Hydrogen can provide a number of benefits for future energy systems. Hydrogen can serve as storage for intermittent renewables or provide grid services. It can replace natural gas in industrial heating processes that are otherwise difficult to decarbonise. Blending hydrogen into the gas grid can reduce the carbon content of the gas as an interim solution. In freight transport by sea and land, hydrogen enables heavier loads over longer distances compared to electric batteries. Fuel cell electric vehicles on the road take no more time to refuel than gasoline. At the point of use, hydrogen fuel cells emit almost no particulates that affect local air quality.

Hydrogen is an energy vector like electricity. It is a means of storing, transmitting, and using energy obtained from other sources such as geothermal energy. It is not an energy source that can be extracted like fossil fuels. It must be produced like electricity. Recently, interest in hydrogen has increased worldwide. The last hydrogen hype was around 20 years ago which lasted until new or unconventional fossil fuel sources were developed and oil prices fell again. This was because the focus at the time was

DOI: 10.1201/9781003382270-18

only on promoting hydrogen mobility applications. The current interest in hydrogen may be different from the false starts of the past. For the first time, a comprehensive system change is being proposed, driven by the significant cost reductions in renewable energy sources.

Because hydrogen must be produced at an energy cost, there is still room for improvement in overall cycle efficiency. The hydrogen infrastructure is not yet mature. Because of its historical legacy, there is a negative public perception regarding safety issues that need to be addressed through careful planning. Hydrogen storage technologies are not yet practical due to the further energy costs associated with pressurisation. The cost of electrolysers and fuel cells must be reduced along their experience curves so that they can be produced in greater quantities in the future. This requires market demand and supply infrastructure, neither of which currently exists. Public targets and subsidies may be needed to encourage these. However, regulatory and policy mechanisms for hydrogen are still emerging around the world.

Today, almost all hydrogen demand comes from petroleum refineries and fertiliser plants. Petroleum refineries account for slightly more than half of this demand. Almost all of the hydrogen for these two industries is produced either from natural gas or, to a lesser extent, from coal. Hydrogen production by methane steam reforming is much less expensive than coal gasification. As for carbon emissions, hydrogen production is currently part of the problem, not the solution. Global emissions are significant and are expected to increase. Therefore, a strong momentum is building up for clean hydrogen production as many countries have recently published their hydrogen road maps and announced their public financing programmes (World Energy Council, 2022).

Geothermal energy is a technically and economically viable option for producing clean hydrogen through electrolysis in countries where geothermal resources are abundant (Mahmoud et al., 2021). Due to its high capacity factors as a baseload supply, geothermal energy can increase the utilisation rates of the electrolysers, thereby reducing their levelised cost of hydrogen production in comparison with intermittent alternatives such as wind and solar. In addition, geothermal fluids can be used to preheat feed water, achieving incremental gains in the overall electrolysis process (Kanoglu et al., 2010). The cost of clean hydrogen production can be reduced at higher feed water temperatures when the electrolysers are coupled with geothermal energy sources (Balta et al., 2009). This section will outline the relevant aspects of geothermal energy with a focus on hybrid geothermal systems as well as the future of the hydrogen economy, covering its supply, demand, and infrastructure.

14.2 GEOTHERMAL ENERGY AND ITS TYPES

The primary source of Earth's natural heat is the radioactive decay of potassium, thorium, and uranium, with heat also generated during the evolution of Earth's core and mantle (Turcotte & Schubert, 2014). Since the Earth's formation, it has been gradually cooling, allowing heat from the planet's interior to reach the surface, where it is released (Jaupart et al., 2007).

Conduction, convection, and radiation are the three modes of heat transfer. The net result of molecular collisions, where kinetic energy is transferred from one molecule to another when they collide, is conductive heat transfer through a medium. Thus, heat is transferred through a medium from a hot region to a cool region and vice versa. Convective heat transfer, on the other hand, involves the movement of a substance (Turcotte & Schubert, 2014). The way heat is distributed in the continental crust is also influenced by convective processes. These convective processes are found primarily within active zones of magmatism and metamorphism and have the potential to create regions of significant heat flow (Ledru & Guillou Frottier, 2010).

Deep sedimentary basins or zones of faulting containing rocks with significant primary or secondary porosity and permeability may allow easy movement of fluids in the crust on a local scale. These types of systems may allow fluid circulation to entrain heat from several kilometres below the surface, amplifying temperature anomalies (Manning & Ingebritsen, 1999).

Geothermal energy extraction techniques and depth affect how much of the energy is usable. Globally, the upper part of the lithosphere warms by 20–30°C/km of depth; this rate of warming is much higher in most known geothermal regions of the Earth. In order to transport the energy to the surface for heat generation, a fluid/steam must usually be present. It can be difficult to find and exploit geothermal resources. This is especially true for high-temperature resources used to generate electricity (Lund, 2023). These resources are often limited to regions of the Earth that have experienced recent volcanic activity, are located on plate boundaries, or are in hot zones of the Earth's crust (Figure 14.1). Heat flow and plate tectonic conditions are closely linked (Elders & Moore, 2016). The development of high-temperature geothermal systems is favoured by the presence of volcanic activity, high heat flow, and stress regimes near divergent and convergent plate boundaries (Moeck, 2014). The geologic setting is directly related to the formation of the geothermal resource (Table 14.1).

There are two ways of using geothermal energy: direct and indirect. Geothermal energy is used indirectly to generate electricity in the binary cycle, dry steam, and

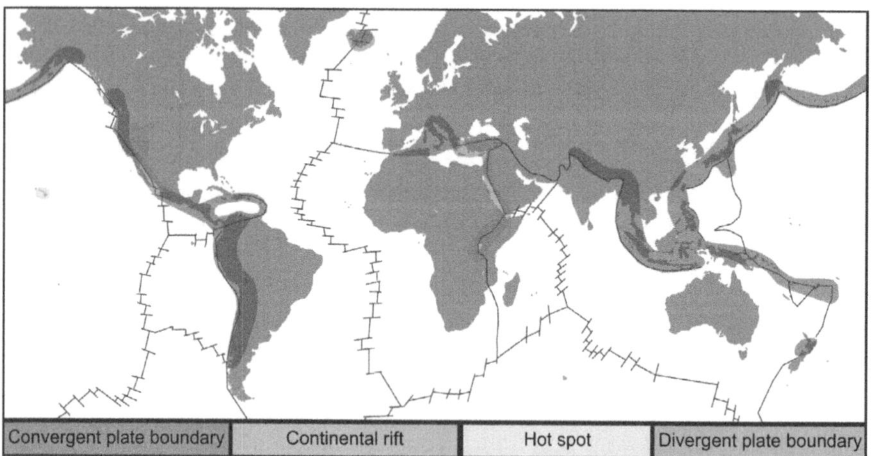

FIGURE 14.1 Areas with high heat flow and geothermal activity (Elders & Moore, 2016; Moeck, 2014).

TABLE 14.1

Geological Characteristics of Geothermal Systems in Different Environments

Geothermal Systems	Tectonic Settings	Reservoir Rock Type	Example
Andesitic volcano	Volcanic arcs at the subduction zone	Andesitic lava flows, volcanoclastic deposit	Indonesia, Philippines
Silicic volcanic	Continental environment	Rhyolite to dacite, pyroclastic deposit	Taupo volcanic zone, New Zealand
Sedimentary hosted	Rift basin	Recent sedimentary deposit	Salton Trough, USA
Extensional tectonics	Continental rift zones and back-arc basin	Mesozoic/palaeozoic basement rocks (sedimentary, metamorphic, intrusive)	Basin and Range, USA
Oceanic spreading centre	Ocean spreading centre	Basalt	Iceland

Source: Adapted from Elders and Moore (2016).

flash steam power plants (Islam et al., 2022). Direct uses include geothermal heat pumps, space heating, greenhouses and ground heating, aquaculture pond, flume heating, agricultural product drying, industrial process heating, bathing and swimming, snowmelt, space cooling, and geothermal tourist parks. Annual use increased rapidly between 1995 and 2020 (Lund & Toth, 2021). In 2020, about 97 terawatt hours (TWh) of geothermal electricity was generated, while about 128 TWh (462 petajoules, PJ) of direct geothermal heat use was recorded (REN21, 2022).

The numerous geothermal technologies provide access to the inherent heat of the Earth (Figure 14.2). With these geothermal technologies, it is possible to extract enough energy from a portion of the Earth's crust to sustain all life on Earth for hundreds of thousands of years. To access and use this energy, a variety of great technologies have been developed over the years (Eavor Geothermal Technology, 2023).

Geo-exchange/residential heat pump systems are very common and can be placed almost anywhere. Soil, groundwater, or surface water is used as heat sources in the geo-exchange/residential heat pump system to produce heating and cooling (Lund, 2023). Given that it offers a high coefficient of performance (COP) of up to four for the indirect heating system and five for the direct heating system, it is widely considered one of the most exceptional heating and cooling systems in both residential and commercial buildings. The fundamental advantage of employing a geo-exchange/residential heat pump is that the subsurface temperature is not affected by the significant changes that the air is susceptible to. Geo-exchange/residential heat pumps are less expensive to operate than conventional heating and cooling systems and do not require big cooling towers (Mustafa Omer, 2008).

Traditional geothermal energy for direct heat is rarely used because it is assumed that there is a high population density in areas where suitable geological conditions exist. Traditional geothermal energy for direct heat generally uses a hot, naturally permeable sedimentary aquifer, although igneous rocks may occasionally be used. Exploration work is required to determine the ideal geological conditions for

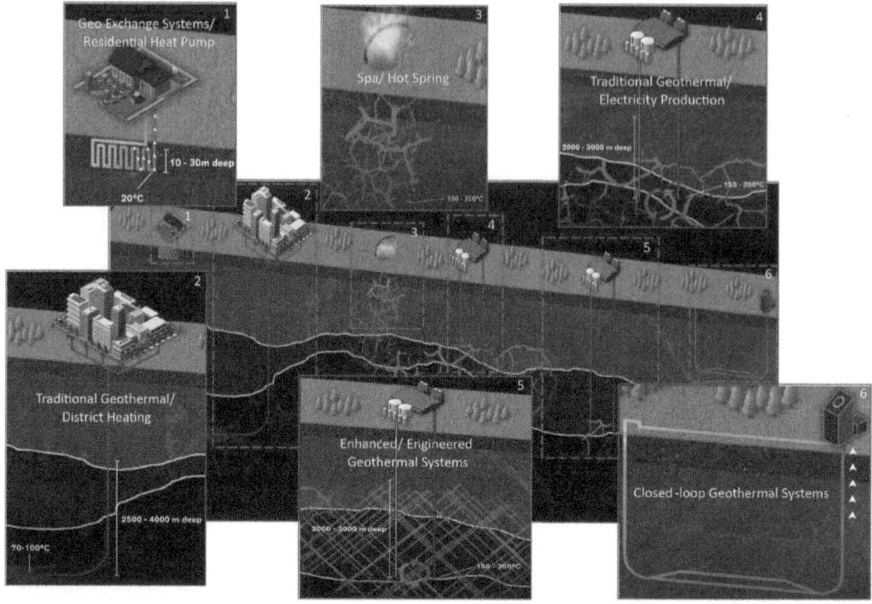

FIGURE 14.2 Geothermal technologies in use. Adapted from Eavor Geothermal Technology, 2023.

heat production, and then, wells are drilled to extract brine. The depth of production wells in hydrothermal and geopressured geothermal reservoirs, where thermofluids are naturally trapped in porous hot rock, ranges from 300 to more than 3,000 m, depending on the reservoir characteristics (Moya et al., 2018).

Spas and hot springs are extremely rare examples of hot water rising to the surface along fault lines from deep down. Fractures, which are produced by these fault zones, are natural passageways that let the water rise from the depths below. Permeability, or the capacity of fluid to move via pore gaps or rock fractures, is the connectedness of these channels (Gallois, 2007).

Traditional geothermal power generation occurs where the temperature and flow from reservoirs are naturally suitable for power generation. In this case, the available resource is present when there is sufficient near-surface water, adequate heat transfers to the system, and sufficient subsurface porosity (Livescu et al., 2023). Wells are drilled to extract brine from the reservoir, and then, a heat-power system is used to convert the heat at the surface into electricity. Convection and thermal conduction work together to transfer heat from the Earth's mantle into shallow porous rock and to fluids located in the pore space of the rock. Higher power plant efficiency is best at temperatures above 225°C (European Geothermal Energy Council, 2020).

Enhanced/engineered geothermal systems (EGSs) are used when fractured zones do not have sufficient permeability to be useful for geothermal energy without intervention. These zones of low permeability can be more penetrated using the emerging technology known as EGS. This allows the permeability to be increased, which improves access to the hot rock. Hydraulic stimulation fracking technology is used for this purpose. However, another type of EGS system uses thermal cooling of reservoirs by continuous pumping of cold water to "induce" fracturing of the geothermal source rock. The goal of EGS is to fracture the rock and inject water into the formation, which flows through the fractures and is then pumped back to the surface for energy conversion, in situations where the rock is hot but not permeable. Untapped geothermal resources in hot, dry rock with little or no permeability or naturally occurring fluids are now achievable thanks to recent technological breakthroughs in deep drilling, logging, and construction, as well as in materials such as cement and well casing (Blackwell et al., 2006; Koelbel & Genter, 2017; Li et al., 2016; Tester et al., 2006; Wendt et al., 2018). Although this technology is still under development, it has the potential to significantly improve the scalability of geothermal energy. EGS is one of the few technologies developed to address the problem of geothermal scalability and could help make geothermal energy "commonplace" rather than extremely rare in the future.

The advanced geothermal (AGS)/closed-loop system is a conduction-based system that significantly increases the scalability of geothermal energy. Conduction systems provide a critical and important differentiator that allows geothermal energy to be harnessed virtually anytime, anywhere on Earth by demonstrating the viability of a closed-loop system for heat recovery through conduction alone. A closed-loop system is a geothermal heat exchange system that is extremely deep and extensive. It provides incredibly predictable and reliable performance, and the system can be tuned to achieve desired results. This means low risk to the system. System efficiency is also greatly improved by taking advantage of the thermosiphon phenomenon, which eliminates the need to lower a parasitic pump load. Geographically, there are no special requirements for a loop system; it can be installed in virtually any location. In addition, the advanced geothermal/closed-loop system is considered a potentially viable geothermal project around the world due to its potential use in inefficient geothermal wells, in coproduction situations at active oil and gas wells, or in regions of the world where the use of hydraulic fracturing is prohibited (Alimonti et al., 2018; Amaya et al., 2020; Elders & Moore, 2016; Gosnold et al., 2015; Livescu & Dindoruk, 2022; Scherer & Inc, 2020).

Supercritical geothermal systems are characterised by extremely high temperatures, and a geothermal reservoir contains supercritical fluid. This means a temperature of at least 374°C and a pressure of 221 bar for pure water (Reinsch et al., 2017). In many places, including Iceland, Japan, Kenya, Mexico, and New Zealand, there are deep volcanic hydrothermal systems that contain such supercritical fluids. Because of the greater energy content of the fluid, such supercritical systems can provide more energy per unit of energy than conventional high-temperature geothermal systems (Fridhleifsson et al., 2014).

14.3 GEOTHERMAL ENERGY IN THE WORLD

The ability of geothermal energy to provide an uninterrupted supply of sustainable renewable energy to communities of varying economic and political backgrounds is well established. Geothermal energy is used directly or indirectly in more than 78

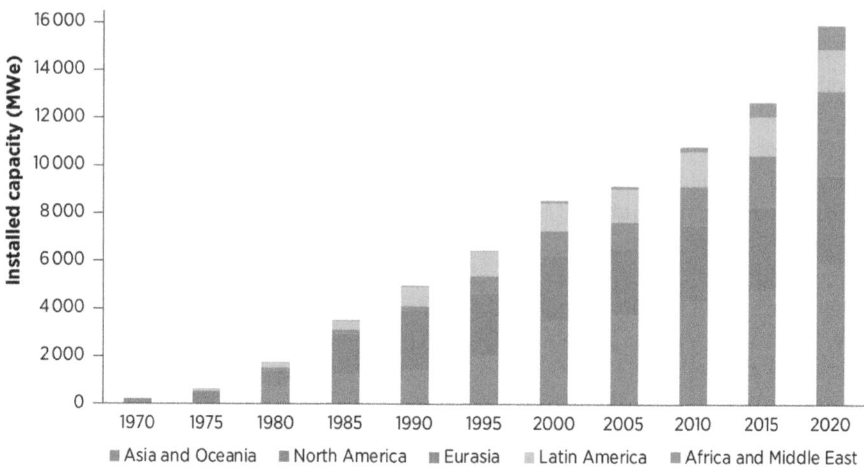

FIGURE 14.3 Growth of installed geothermal electricity capacity by region (IRENA & IGA, 2023).

countries, whether for electricity generation or for domestic, industrial, agricultural, or residential applications (Islam et al., 2022).

Geothermal power plants are with a capacity of about 16 GWe in 2020, up from 200 MWe in the early 1950s (Figure 14.3). The oil crises of 1973 and 1980/1981 contributed to a large expansion of geothermal capacity in the 1970s and 1980s. The development of many alternative power sources, including geothermal, was triggered by the sharp increase in oil costs. Geothermal energy, available locally, played a key role in these advances by enabling countries to reduce their dependence on imported fossil fuels for power generation. Since 2000, installed geothermal power generation capacity has increased by an average of 3% annually; Indonesia, Kenya, Turkey, and the United States have all made significant investments. Despite this expansion, geothermal accounted for only 0.5% of the global renewable electricity market in 2022 (IRENA & IGA, 2023).

Geothermal power and heat generation are highly concentrated in certain countries and concentrated in strategic geographic areas within regions. Turkey (0.9 GW additional), Indonesia (0.7 GW), Kenya (0.2 GW), and the United States (0.2 GW) were the top markets for reported capacity additions during 2016–2021, followed by Iceland, Chile, Japan, New Zealand, Costa Rica, and Mexico (Figure 14.4).

Geothermal power generation capacity increased by 0.3 gigawatts (GW) in 2021, bringing total installed capacity to about 14.5 GW. This is more than the increase in 2020, but less than the five-year average of 0.5 GW since 2016 (REN21, 2022). Today's installed geothermal capacity is dominated by the United States with about 3.7 GW, followed by Indonesia (2.1GW), the Philippines (1.9 GW), Turkey (1.7 GW), and New Zealand.

Considering the net-zero emission's (NZE's) ten-year total energy supply scenarios, it is seen that the transformation towards low-emission energy sources such as

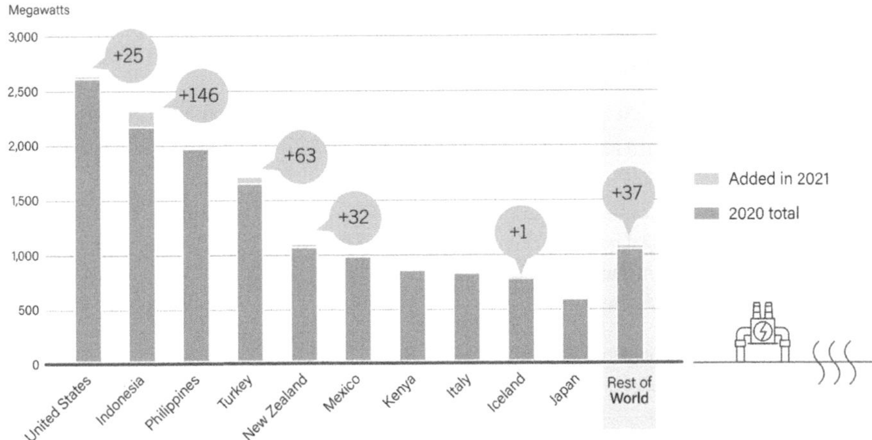

FIGURE 14.4   Geothermal power capacity and additions, top 10 countries and rest of world, 2021 (REN21, 2022).

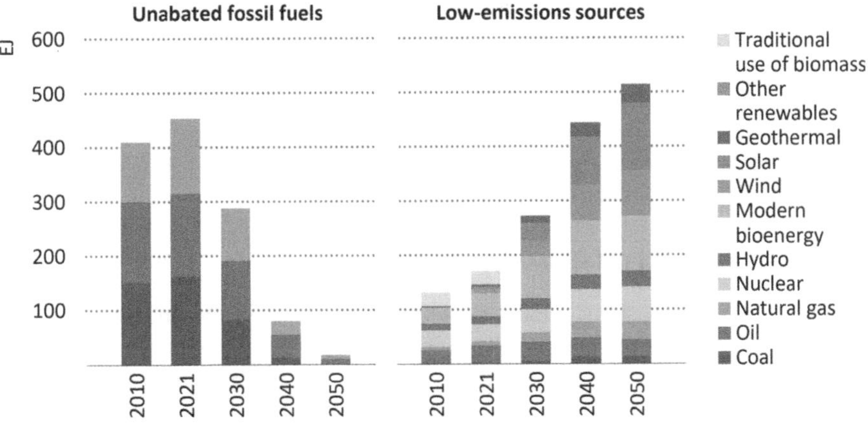

FIGURE 14.5   Total energy supply of unabated fossil fuels and low emissions (International Energy Agency, 2022).

geothermal is inevitable (Figure 14.5). The amount of low-emission supply sources is expected to rise by roughly 125 exajoules (EJ) between 2021 and 2030 (International Energy Agency, 2022).

Three entirely updated scenarios of the World Energy Outlook 2022 offer a framework for considering the energy future and examining the effects of various policy decisions, investment patterns, and technological developments:

- **Stated Policies Scenario:** The scenario examines where this takes the energy industry rather than focusing on what governments claim they will do but rather on what they are actually doing to meet the goals they have set.

TABLE 14.2
World Geothermal Energy Sector

Geothermal	Scenario	2010	2020	2021	2030	2040	2050
Generation (TWh)	Stated policies	–	95	97	183	335	458
	Announced pledges	68	95	97	237	479	686
	NZEs by 2050	68	95	97	312	655	857
Capacity (GW)	Stated policies	11	15	16	28	50	66
	Announced pledges	11	15	16	37	72	102
	NZEs by 2050	11	15	16	50	98	126

Source: Adapted from International Energy Agency (2022).

- **Announced Pledges Scenario:** The scenario looks at where the energy sector would be if all recently stated energy and climate commitments, including NZE's pledges as well as commitments in areas like energy access, were fully and promptly carried out.
- **NZE by 2050 Scenario:** The scenario shows how to fulfil the key energy-related UN Sustainable Development Goals while restoring stability to the global average temperature at 1.5°C.

Under the above scenarios, it appears that the rate of increase in geothermal generation (TWh) and geothermal capacity (GW) to date is insufficient. To achieve the average scenario targets, geothermal production and geothermal capacity should increase at least twofold by 2030 (Table 14.2).

14.4 HYBRID GEOTHERMAL SYSTEMS

Hybrid geothermal systems (HGSs) may be referred to in the literature as "multisystem hybrid systems" or "combined systems." They may be a hybrid system of two geothermal systems, such as EGS-AGS, and renewable energy sources, such as solar photovoltaic geothermal, hydrogen production geothermal, or fossil fuels and renewables, such as natural gas geothermal. Hybrid geothermal systems can also be considered, which provide other benefits in addition to power generation, such as carbon capture geothermal, desalination geothermal, and mineral extraction geothermal. To increase efficiency, improve performance, or provide additional energy services, hybrid geothermal systems integrate geothermal energy with other renewable energy sources or technologies (DiPippo, 2016; Livescu et al., 2023; Wendt et al., 2018).

Geothermal hybrid systems not only have advantages such as low energy consumption, lower electricity costs, support of higher peak loads, low investment costs, low operating costs, no soil contamination, and smaller required borehole area, but also have disadvantages such as higher complexity, control problems, and connection problems (Olabi et al., 2020).

Geothermal hybrid systems can be classified based on several factors, including the types of energy sources integrated, the applications they serve, and the specific technologies used.

14.4.1 Geothermal–Electricity Hybrid Systems

Geothermal–Photovoltaic Hybrid System: Geothermal energy is used as the main energy source in this hybrid system, providing a constant baseload of heat or steam for electricity generation. Solar photovoltaic panels are incorporated into the system to supplement the energy output during periods of higher demand or when more electricity generation is required (Olabi et al., 2020).

Geothermal–Biomass Hybrid System: In this system, electricity is generated by combining a geothermal power plant with a biomass plant. The baseload is supplied by the geothermal plant, while the biomass plant provides additional power during periods of high demand or when geothermal resources are depleted. The hybrid system increases the reduction of greenhouse gas emissions and the reliability of electricity generation (Thain & Dipippo, 2015).

Geothermal–Wind Hybrid System: This hybrid system combines wind turbines and geothermal energy. While wind turbines generate electricity when the wind is strong, geothermal systems provide a steady supply of energy. The combination ensures more consistent and reliable renewable energy generation (Boubaker et al., 2013).

14.4.2 Geothermal–Heat Pump Hybrid Systems

Geothermal–Heat Recovery Ventilation (HRV) Hybrid System: Geothermal heat recovery ventilation systems integrate HRV systems with geothermal heat pumps. By extracting heat from the exhaust air of the HRV system, the geothermal heat pump reduces the energy required to process fresh air. This hybrid system improves both indoor air quality and energy savings (Liu et al., 2023).

Geothermal–Solar Hybrid System: To meet heating and cooling needs, this system combines solar thermal collectors and geothermal heat pumps. The water or working fluid is heated by solar collectors before entering the geothermal heat pump, reducing the energy demand for heating or cooling. The hybrid system can reduce operating costs and increase overall efficiency (Li et al., 2020).

Hybrid System for Geothermal Cooling: In this system, geothermal cooling is combined with advanced cooling technologies such as adsorption systems or absorption chillers. The geothermal resource provides a source of low-temperature heat for cooling, increasing the effectiveness of the cooling system and reducing energy consumption (Ashwood & Bharathan, 2011).

14.4.3 Geothermal–Hydrogen Hybrid Systems

Geothermal–Electrolysis–Hydrogen Hybrid System: This hybrid system combines the generation of hydrogen with geothermal energy. Geothermal energy is used to generate electricity, which is then converted to hydrogen through electrolysis. Hydrogen is a clean fuel that can be used and stored in a variety of ways (Shah et al., 2022).

Geothermal Fuel Cell Hybrid System: Geothermal energy is used as a heat source to power a fuel cell that converts thermal energy into clean electrical energy through an electrochemical process (Alirahmi et al., 2021).

Karayel et al. (2020) mention the effective use of geothermal energy for hydrogen production. They found that geothermal resources with temperatures above 170°C are suitable for hydrogen production with high-temperature electrolysis systems and presented the additional hydrogen productions.

14.5   HYDROGEN SUPPLY

There are two main technologies to produce hydrogen: electrolysis and steam methane reforming. The latter is a mature process which delivers almost all of the current production everywhere. Under heat, methane reacts with steam when in contact with a catalyst to produce carbon monoxide and hydrogen. So, it does have a large carbon footprint. Electrolysis, on the other hand, is the only viable option to produce sustainable hydrogen from renewable electricity. It splits water into hydrogen and oxygen using direct electric current. Hydrogen is stored while oxygen can be either stored or released into the air. In order to scale up the hydrogen production in the future, new renewable power plants must be deployed, accounting for the cycle efficiency limits and the extra grid capacity requirements.

Today, less than 1% of the global hydrogen is actually produced by electrolysis (Figure 14.6). That is limited to markets like electronics where the required high-purity hydrogen is obtained from water electrolysis or to chlorine production where hydrogen comes as a by-product of brine electrolysis. Incidentally, not distinguishing this by-product hydrogen is a frequent source of statistical miscommunication. Almost a century ago, hydropower used to be a major source of hydrogen production by electrolysis. But it has been displaced by the commercialisation of steam methane reforming, purely on economic grounds. Even today, unsubsidised clean hydrogen from renewables would be two to three times more expensive than producing hydrogen from fossil fuels. That is the challenge to overcome.

Virtually zero-carbon hydrogen can be produced in electrolysers that run on dedicated renewables like geothermal energy. Grid-connected electrolysers, on the other hand, can operate for longer durations, reducing the cost of hydrogen. However, grid electricity may include electricity generated by fossil fuel plants. For hydrogen

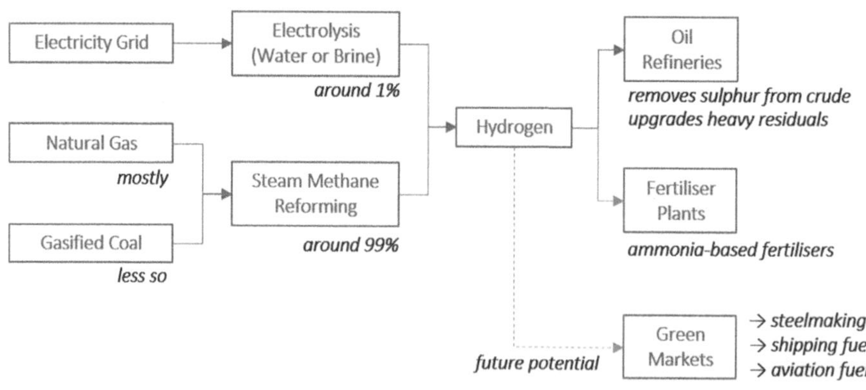

FIGURE 14.6   Global supply and demand of hydrogen.

produced from electrolysis to have lower overall emissions than hydrogen from steam methane reforming, the grid must be sufficiently decarbonised. Yet most countries fail to achieve these criteria with their current grid mixes. Hence, the investments in electrolysis and renewable generation need to be made in conjunction. Countries that lack the required renewable resources potential may import hydrogen and hydrogen-based products through ports or pipelines (IRENA, 2022).

Electrolyser plants need to run on purified and demineralised water in order to avoid damaging the stack components. Therefore, water treatment systems are usually integrated into the electrolysis process even when there is an access to potable water sources. Water is also needed for the cooling of the plant. In water-stressed coastal regions, desalination may need to be deployed. Unless desalinated, the direct use of seawater causes corrosive damage and results in the production of chlorine. Typical water treatment designs that use reverse osmosis do not lead to major cost uplifts on the total electrolysis costs. Nevertheless, water availability is an important consideration for site selection (Figure 14.7).

For clean hydrogen to become more affordable, there are four main factors determining the clean hydrogen production costs: the electrolyser capital costs, their conversion rates, their utilisation rates, and the cost of renewable electricity used to operate them. The levelised cost of renewable electricity has fallen substantially during the last decade. While the wind turbines benefited from economies of scale, the solar photovoltaic costs have fallen along their learning curves. With a greater market uptake in the future, the electrolyser capital costs can also be expected to fall (LAZARD, 2023). Today, there is more and cheaper renewable electricity to operate electrolysers than ever before. With a greater system integration in the future, the electrolyser utilisation rates will also increase. Where possible, coupling the electrolysers with geothermal plants can yield especially high utilisation rates.

The electrolyser conversion rates will remain as the dominant cost factor to be addressed. Clean hydrogen incurs energy losses at each stage of its value chain. At the current electrolyser efficiencies, about one-third of the useful energy in electrolysis

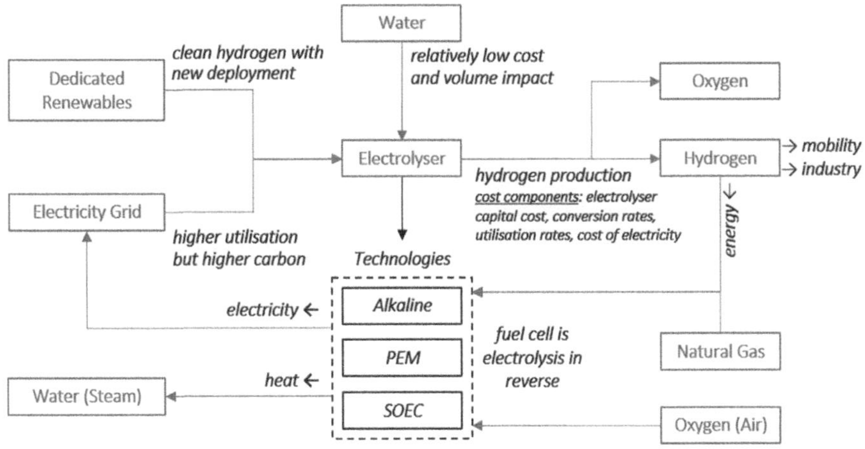

FIGURE 14.7 Energy and material flows for clean hydrogen technologies.

is lost. That is an energy cost penalty in comparison with direct electrification. Preheating the water with geothermal fluid can increase the conversion efficiency of the electrolysers which would be run on electricity also provided by geothermal energy. From an overall system sustainability viewpoint, clean hydrogen production must not hinder the use of renewable electricity that would be more effective and efficient elsewhere. Clean hydrogen needs to be produced from renewable electricity that would not be otherwise consumed and capacity that would not be otherwise commissioned. Disregarding this principle of additionality in one place could lead to an increase in fossil fuel consumption in another place.

Cost reductions can be achieved through the commissioning of new electrolysers that are larger than the operational modules by an order of magnitude, to supersede the steam methane reforming units. Of all electrolyser capital costs, the cell stack represents around half of the electrolyser costs. Most of the remaining capital cost is due to the power electronics, gas-conditioning, and plant components. In order to inject the produced hydrogen into gas transmission grids, additional compressors will be needed especially for alkaline electrolysers which means additional cost. Electrolyser assembly is still mostly done manually due to the complexity involved with its automation. That would require a lot of upfront capital which may only be justified with a demand uptake for electrolysers.

Electrolysers have been developing with incremental steps rather than technological breakthroughs. There are three main types of electrolyser technologies: alkaline, proton exchange membrane (PEM), and solid oxide electrolyser cell (SOEC). The first two are currently mature enough for large-scale implementation. As the oldest and most established technology, alkaline electrolysers avoid any precious metals and therefore are relatively cheaper than the other two in terms of the capital cost. For alkaline electrolysers, there will be no long-term constraints regarding primary nickel reserves. However, if their operational performances are compared, alkaline electrolysers suffer from greater efficiency drops due to the variable hydrogen output under variable renewable electricity loads.

Another established technology, PEM electrolysers, maintains their rated efficiency far better under fluctuating loads coming from renewables due to their short response times for switching on and off. Also, their compact design makes them more suitable for decentralised plants. However, their design uses expensive and trademarked membrane materials as well as expensive iridium and platinum catalysts, whose distribution is quite uneven across the Earth. Recent demonstration projects tend to favour PEM electrolysers with the assumption that they will have a higher potential for cost reduction being less mature and less widely deployed than alkaline electrolysers. Also, the lifetime of PEM electrolysers is currently shorter than that of alkaline electrolysers which can be improved.

SOEC electrolysers are the least developed of all three technologies. However, they also promise the highest efficiencies and gradually beginning to enter the market. One key advantage of SOEC electrolysers is that they can be operated in reverse mode as a fuel cell. In a fuel cell, hydrogen and oxygen are converted into electricity, with water steam as the only emission. This increases the overall utilisation rate considerably and reduces the production costs for hydrogen and electricity. Fuel cells can be used to deliver power to remote locations, micro-grids, construction sites, port facilities, and the armed forces. The key challenge hindering the commercialisation of the SOEC electrolysers is the degradation of their materials as a consequence of the high operating temperatures.

14.6 HYDROGEN DEMAND

In tandem with the process of electrifying the existing energy system, the electricity generation mix also needs to accommodate significantly more renewables, almost entirely phasing out fossil fuels. Using hydrogen may be an unavoidable route for certain sectors with high energy density requirements. What is common to these sectors is that they are all hard to electrify and therefore hard to abate. Steel, glass, ceramics, petrochemicals, fertilisers, long-haul road, and maritime transport can be listed among the major hard-to-abate emissions sources (International Energy Agency, 2023). Providing a link between the growing and sustainable renewable electricity generation sources and the hard-to-electrify sectors as a decarbonisation strategy is termed sector coupling. Conversely, there is a limited scope of using hydrogen as an energy vector in the electricity and heat sectors.

Within the electricity sector, hydrogen can play a role in carbon reduction, capacity utilisation, and constraint management. Hydrogen has been proposed as a flexible source of potentially low-carbon electricity that can capture the curtailed energy under high penetration levels of variable renewables, such as wind and solar. At the grid scale, hydrogen can be used for seasonal storage, complementary to the pumped-storage hydropower plants. In more de centralised modes, hydrogen can replace the peaking natural gas plants and act as a backup generator. Due to their flexible operation, electrolysers can participate in demand-side management by reacting to the signals from spot electricity prices or grid carbon intensity. Utilisation rates of such electrolysers need to be high enough to justify their investment (Figure 14.8).

Heat is a sector that is harder to abate than electricity, for both domestic and non-domestic loads. Hydrogen is unlikely to completely replace the natural gas on the mains supply, though. That is because the price of hydrogen is likely to remain higher than natural gas on a commercial basis. Blending and co-firing hydrogen in

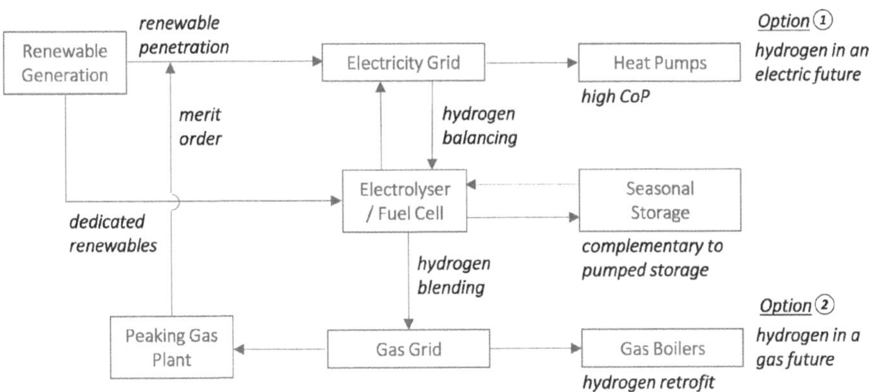

FIGURE 14.8 Hydrogen for energy in the future electricity and heating systems.

regional gas networks can have a marginal contribution in reducing carbon emissions because hydrogen has a very low energy density by volume and there is a volumetric upper limit due to safety issues in dispersed hydrogen applications in buildings. Yet the incumbent gas networks may opt for a transitional role to avoid becoming stranded assets. In the long term, heating can be decarbonised by electrification through more heat pumps plus better energy efficiency measures. Clean hydrogen can be allocated to other sectors with a comparative advantage.

Despite the steady progress in battery electric vehicles recently, the long-distance and heavy-haul freight transportation challenges remain to be answered. The energy density of hydrogen fuel cells is higher than batteries on a gravimetric basis. Yet hydrogen is bulky due to its very low energy density per volume. In that regard, clean hydrogen via ammonia conversion may have an advantage over batteries for transporting heavy loads over long distances on land and especially in maritime international trade. Ships require energy not only for propulsion but also for on-board electricity generation as well as on-board heating and cargo operations. In its historical evolution, shipping may need to skip out natural gas and directly move from heavy fuel oil to clean ammonia. Vessels typically have long lifespans so the procured designs will need to be adaptable to hydrogen retrofits in the near future (DNV, 2022).

On land, hydrogen is likely to be used as a transport fuel only where batteries fall short of dominating. Hydrogen fuel cells may be the only viable solution for decarbonising the long-distance and heavy-haul trucks as direct electrification by batteries proves to be more challenging. Fuel cells may also be suitable for materials handling in warehouses and port facilities. Regarding public and private transport in cities, there is less room for fuel cells, mainly due to the path dependency of the mobility sector on battery technologies. Fundamentally, hydrogen vehicles require more renewable electricity to run, compared to battery electric vehicles. Lately, there has been a dramatic fall in battery prices and an increase in the deployment rate of the charging infrastructure for electric passenger road transport.

The flexibility of hydrogen lends itself to a broad range of applications in the industrial sectors, some of which are existing and some potential. Hydrogen is already used as a chemical feedstock in oil refineries and fertiliser plants. Currently, this hydrogen demand is met almost entirely by fossil fuels. Under new regulations, these existing industries can act as anchor loads for the wider transition towards clean hydrogen. For instance, the use of clean hydrogen in petrochemical industry can act as an enabler to replace the regular trailer trucks with hydrogen trucks that transport oil and fertilisers, which will further unlock the investment potential for the hydrogen refuelling infrastructure.

Due to its additional distribution costs, hydrogen production is better suited for large-scale central industrial hubs which can aggregate the demand for hydrogen, thereby creating a virtuous circle of supply and demand. These industrial hydrogen hubs will be usually co-located with port facilities for bunkering and exporting purposes. The complementary hydrogen demand patterns from agriculture, industry, and transport activities in the port hinterlands can increase the feasibility of centralised electrolysers. Ports will allow not only the trade of hydrogen feedstock itself but also hydrogen-based products such as green steel, glass, and ceramics that will be exempt from the carbon border taxes (Figure 14.9).

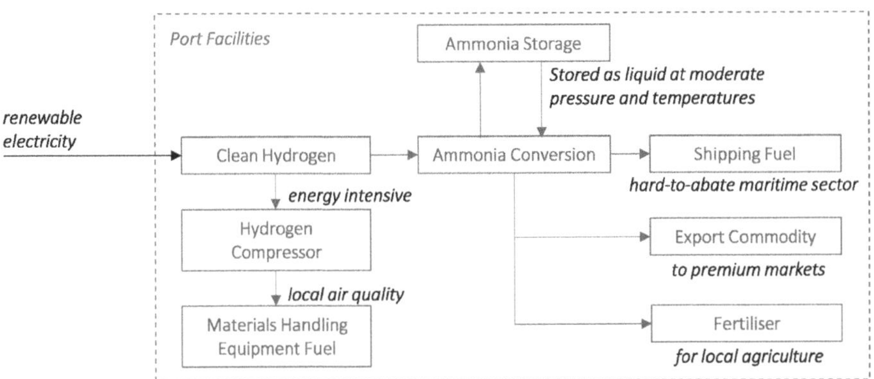

FIGURE 14.9 Hydrogen for transport in future ports and shipping.

Steelmaking is a heavy industry that requires extreme heat to separate, or reduce, the iron oxide mineral, lime, and coke. Steel can be produced mainly by two alternative processes: blast furnace to basic oxygen furnace (BF-BOF) and electric arc furnace (EAF). The second route uses either scrap metal or direct-reduced iron (DRI) as the input. In the BF-BOF route, hydrogen can reduce the carbon footprint of steel production by replacing the coking coal as the reducing agent or it can be co-fed into the blast furnace. In the EAF route, clean hydrogen can be injected instead of natural gas in the DRI stage. The BF-BOF route with hydrogen injection is expected to provide a decarbonisation solution in the short term. However, if the steelmaking industry is to fully decarbonise, large-scale DRI-EAF deployment with hydrogen injection is widely considered as the only viable solution (Figure 14.10).

14.7 HYDROGEN INFRASTRUCTURE

Hydrogen technologies are a part of a wider infrastructure that covers everything from the feedstock and production, through distribution and storage, to the end-use markets. The use of hydrogen will require its own integrated assets in the future. A decarbonised energy system based only on electricity would be much more flow-based without hydrogen and therefore be vulnerable to the disruptions of supply in real time. As a chemical energy vector, hydrogen can make the energy system more stock-based and therefore more resilient. Hydrogen can be stored and transported in a stable way, similar to other chemical energy vectors such as coal, oil, and natural gas. Alternatively, it can be combined with other elements such as carbon and nitrogen to synthesise hydrogen-based fuels.

One of the challenges of integrating hydrogen in the energy system is its storage infrastructure. Due to its low volumetric energy density, storing hydrogen requires considerably more space or energy compared to other fossil fuels. On average, three to four times storage infrastructure would be required to hypothetically replace the entire natural gas infrastructure today. This implies either large volumes such as salt caverns where the local geology permits or the application of compression technologies and pressure tanks. While it offers a viable alternative for seasonal storage, the

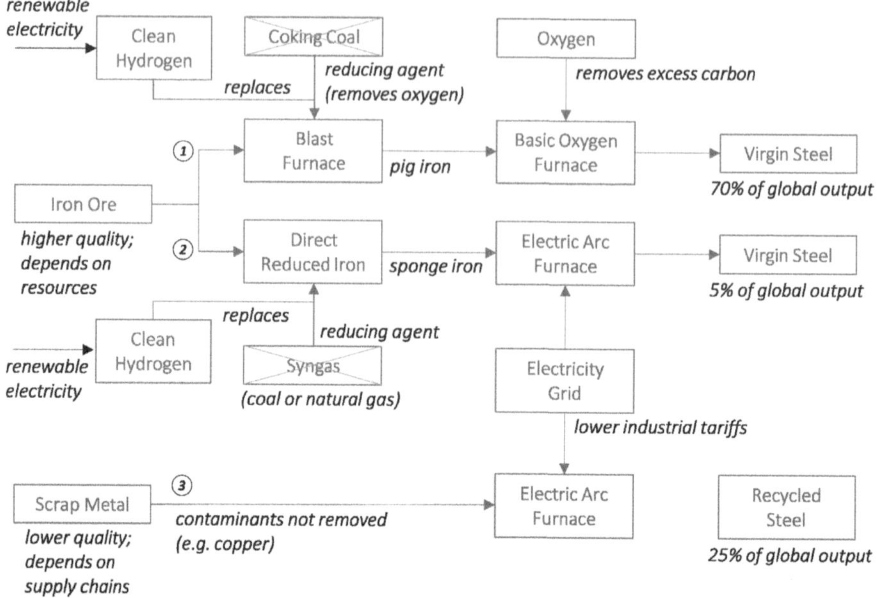

FIGURE 14.10   The role of hydrogen in different steelmaking routes.

roundtrip electrical efficiency of hydrogen storage is low, meaning that the final electricity output has a cost premium. Therefore, electricity generation from hydrogen will be practical only when the electricity price exceeds this cost (Energy Transitions Commission, 2021).

Hydrogen can also be chemically converted into ammonia which is then stored in liquid form at modest temperatures and pressures. This option is suitable for coastal areas where ammonia can be directly used as shipping fuel, sold as fertiliser, or stored at port facilities. Ammonia is already produced through a combination of steam methane reforming and the Haber–Bosch process, causing significant carbon emissions. The synthesis of ammonia can be completely electrified, eliminating the steam methane reforming stage. In general, these methods are called Power-to-X by which synthetic fuels are produced with renewable electricity. Ammonia cannot be decomposed back into hydrogen with any known industrial process today, and its toxicity requires high safety standards for storage and handling.

The majority of hydrogen is used on site today and not distributed as transporting the hydrogen gas consumes more than one-tenth of its useful energy. Trailer trucks, pipelines, and cargo ships are all possible modes to distribute hydrogen depending on the volumes and distances involved. At low volumes and short distances, transporting compressed hydrogen with trucks can be viable before it starts to erode the carbon abatement benefits. For high volumes and if the end-use is also hydrogen, pipelines are cheaper than shipping with its reconversion costs. Shipping high volumes of hydrogen through ammonia conversion is best suited for longer distances as the global shipping of ammonia-based fertilisers is already a well-established operation. Ammonia is easier to transport than LNG, and its liquefaction require-ments are similar to LPG, allowing for the same vessels to be repurposed.

As a general rule, moving the same amount of energy in gas pipelines takes much less infrastructure than moving it in electricity transmission lines. Using the existing gas pipelines to distribute hydrogen will be an order of magnitude cheaper than the cost of transmitting the same energy as electricity across high-voltage power lines. The spatial footprint requirement of the pipelines will be almost two orders of mag-nitude less than the power lines. Hydrogen distribution pipelines can also be used for line packing which is an indirect short-term storage technique for the peak demand times, akin to the current practice in the pressurised natural gas pipelines. Over very long distances, however, the cost of the HVDC power lines will decrease more steeply than the cost of gas pipelines due to the need for compressing the hydrogen gas at regular intervals (Figure 14.11).

Safety regulations for hydrogen need to define the technical risks as well as the measures to mitigate them effectively. When transmitted through pipelines, hydrogen has a high propensity to leak through the microscopic cracks in certain iron and steel pipes while also causing their embrittlement. Hydrogen is extremely flammable with a very high flame velocity. This flame is invisible while the gas itself is colourless and odourless, which makes leak and fire detection quite difficult. Yet it is important to note that hydrogen is a non-toxic gas. The extreme flammability is also partly bal-anced out by its tendency to dissipate quickly due to its high buoyancy and diffusivity. All safety issues are considered; hydrogen is likely to develop as a business-to-busi-ness market rather than a retail market (Bloomberg New Energy Finance, 2020).

In contrast, ammonia is a toxic chemical that requires stringent safety procedures. It is also flammable, corrosive, and prone to leak in the gaseous state. However, these

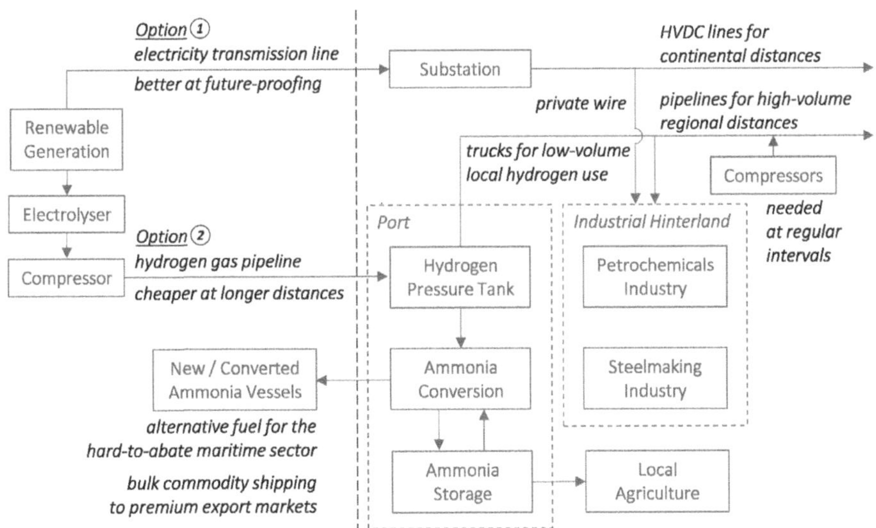

FIGURE 14.11 Future infrastructure options for hydrogen storage and transport.

leaks are quite easy to detect due to its pungent smell. Heavy industries have a long and extensive experience of handling both hydrogen and ammonia. Existing protocols for the safe use of hydrogen and ammonia can be extended to cover new types of site-specific infrastructure such as electrolysers or refuelling stations. The public perception regarding the health and safety of hydrogen is also likely to improve due to the increased familiarity with it, analogous to the risks associated with already established fuels such as oil and natural gas.

These issues are likely to be resolved as clean hydrogen production grows on a global scale. The grid carbon intensity in most countries is currently too high to produce clean hydrogen from electrolysis. This calls for further deployment of renewables either dedicated to electrolysers or connected to the grids. Countries with high potential in baseload renewables such as geothermal and hydropower will have an opportunity to produce and export clean hydrogen or its related products at very competitive costs. The market demand for hydrogen products will inform the optimum operation strategies for these plants.

While the uncertainty around the optimal role of hydrogen remains, there is a renewed interest from the private sector as a result of these steps. A diverse set of companies from the energy, industry, and transport sectors create a wide support base for hydrogen, all looking for a valid business case. Their plans for a hydrogen transition rest upon the overarching public motives of energy security, long-term affordability, renewable capacity expansion, global carbon emissions, and local air quality. With these in mind, hydrogen investments need to be targeted towards areas where they add the most value.

14.8 CONCLUSION

It is possible to generate electricity from geothermal energy and then hydrogen by electrolysis. In addition, hydrogen can also be produced directly from geothermal gas or by chemical processes. Geothermal resources are large enough to provide enough electricity to produce hydrogen for various sectors such as public transportation. Geothermally driven electrolysers can produce hydrogen at similar cost ranges to the current gasoline prices. Geothermal gas emissions from power plants contain pure hydrogen and hydrogen sulphide. Geothermal energy can be used to support hydrogen electrolysis through the use of steam turbines and high-temperature electrolysis or through solid-state methods that can use much lower-temperature resources. When metal hydrides are used, geothermal energy can be used to support storage management. Hydrides can also be used to transport thermal energy over long distances as hydrogen gas for heating and cooling purposes (Arnason & Sigfusson, 2007).

Electrolysers that run on geothermal electricity can be utilised more effectively due to the high capacity factors of geothermal power plants. Geothermal fluids can also be used to preheat the feed water to achieve incremental gains in the conversion efficiency of the electrolysers. Overall, the cost of hydrogen production from geothermal energy can be lower than the electrolysis driven by either wind or solar energy (Ghazvini et al., 2019). Moreover, geothermal electricity can be used to power the compressors required for the underground storage of gaseous hydrogen and its distribution via pipelines. Alternatively, geothermal heat can be used to precool gaseous hydrogen before its chemical conversion into liquid ammonia for shipping and handling at more modest pressure and temperatures (Yılmaz & Kanoglu, 2017). It is possible that many communities around the world could benefit from linking their geothermal resources to the development of a hydrogen economy.

REFERENCES

Alimonti, C., Soldo, E., Bocchetti, D., & Berardi, D. (2018). The wellbore heat exchangers: A technical review. *Renewable Energy*, *123*, 353–381.

Alirahmi, S. M., Assareh, E., Chitsaz, A., Ghazanfari Holagh, S., & Jalilinasrabady, S. (2021). Electrolyzer-fuel cell combination for grid peak load management in a geothermal power plant: Power to hydrogen and hydrogen to power conversion. *International Journal of Hydrogen Energy*, *46*(50), 25650–25665. https://doi.org/10.1016/j.ijhydene.2021.05.082

Amaya, A., Scherer, J., Muir, J., Patel, M., & Higgins, B. (2020). Greenfire energy closed-loop geothermal demonstration using supercritical carbon dioxide as working fluid. In *Proceedings of the 45th Workshop on Geothermal Reservoir Engineering*, Stanford, CA, pp. 10–12.

Arnason, B., & Sigfusson, T. (2007). *Application of Geothermal Energy to Hydrogen Production and Storage*. https://www.researchgate.net/publication/237613011

Ashwood, A., & Bharathan, D. (2011). Hybrid cooling systems for low-temperature geothermal power production. *Transactions—Geothermal Resources Council*, *35* 2(September), 1275–1279. https://doi.org/10.2172/1009690

Balta, T. M., Dincer, I., & Hepbasli, A. (2009). Thermodynamic assessment of geothermal energy use in hydrogen production. *International Journal of Hydrogen Energy*, *34*, 2925–2939.

Blackwell, D. D., Negraru, P. T., & Richards, M. C. (2006). Assessment of the enhanced geothermal system resource base of the United States. *Natural Resources Research*, *15*, 283–308.

Bloomberg New Energy Finance. (2020). *Hydrogen Economy Outlook: Key Messages*. https://data.bloomberglp.com/professional/sites/24/BNEF-Hydrogen-Economy-Outlook-Key-Messages-30-Mar-2020.pdf

Boubaker, K., Colantoni, A., Allegrini, E., Longo, L., Di Giacinto, S., & Biondi, P. (2013). Short-term perspectives for hybrid wind/solar/geothermal renewable energy. *International Journal of Renewable Energy Research*, *3*(2), 436–440.

DiPippo, R. (2016). Combined and hybrid geothermal power systems. In *Geothermal Power Generation: Developments and Innovation*, pp. 391–420. https://doi.org/10.1016/B978-0-08-100337-4.00014-0

DNV. (2022). *Energy Transition Outlook: Hydrogen Forecast to 2050*. https://www.dnv.com/focus-areas/hydrogen/forecast-to-2050.html

Eavor Geothermal Technology. (2023). *Eavor Geothermal Technology*. https://www.eavor.com/technology/

Elders, W. A., & Moore, J. N. (2016). Geology of geothermal resources. In *Geothermal Power Generation: Developments and Innovation*, pp. 7–32. https://doi.org/10.1016/B978-0-08-100337-4.00002-4

Energy Transitions Commission. (2021). *Making the Hydrogen Economy Possible*. https://www.energy-transitions.org/publications/making-clean-hydrogen-possible/

European Geothermal Energy Council. (2020). *Sectoral Integration: The Multiple Benefits of Geothermal Energy* (Issue April). https://doi.org/10.18356/bf78500e-en

Fridhleifsson, G. Ó., Elders, W. A., & Albertsson, A. (2014). The concept of the Iceland deep drilling project. *Geothermics*, *49*, 2–8.

Gallois, R. (2007). The formation of the hot springs at Bath Spa, UK. *Geological Magazine*, *144*(4), 741–747. https://doi.org/10.1017/S0016756807003482

Ghazvini, M., Sadeghzadeh, M., Ahmadi, M. H., Moosavi, S., & Pourfayaz, F. (2019). Geothermal energy use in hydrogen production: A review. *International Journal Energy Research*, 1–29. https://doi.org/10.1002/er.4778

Gosnold, W., Crowell, A., Nordeng, S., & Mann, M. (2015). Co-produced and low-temperature geothermal resources in the Williston Basin. *Geothermal Resources Council—Transactions*, *39*, 653–660.

International Energy Agency. (2022). *International Energy Agency (IEA) World Energy Outlook 2022*. https://www.iea.org/reports/world-energy-outlook-2022

International Energy Agency. (2023). *Towards Hydrogen Definitions Based on Their Emissions Intensity*. https://www.iea.org/reports/towards-hydrogen-definitions-based-on-their-emissions-intensity

IRENA. (2022). *Geopolitics of the Energy Transformation: The Hydrogen Factor*. https://www.irena.org/publications/2022/Jan/Geopolitics-of-the-Energy-Transformation-Hydrogen

IRENA, & IGA. (2023). *Global Geothermal Market and Technology Assessment*. www.irena.org

Islam, M. T., Nabi, M. N., Arefin, M. A., Mostakim, K., Rashid, F., Hassan, N. M. S., Rahman, S. M. A., McIntosh, S., Mullins, B. J., & Muyeen, S. M. (2022). Trends and prospects of geothermal energy as an alternative source of power: A comprehensive review. *Heliyon*, *8*(12), e11836. https://doi.org/10.1016/J.HELIYON.2022.E11836

Jaupart, C., Labrosse, S., Lucazeau, F., & Mareschal, J. C. (2007). 7.06-temperatures, heat and energy in the mantle of the earth. *Treatise on Geophysics*, *7*, 223–270.

Kanoglu, M., Bolatturk, A., & Yilmaz, C (2010). Thermodynamic analysis of models used in hydrogen production by geothermal energy. *International Journal of Hydrogen Energy*, *35*, 8783–8791.

Karayel, K., Javani, N., & Dincer, I. (2020). Effective use of geothermal energy for hydrogen production: A comprehensive application, *Energy*, *249*, 123597.

Koelbel, T., & Genter, A. (2017). *Enhanced Geothermal Systems: The Soultz-sous-Forêts Project*, pp. 243–248. https://doi.org/10.1007/978-3-319-45659-1_25

LAZARD. (2023). *Levelised Cost of Hydrogen Analysis: Version 3.0*. https://www.lazard.com/research-insights/2023-levelized-cost-of-energyplus/

Ledru, P., & Guillou Frottier, L. (2010). Reservoir definition. In *Geothermal Energy Systems: Exploration, Development, and Utilization*, pp. 1–36, Wiley.

Li, K., Liu, C., Jiang, S., & Chen, Y. (2020). Review on hybrid geothermal and solar power systems. *Journal of Cleaner Production*, *250*, 119481. https://doi.org/10.1016/j.jclepro.2019.119481

Li, T., Shiozawa, S., & McClure, M. W. (2016). Thermal breakthrough calculations to optimize design of a multiple-stage enhanced geothermal system. *Geothermics*, *64*, 455–465.

Liu, Z., Xie, M., Zhou, Y., He, Y., Zhang, L., Zhang, G., & Chen, D. (2023). A state-of-the-art review on shallow geothermal ventilation systems with thermal performance enhancement system classifications, advanced technologies and applications. *Energy and Built Environment*, *4*(2), 148–168.

Livescu, S., & Dindoruk, B. (2022). Subsurface technology pivoting from oil and gas to geothermal resources: A sensitivity study of well parameters. *SPE Production & Operations*, *37*(2), 280–294.

Livescu, S., Dindoruk, B., Schulz, R., Boul, P., Kim, J., & Wu, K. (2023). Geothermal and electricity production: Scalable geothermal concepts in the future of geothermal in Texas: Contemporary prospects and perspectives. *The University of Texas Faculty/Researcher Works*. 25–46 https://doi.org/10.26153/tsw/44083

Lund, J. W. (2023). Geothermal energy. In *Encyclopedia Britannica*. https://www.britannica.com/science/geothermal-energy

Lund, J. W., & Toth, A. N. (2021). Direct utilization of geothermal energy 2020 worldwide review. *Geothermics*, *90*, 101915. https://doi.org/10.1016/J.GEOTHERMICS.2020.101915

Mahmoud, M. Ramadan M., Naher, S., Pullen, K., Abdelkareem, M.A., & Olabi, A. (2021). A review of geothermal energy-driven hydrogen production systems. *Thermal Science and Engineering Progress*, *22*, 100854. https://doi.org/10.1016/j.tsep.2021.100854

Manning, C. E., & Ingebritsen, S. E. (1999). Permeability of the continental crust: Implications of geothermal data and metamorphic systems. *Reviews of Geophysics*, *37*(1), 127–150.

Moeck, I. S. (2014). Catalog of geothermal play types based on geologic controls. *Renewable and Sustainable Energy Reviews*, *37*, 867–882.

Moya, D., Aldás, C., & Kaparaju, P. (2018). Geothermal energy: Power plant technology and direct heat applications. *Renewable and Sustainable Energy Reviews*, *94*, 889–901. https://doi.org/10.1016/J.RSER.2018.06.047

Mustafa Omer, A. (2008). Ground-source heat pumps systems and applications. *Renewable and Sustainable Energy Reviews*, *12*(2), 344–371. https://doi.org/10.1016/J.RSER. 2006.10.003

Olabi, A. G., Mahmoud, M., Soudan, B., Wilberforce, T., & Ramadan, M. (2020). Geothermal based hybrid energy systems, toward eco-friendly energy approaches. *Renewable Energy*, *147*, 2003–2012. https://doi.org/10.1016/j.renene.2019.09.140

Reinsch, T., Dobson, P., Asanuma, H., Huenges, E., Poletto, F., & Sanjuan, B. (2017). Utilizing supercritical geothermal systems: a review of past ventures and ongoing research activities. *Geothermal Energy*, *5*(1). https://doi.org/10.1186/s40517-017-0075-y

REN21. (2022). Renewables 2022 global status. In *Global Status Report for Buildings and Construction*: *Towards a Zero-emission, Efficient and Resilient Buildings and Construction Sector*. https://www.ren21.net/gsr-2022/

Scherer, J. A., & Inc, G. E. (2020). *Closed-Loop Geothermal Demonstration Project*: *Confirming Models for Large-Scale, Closed-Loop Geothermal Projects in California*: *Consultant Report*. California Energy Commission.

Shah, M., Prajapati, M., Yadav, K., & Sircar, A. (2022). A review of the geothermal integrated hydrogen production system as a sustainable way of solving potential fuel shortages. *Journal of Cleaner Production*, *380*(P1), 135001. https://doi.org/10.1016/j. jclepro.2022.135001

Tester, J. W., Anderson, B. J., Batchelor, A. S., Blackwell, D. D., DiPippo, R., Drake, E. M., Garnish, J., Livesay, B., Moore, M. C., Nichols, K., & others. (2006). The Future of Geothermal Energy, p. 358. Massachusetts Institute of Technology.

Thain, I., & Dipippo, R. (2015). Hybrid geothermal-biomass power plants: Applications, designs and performance analysis. In *Proceedings World Geothermal Congress*, April, 19–25, Melbourne, Australia.

Turcotte, D. L., & Schubert, G. (2014). *Geodynamics*, 3rd Edition. Cambridge University Press.

Yılmaz, C., & Kanoglu, M., (2017). Investigation of hydrogen production cost by geothermal energy. *International Advanced Researches and Engineering Journal 1*(1): 5–10.

Wendt, D. S., Neupane, G., Davidson, C. L., Zheng, R., & Bearden, M. A. (2018). *GeoVision Analysis Supporting Task Force Report*: *Geothermal Hybrid Systems* (No. INL/ EXT-17-42891; PNNL-27386). Idaho National Lab.(INL), Idaho Falls, ID (United States); Pacific Northwest National Lab.(PNNL), Richland, WA (United States)

World Energy Council. (2022). *Regional Insights into Low-Carbon Hydrogen Scale Up*. https:// www.worldenergy.org/publications/entry/regional-insights-low-carbon-hydrogen- scale-up-world-energy-council

Index

Note: **Bold** page numbers refer to tables and *italic* page numbers refer to figures.